Selected Titles in This Series

197 **Kenji Ueno,** Algebraic geometry 2: Sheaves and cohomology, 2001
196 **Yu. N. Lin'kov,** Asymptotic statistical methods for stochastic processes, 2001
195 **Minoru Wakimoto,** Infinite-dimensional Lie algebras, 2001
194 **Valery B. Nevzorov,** Records: Mathematical theory, 2001
193 **Toshio Nishino,** Function theory in several complex variables, 2001
192 **Yu. P. Solovyov and E. V. Troitsky,** C^*-algebras and elliptic operators in differential topology, 2001
191 **Shun-ichi Amari and Hiroshi Nagaoka,** Methods of information geometry, 2000
190 **Alexander N. Starkov,** Dynamical systems on homogeneous spaces, 2000
189 **Mitsuru Ikawa,** Hyperbolic partial differential equations and wave phenomena, 2000
188 **V. V. Buldygin and Yu. V. Kozachenko,** Metric characterization of random variables and random processes, 2000
187 **A. V. Fursikov,** Optimal control of distributed systems. Theory and applications, 2000
186 **Kazuya Kato, Nobushige Kurokawa, and Takeshi Saito,** Number theory 1: Fermat's dream, 2000
185 **Kenji Ueno,** Algebraic Geometry 1: From algebraic varieties to schemes, 1999
184 **A. V. Mel'nikov,** Financial markets, 1999
183 **Hajime Sato,** Algebraic topology: an intuitive approach, 1999
182 **I. S. Krasil'shchik and A. M. Vinogradov, Editors,** Symmetries and conservation laws for differential equations of mathematical physics, 1999
181 **Ya. G. Berkovich and E. M. Zhmud',** Characters of finite groups. Part 2, 1999
180 **A. A. Milyutin and N. P. Osmolovskii,** Calculus of variations and optimal control, 1998
179 **V. E. Voskresenskiĭ,** Algebraic groups and their birational invariants, 1998
178 **Mitsuo Morimoto,** Analytic functionals on the sphere, 1998
177 **Satoru Igari,** Real analysis—with an introduction to wavelet theory, 1998
176 **L. M. Lerman and Ya. L. Umanskiy,** Four-dimensional integrable Hamiltonian systems with simple singular points (topological aspects), 1998
175 **S. K. Godunov,** Modern aspects of linear algebra, 1998
174 **Ya-Zhe Chen and Lan-Cheng Wu,** Second order elliptic equations and elliptic systems, 1998
173 **Yu. A. Davydov, M. A. Lifshits, and N. V. Smorodina,** Local properties of distributions of stochastic functionals, 1998

(*Continued in the back of this publication*)

Translations of
MATHEMATICAL MONOGRAPHS

Algebraic Geometry 2
Sheaves and Cohomology

Kenji Ueno

Translated by Goro Kato

IWANAMI SERIES IN MODERN MATHEMATICS

AMERICAN MATHEMATICAL SOCIETY
Providence, Rhode Island

Editorial Board

Shoshichi Kobayashi (Chair)
Masamichi Takesaki

代数幾何 2
層とコホモロジー

DAISŪ KIKA (ALGEBRAIC GEOMETRY 2)
by Kenji Ueno
with financial support
from the Japan Association for Mathematical Sciences

Copyright © 1997 by Kenji Ueno
Originally published in Japanese
by Iwanami Shoten, Publishers, Tokyo, 1997
Translated from the Japanese by Goro Kato

2000 *Mathematics Subject Classification.* Primary 14–01, 14F99.

ABSTRACT. This is the second of three books by the author aimed at introducing the reader to Grothendieck's scheme theory as a method of studying algebraic geometry. This book contains definitions and results related to coherent schemes, proper and projective morphisms, and cohomology of sheaves on schemes. As in the first book, the author includes many examples and problems illustrating the topics discussed in the main text.

The book is aimed at graduate and upper-level undergraduate students who want to learn modern algebraic geometry.

Library of Congress Cataloging-in-Publication Data

Ueno, Kenji, 1945–
  [Daisū kika. English]
  Albegraic geometry / Kenji Ueno ; translated by Goro Kato.
    p. cm. — (Translations of mathematical monographs, ISSN 0065-9382 ; v. 185) (Iwanami series in modern mathematics)
  Includes index.
  contents: 1. From algebraic varieties to schemes
  ISBN 0-8218-0862-1 (v. 1 : pbk. : acid-free)| v. 2 ISBN 978-0-8218-1357-7
  1. Geometry, Algebraic. I. Title. II. Series. III. Series: Iwanami series in modern mathematics.
QA564.U3513   1999
516.3′5—dc21   99-22304
                                                                                                          CIP

© 2001 by the American Mathematical Society. All rights reserved.
The American Mathematical Society retains all rights
except those granted to the United States Government.
Printed in the United States of America.
Reprinted by the American Mathematical Society, 2022.

∞ The paper used in this book is acid-free and falls within the guidelines
established to ensure permanence and durability.
Information on copying and reprinting can be found in the back of this volume.
Visit the AMS home page at URL: http://www.ams.org/
  13 12 11 10 9 8 7 6 5 4   27 26 25 24 23 22

Contents

Chapter 4. Coherent Sheaves	1
4.1. Exact Sequence of Sheaves	2
4.2. Quasicoherent Sheaves and Coherent Sheaves	16
4.3. Direct Image and Inverse Image	36
4.4. Schemes and Quasicoherent Sheaves	44
Summary	49
Exercises	50
Chapter 5. Proper and Projective Morphisms	53
5.1. Proper Morphisms	53
5.2. Quasicoherent Sheaves over a Projective Scheme	67
5.3. Projective Morphisms	91
Summary	106
Exercises	107
Chapter 6. Cohomology of Coherent Sheaves	111
6.1. Cohomology of Sheaves	111
6.2. Cohomology of a Projective Scheme	138
6.3. Higher Direct Image	153
Summary	158
Exercises	159
Solutions to Problems	161
Chapter 4	161
Chapter 5	166
Chapter 6	170
Solutions to Exercises	173
Chapter 4	173
Chapter 5	177
Chapter 6	181
Index	183

CHAPTER 4

Coherent Sheaves

In this chapter we will discuss the most important concept in algebraic geometry: coherent sheaves. Kiyoshi Oka introduced the concept of an ideal with indeterminate domains (that is, the stalk $\mathcal{O}_{X,x}$ at x of a sheaf $\mathcal{O}_X$ of holomorphic functions) and discovered the important properties of the ideal of indeterminate domains. H. Cartan recognized that Oka's work essentially coincided with the notion of Leray's sheaf. Consequently, by introducing the concept of a coherent sheaf, Cartan expressed Oka's results as the coherency of the sheaf of holomorphic functions. Cartan and J.-P. Serre reinterpreted the main results in the theory of holomorphic functions of several complex variables in terms of coherent sheaves. Grothendieck's scheme theory is the ultimate result of Serre's plan. These historical events indicate how important coherent sheaves are. For the applications of coherent sheaves to schemes, we find it more convenient to generalize the notion of a coherent sheaf to that of a quasicoherent sheaf. Following Grothendieck, we will begin with the theory of quasicoherent sheaves. Note that sheaves which are neither coherent nor quasicoherent play an important role in algebraic geometry.

We have briefly described the theory of sheaves. In this chapter we will establish the fundamental properties of sheaves on schemes in detail. Sheaf theory requires one entire book for its full treatment. Consequently, books on algebraic geometry cover sheaf theory only by giving necessary definitions, and then proceed to the next topic. Our intention is to describe sheaves in as much detail as possible. It is often the case that a lack of understanding of sheaf theory causes students to have difficulties grasping algebraic geometry. In order to understand sheaf theory, it is important to witness sheaves in action. For this purpose, we will provide many examples.

The sheaf induced by an R-module M over $\operatorname{Spec} R$ is denoted as either $\widetilde{M}$ or $(M)^\sim$. When the description for M is long, we will use the latter notation.

4.1. Exact Sequence of Sheaves

A homomorphism of sheaves has been described in §2.3(a). We will discuss it fully in this section. We will define the kernel, the image, and the cokernel of a sheaf homomorphism, and show that the notion of sheaves naturally generalizes that of additive groups (abelian groups). All of our sheaves or presheaves are assumed to be sheaves or presheaves of additive groups.

(a) **Sheafification of Presheaves**

We will review the construction of a sheaf of a presheaf $\mathcal{G}$ over a topological space X (see Exercise 2.5). Define the stalk $\mathcal{G}_x$ of $\mathcal{G}$ at $x \in X$ as follows:

$$\mathcal{G}_x = \varinjlim_{x \in U} \mathcal{G}(U).$$

The right-hand side is the inductive limit (direct limit) over $\mathcal{U}_x$ of all the open sets containing x where the order $U < V$ is defined by $V \subset U$. Let $^a\mathcal{G}(U)$ be the collection (totality) of maps s from U to $\bigcup_{x \in U} \mathcal{G}_x$ satisfying the following:

(1) $s(x) \in \mathcal{G}_x$ for $x \in U$.

(2) For each $x \in U$, one can choose an open set $V \subset U$ containing x, and $t \in \mathcal{G}(V)$ so that the germ $t_y \in \mathcal{G}_y$ of t at an arbitrary point y in V coincides with $s(y)$. Namely, define

(4.1)
$$^a\mathcal{G}(U) = \left\{ \{s(x)\} \in \prod_{x \in U} \mathcal{G}_x \;\middle|\; \begin{array}{l} \text{an open neighborhood } V \subset U \text{ and} \\ t \in \mathcal{G}(U) \text{ can be chosen so that} \\ t_y = s(y),\ y \in V \end{array} \right\}.$$

The restriction map $\rho_{V,U} : {}^a\mathcal{G}(U) \to {}^a\mathcal{G}(V)$ is defined by restricting $\{s(x)\}_{x \in U}$ to $\{s(y)\}_{y \in V}$. Then $^a\mathcal{G}$ is a sheaf of additive groups over X.

PROBLEM 1. Prove that $^a\mathcal{G}$ is a sheaf of additive groups.

PROBLEM 2. Prove that the sheaf $^a\mathcal{G}$ coincides with the sheaf $\widetilde{\mathcal{G}}$ as defined in Exercise 2.5.

For presheaves $\mathcal{G}$ and $\mathcal{H}$ of additive groups over a topological space X, a *homomorphism* $\varphi : \mathcal{G} \to \mathcal{H}$ is defined as follows. For each open set U in X, a homomorphism $\varphi_U : \mathcal{G}(U) \to \mathcal{H}(U)$ is defined satisfying the compatibility condition, i.e., for open sets $V \subset U$ the diagram

$$\begin{array}{ccc} \mathcal{G}(U) & \xrightarrow{\varphi_U} & \mathcal{H}(U) \\ {\scriptstyle \rho^{\mathcal{G}}_{V,U}} \downarrow & & \downarrow {\scriptstyle \rho^{\mathcal{H}}_{V,U}} \\ \mathcal{G}(V) & \xrightarrow{\varphi_V} & \mathcal{H}(V) \end{array}$$

commutes. Namely, φ is a natural transformation from $\mathcal{G}$ to $\mathcal{H}$ (see §3.1). In particular, when either $\mathcal{G}$ or $\mathcal{H}$ is a sheaf, we can consider a presheaf homomorphism from $\mathcal{G}$ to $\mathcal{H}$. Then a sheaf homomorphism coincides with a presheaf homomorphism. Let $\mathrm{Hom}_{\mathrm{sheaf}}(\mathcal{G}, \mathcal{H})$ be the totality of homomorphisms from a sheaf $\mathcal{G}$ to a sheaf $\mathcal{H}$, and let $\mathrm{Hom}_{\mathrm{presheaf}}(\mathcal{G}, \mathcal{H})$ be the totality of homomorphisms from a presheaf $\mathcal{G}$ to a presheaf $\mathcal{H}$. That is, we have

$$\mathrm{Hom}_{\mathrm{sheaf}}(\mathcal{G}, \mathcal{H}) = \mathrm{Hom}_{\mathrm{presheaf}}(\mathcal{G}, \mathcal{H}).$$

Note also that $\mathrm{Hom}_{\mathrm{presheaf}}(\mathcal{G}, \mathcal{H})$ and $\mathrm{Hom}_{\mathrm{sheaf}}(\mathcal{G}, \mathcal{H})$ are additive groups. Namely, for $\varphi : \mathcal{G} \to \mathcal{H}$ and $\psi : \mathcal{G} \to \mathcal{H}$, and for each open set U of X, define $\varphi + \psi$ as follows. For $s \in \mathcal{G}(U)$,

$$(\varphi + \psi)_U(s) \equiv \varphi_U(s) + \psi_U(s).$$

The zero element is the zero map 0, i.e., for each open set U and $s \in \mathcal{G}(U)$ the zero map is a homomorphism satisfying

$$O_U(s) = 0.$$

A presheaf (or sheaf) homomorphism $\varphi : \mathcal{G} \to \mathcal{H}$ is said to be an *isomorphism* if for each open set U of X, φ_U is an isomorphism. Then presheaves (or sheaves) $\mathcal{G}$ and $\mathcal{H}$ are said to be *isomorphic*.

For a presheaf $\mathcal{G}$ over a topological space X, we constructed a sheaf ${}^a\mathcal{G}$. Then, for an open set U in X, one can define a homomorphism of additive groups

(4.2)
$$\begin{aligned} \theta_U : \mathcal{G}(U) &\to {}^a\mathcal{G}(U), \\ t &\mapsto \{t_x\}_{x \in U}, \end{aligned}$$

where t_x is the germ of $t \in \mathcal{G}(U)$ at x. From the definition (4.1) and the definition of the restriction map of ${}^a\mathcal{G}$, it is clear that (4.2) defines a presheaf homomorphism $\theta : \mathcal{G} \to {}^a\mathcal{G}$.

The following proposition characterizes the sheaf ${}^a\mathcal{G}$ and the homomorphism $\theta : \mathcal{G} \to {}^a\mathcal{G}$.

PROPOSITION 4.1. (i) *For presheaves $\mathcal{G}$ and $\mathcal{H}$ of additive groups over a topological space X, the map determined by the sheaf $^a\mathcal{G}$ of additive groups and the homomorphism of presheaves in* (4.2)

(4.3)
$$\mathrm{Hom}_{\mathrm{sheaf}}(^a\mathcal{G}, \mathcal{H}) \to \mathrm{Hom}_{\mathrm{presheaf}}(\mathcal{G}, \mathcal{H}),$$
$$\varphi \mapsto \varphi \circ \theta,$$

is an isomorphism of additive groups. Conversely, for a presheaf $\mathcal{G}$ of additive groups, the sheaf $^a\mathcal{G}$ and the homomorphism of presheaves $\theta : \mathcal{G} \to {}^a\mathcal{G}$ are determined uniquely up to isomorphisms.

(ii) *When $\mathcal{G}$ is a sheaf, we have $^a\mathcal{G} = \mathcal{G}$.*

PROOF. By the definition of the map in (4.3), the map is a homomorphism of additive groups. For $^a\mathcal{G}$ as constructed above, we define a map

(4.4) $$\mathrm{Hom}_{\mathrm{presheaf}}(\mathcal{G}, \mathcal{H}) \to \mathrm{Hom}_{\mathrm{sheaf}}(^a\mathcal{G}, \mathcal{H}).$$

For a given homomorphism of presheaves $\psi : \mathcal{G} \to \mathcal{H}$, we will show that for an open set U, one can construct a homomorphism of additive groups $^a\psi_U : {}^a\mathcal{G}(U) \to \mathcal{H}(U)$ making the diagram

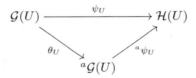

commutative. From (4.1), for $s = \{s(x)\} \in {}^a\mathcal{G}(U)$, one can find an open covering $\{V_\lambda\}_{\lambda \in \Lambda}$ of U and $t_\lambda \in \mathcal{G}(V_\lambda)$ so that $s(y) \equiv t_{\lambda_y}$, $y \in V_\lambda$. Let $\tilde{t}_\lambda = \varphi_{V_\lambda}(t_\lambda)$. Then we have $\tilde{t}_\lambda \in \mathcal{H}(V_\lambda)$. For $V_{\lambda\mu} = V_\lambda \cap V_\mu \neq \varnothing$, we have $\rho_{V_{\lambda\mu},V_\lambda}(\tilde{t}_\lambda) = \rho_{V_{\lambda\mu},V_\mu}(\tilde{t}_\mu)$ since $\rho_{V_{\lambda\mu},V_\lambda}(t_\lambda) = \rho_{V_{\lambda\mu},V_\mu}(t_\mu)$. Since $\mathcal{H}$ is a sheaf, property (F2) in §2.2 implies that $\tilde{t} \in \mathcal{H}(U)$ exists satisfying $\rho_{V_\lambda,U}(\tilde{t}) = t_\lambda$, $\lambda \in \Lambda$. Furthermore, from (F1) such a $\tilde{t}$ is uniquely determined. Then define $^a\psi_U(s) = \tilde{t}$. For $t \in \mathcal{G}(U)$, let $s = \{t_x\} \in {}^a\mathcal{G}(U)$. We have $\theta_U(t) = s$ and $\tilde{t} = \psi_U(t)$. Namely, we obtain $^a\psi_U \circ \theta_U = \psi_U$. Consequently, the above diagram is commutative. Therefore, $^a\psi \circ \theta = \psi$ implies that the map (4.3) is surjective.

Conversely, suppose $\psi = \varphi \circ \theta, \varphi \in \mathrm{Hom}_{\mathrm{sheaf}}(^a\mathcal{G}, \mathcal{H})$. Then from the construction of $^a\psi$ in the above we get $^a\psi = \varphi$. Hence, the map (4.3) is injective, and so it is an isomorphism. (The map (4.4) is the inverse map of (4.3).)

Assume that for a sheaf $\mathcal{F}$ and for a presheaf homomorphism $\eta : \mathcal{G} \to \mathcal{F}$, the map
$$\text{Hom}_{\text{sheaf}}(\mathcal{F}, \mathcal{H}) \to \text{Hom}_{\text{presheaf}}(\mathcal{G}, \mathcal{H}),$$
$$\varphi \mapsto \varphi \circ \eta,$$
is an isomorphism, where $\mathcal{H}$ is an arbitrary sheaf. In particular, for $\mathcal{H} = {}^a\mathcal{G}$ and $\theta : \mathcal{G} \to {}^a\mathcal{G}$, a sheaf homomorphism $\varphi : \mathcal{F} \to {}^a\mathcal{G}$ is uniquely determined to satisfy $\theta = \varphi \circ \eta$. In (4.3) let $\mathcal{H} = \mathcal{F}$. Then a sheaf homomorphism $\psi : {}^a\mathcal{G} \to \mathcal{F}$ is uniquely determined to satisfy $\eta = \psi \circ \theta$. We have $\theta = \varphi \circ \eta = \theta \circ (\psi \circ \theta) = (\varphi \circ \psi) \circ \theta$, which implies that, for a sheaf homomorphism $\varphi \circ \psi : {}^a\mathcal{G} \to {}^a\mathcal{G}$, if $\mathcal{H} = {}^a\mathcal{G}$ in (4.3), $\varphi \circ \psi$ corresponds to $\theta \in \text{Hom}_{\text{presheaf}}(\mathcal{G}, \mathcal{H})$. On the other hand, $\text{id}_{{}^a\mathcal{G}}$ also corresponds to θ in (4.3). Since (4.3) is an isomorphism, we have $\varphi \circ \psi = \text{id}_{{}^a\mathcal{G}}$. Furthermore, $\eta = \psi \circ \theta = \psi \circ (\varphi \circ \eta) = (\psi \circ \varphi) \circ \eta$ implies that the isomorphism (4.5) provides $\psi \circ \varphi = \text{id}_{\mathcal{F}}$ as well. Therefore, $\psi : {}^a\mathcal{G} \to \mathcal{F}$ is an isomorphism, and $\eta = \psi \circ \theta$ implies that $({}^a\mathcal{G}, \theta)$ is uniquely determined up to an isomorphism to satisfy (4.3).

(ii) If $\mathcal{G}$ is a sheaf, then by properties (F1) and (F2), the definition of ${}^a\mathcal{G}$ in (4.1) implies ${}^a\mathcal{G}(U) = \mathcal{G}(U)$. □

As was mentioned in Exercise **2.5**, ${}^a\mathcal{G}$ is said to be *sheafification* of a presheaf $\mathcal{G}$. The above proposition characterizes the sheafification by the universal mapping property.

EXAMPLE 4.2. For a presheaf $\mathcal{G}$ and its sheafification ${}^a\mathcal{G}$ over a topological space X, their stalks $\mathcal{G}_x$ and ${}^a\mathcal{G}_x$ at x coincide.

PROOF. For an open set U containing $x \in X$ and $s \in {}^a\mathcal{G}(U)$, one can find $x \in V \subset U$ and $t \in \mathcal{G}(V)$ so that $\rho_{V,U}(s) = \theta_U(t)$. Hence, we get $s_x = t_x$. □

(b) **Kernels and Cokernels of Sheaf Homomorphisms**

For a given homomorphism of sheaves of additive groups $\varphi : \mathcal{G} \to \mathcal{H}$ over a topological space X, we will define the kernel and the cokernel of φ as sheaves of additive groups. We will begin with the kernel, which is simpler than the cokernel.

EXERCISE 4.3. Let $\varphi : \mathcal{G} \to \mathcal{H}$ be a sheaf homomorphism over a topological space X. For an open set U of X, define

(4.5) $$\mathcal{F}(U) = \{s \in \mathcal{G}(U) | \varphi_U(s) = 0\}.$$

Then $\mathcal{F}$ is a sheaf of additive groups over X.

PROOF. For open sets $V \subset U$, let $\rho^{\mathcal{G}}_{V,U}$ and $\rho^{\mathcal{H}}_{V,U}$ be the restriction maps of sheaves $\mathcal{G}$ and $\mathcal{H}$, respectively. From the definition of a sheaf homomorphism, we obtain the commutative diagram

$$\begin{array}{ccc} \mathcal{G}(U) & \xrightarrow{\varphi_U} & \mathcal{H}(U) \\ \rho^{\mathcal{G}}_{V,U} \downarrow & & \downarrow \rho^{\mathcal{H}}_{V,U} \\ \mathcal{G}(V) & \xrightarrow{\varphi_V} & \mathcal{H}(V) \end{array}$$

Then we have $\rho^{\mathcal{G}}_{V,U}(\mathcal{F}(U)) \subset \mathcal{F}(V)$. Define the restriction map $\rho_{V,U}$ for $\mathcal{F}$ as the restriction of $\rho^{\mathcal{G}}_{V,U}$ to $\mathcal{F}(U)$. Then $\mathcal{F}$ is a presheaf of additive groups over X.

We will show that $\mathcal{F}$ satisfies the sheaf properties (F1) and (F2). Since $\mathcal{G}$ is a sheaf, $\mathcal{F}$ clearly satisfies (F1). For an open set U in X, let $\{U_j\}_{j \in I}$ be an open covering of U. Suppose that for $U_{jk} = U_j \cap U_k \neq \varnothing$ we have $\rho_{U_{jk}, U_j}(s_j) = \rho_{U_{jk}, U_k}(s_k)$ for $s_j \in \mathcal{F}(U_j)$, $j \in I$. By regarding $s_j \in \mathcal{G}(U_j)$, there is $s \in \mathcal{G}(U)$ such that $\rho^{\mathcal{G}}_{U_j, U}(s) = s_j$, $j \in I$. Let $t = \varphi_U(s)$ and $t_j = \rho^{\mathcal{H}}_{U_j, U}(t)$. Then $t_j = \varphi_{U_j}(\rho^{\mathcal{G}}_{U_j, U}(s)) = \varphi_{U_j}(s_j) = 0$. Since $\mathcal{H}$ is a sheaf, we get $t = 0$. Therefore $s \in \mathcal{F}(U)$, satisfying (F2). □

The sheaf $\mathcal{F}$ defined by (4.5) is said to be the *kernel* of $\varphi : \mathcal{G} \to \mathcal{H}$ and is denoted by $\operatorname{Ker} \varphi$.

In general, for sheaves $\mathcal{F}$ and $\mathcal{G}$, if $\mathcal{F}(U)$ is an additive subgroup of $\mathcal{G}(U)$, and if the restriction map $\rho_{V,U} : \mathcal{F}(U) \to \mathcal{F}(V)$ is obtained by restricting $\rho^{\mathcal{G}}_{V,U}$ to $\mathcal{F}(U)$, then $\mathcal{F}$ is said to be a *subsheaf* of $\mathcal{G}$. The above $\operatorname{Ker} \varphi$ is an example of a subsheaf of $\mathcal{G}$. For a sheaf homomorphism $\varphi : \mathcal{G} \to \mathcal{H}$, we will see whether

$$\mathcal{I}(U) = \operatorname{Im} \varphi_U = \{\varphi_U(s) \in \mathcal{H}(U) | s \in \mathcal{G}(U)\}$$

is a subsheaf of $\mathcal{H}$ or not, where U is an open set of X. For $V \subset U$ and for $t \in \mathcal{I}(U)$, there is an element $s \in \mathcal{G}(U)$ satisfying $t = \varphi_U(s)$. Hence, we get

$$\rho^{\mathcal{H}}_{V,U}(t) = \rho^{\mathcal{H}}_{V,U}(\varphi_U(s)) = \varphi_V(\rho^{\mathcal{G}}_{V,U}(s)) \in \mathcal{I}(V).$$

Namely, $\mathcal{I}$ becomes a presheaf whose restriction map is obtained from that of $\mathcal{H}$. We will examine (F1) and (F2) for $\mathcal{I}$. For an open covering $\{U_j\}_{j \in J}$ of U, if $t \in \mathcal{I}(U)$ satisfies $\rho_{U_i, U}(t) = 0$, then as an element of $\mathcal{H}(U)$ we have $t = 0$. Hence, as an element of $\mathcal{I}(U)$, we get $t = 0$. Namely, $\mathcal{I}$ satisfies (F1).

4.1. EXACT SEQUENCE OF SHEAVES

Suppose, for an open covering $\{U_j\}_{j\in J}$ of U, that $t_j \in \mathcal{I}(U)$, $j \in J$, satisfy $\rho_{U_{jk},U_j}(t_j) = \rho_{U_{jk},U_k}(t_k)$, where $U_{jk} = U_j \cap U_k \neq \varnothing$. Since $\mathcal{H}$ is a sheaf, by regarding $t_j \in \mathcal{H}(U_j)$, $j \in J$, there is $t \in \mathcal{H}(U)$ satisfying $\rho^{\mathcal{H}}_{U_j,U}(t) = t_j$. Is $t \in \mathcal{I}(U)$? Namely, we ask whether there exists $s \in \mathcal{G}(U)$ satisfying $\varphi_U(s) = t$.

Since $t_j \in \mathcal{I}(U_j)$, there exists $s_j \in \mathcal{G}(U_j)$ such that $t_j = \varphi_{U_j}(s_j)$. Notice that s_j is not uniquely determined. If $t_j = \varphi_{U_j}(s'_j)$, $s'_j \in \mathcal{G}(U_j)$, then $\varphi_{U_j}(s_j - s'_j) = 0$, i.e., $s_j - s'_j \in (\text{Ker}\,\varphi)(U_j)$. That is, for any $u_j \in (\text{Ker}\,\varphi)(U_j)$, we have $t_j = \varphi_{U_j}(s_j) = \varphi_{U_j}(s_j + u_j)$. For each $j \in J$, choose one element $s_j \in \mathcal{G}(U_j)$ satisfying $t_j = \varphi_{U_j}(s_j)$. Then we would like to choose $u_j \in (\text{Ker}\,\varphi)(U_j)$ so that $\tilde{s}_j = s_j + u_j$ may satisfy $\rho^{\mathcal{G}}_{U_{jk},U_j}(\tilde{s}_j) = \rho^{\mathcal{G}}_{U_{jk},U_k}(\tilde{s}_k)$ for $U_{jk} = U_j \cap U_k \neq \varnothing$. If such a choice is possible, then there is $\tilde{s} \in \mathcal{G}(U)$ such that $\rho^{\mathcal{G}}_{U_j,U}(\tilde{s}) = \tilde{s}_j$, and $t_j = \varphi_{U_j}(\tilde{s}_j)$ would give $t = \varphi_U(\tilde{s})$. Thus, the question is if such $\{u_j\}_{j\in J}$ exist or not. This problem depends upon the sheaf $\text{Ker}\,\varphi$. Let us explain this situation. For $U_{jk} = U_j \cap U_k \neq \varnothing$, let

(4.6) $$s_{jk} = \rho_{U_{jk},U_k}(s_k) - \rho_{U_{jk},U_j}(s_j) \in \mathcal{G}(U_{jk}).$$

Then $t_j = \varphi_{U_j}(s_j), t_k = \varphi_{U_k}(s_k)$ and $\rho^{\mathcal{H}}_{U_{jk},U_j}(t_j) = \rho^{\mathcal{H}}_{U_{jk},U_k}(t_k)$ imply $s_{jk} \in (\text{Ker}\,\varphi)(U_{jk})$. Therefore, for $\{s_{jk}\}$ the question becomes the following. Are there $u_j \in (\text{Ker}\,\varphi)(U_j)$, $j \in J$, such that we get

$$s_{jk} = u_k - u_j, \quad U_{jk} \neq \varnothing?$$

Namely, it is a problem about the sheaf $\text{Ker}\,\varphi$. We have that the s_{jk} satisfy

$$s_{kj} = -s_{jk}$$

and for $U_{ijk} = U_i \cap U_j \cap U_k \neq \varnothing$

(4.7) $$\rho_{U_{ijk},U_{ij}}(s_{ij}) + \rho_{U_{ijk},U_{jk}}(s_{jk}) + \rho_{U_{ijk},U_{ki}}(s_{ki}) = 0.$$

Then $\{s_{jk}\}$ is said to be a *one-cocycle*, which will play an important role for *sheaf cohomology*. We will return to sheaf cohomology in Chapter 6. The notion of the image of a sheaf homomorphism has brought us into sheaf cohomology, indicating that sheaf cohomology theory is a naturally needed concept.

A natural example of a presheaf which does not satisfy (F2) is a cokernel. The sheaf associated to the presheaf $\mathcal{I}$ is denoted as $\text{Im}\,\varphi$, and $\text{Im}\,\varphi$ is said to be the *image* of $\varphi : \mathcal{G} \to \mathcal{H}$.

PROBLEM 3. Prove that the s_{jk}, defined by (4.6), satisfy (4.7).

We will define the cokernel of a homomorphism $\varphi : \mathcal{G} \to \mathcal{H}$. For each open set U of X, define

(4.8) $$\mathcal{C}(U) = \operatorname{Coker} \varphi_U = \mathcal{H}(U)/\operatorname{Im} \varphi_U,$$

which is clearly a presheaf of additive groups. The restriction map $\rho_{V,U}$ is induced naturally from the restriction map $\rho^{\mathcal{H}}_{V,U}$, i.e.,

$$\rho_{V,U}(t \bmod \varphi_U(\mathcal{G}(U))) = \rho^{\mathcal{H}}_{V,U}(t) \bmod \varphi_V(\mathcal{G}(V)).$$

The following example will show that $\mathcal{C}$ need not be a sheaf.

EXAMPLE 4.4. Let $\mathbb{P}^1_k$ be the one-dimensional projective space over a field k (see Example 2.31). By Definition (c) of §2.3, $\mathbb{P}^1_k = \operatorname{Proj} k[x_0, x_1]$. Let $x = x_1/x_0$ and $y = x_0/x_1$. Then affine lines

$$U_0 = \operatorname{Spec} k[x] \quad \text{and} \quad U_1 = \operatorname{Spec} k[y]$$

form an open covering for $\mathbb{P}^1_k$. The points, denoted as $\mathfrak{p}_0$ and $\mathfrak{p}_\infty$, are determined by homogeneous ideals $\mathfrak{p}_0 = (x_1)$ and $\mathfrak{p}_\infty(x_0)$ in $k[x_0, x_1]$, respectively. Then $\mathfrak{p}_0$ is contained in U_0, which is the point on $\operatorname{Spec} k[x]$ determined by the ideal (x) of $k[x]$. Namely, $\mathfrak{p}_0$ is the origin of the affine line $\operatorname{Spec} k[x]$. On the other hand, we have $\mathfrak{p}_\infty \in U_\infty$, which is the origin of the affine line $\operatorname{Spec} k[y]$. Define a subsheaf $\mathcal{J}$ of $\mathcal{O}_{\mathbb{P}^1_k}$ as follows:

$$\mathcal{J}(U) = \begin{cases} \mathcal{O}_{\mathbb{P}^1_k}(U) \text{ if } \mathfrak{p}_0, \mathfrak{p}_\infty \notin U, \\ \left\{ s \in \mathcal{O}_{\mathbb{P}^1_k}(U) \;\middle|\; \begin{array}{l} s(\mathfrak{p}_0) = 0 \text{ for } \mathfrak{p}_0 \in U, \\ s(\mathfrak{p}_\infty) = 0 \text{ for } \mathfrak{p}_\infty \in U \end{array} \right\} \end{cases}.$$

It is clear that $\mathcal{J}$ is a subsheaf of $\mathcal{O}_{\mathbb{P}^1_k}$. A natural homomorphism $\iota : \mathcal{J} \to \mathcal{O}_{\mathbb{P}^1_k}$ is induced by $\mathcal{J}(U) \subset \mathcal{O}_{\mathbb{P}^1_k}(U)$. Notice also that for $U_0 = \operatorname{Spec} k[x]$ we get

$$\mathcal{O}_{\mathbb{P}^1_k}(U_0) = k[x] \quad \text{and} \quad \mathcal{J}(U_0) = (x),$$

and for $U_1 = \operatorname{Spec} k[y]$, we get

$$\mathcal{O}_{\mathbb{P}^1_k}(U_1) = k[y] \quad \text{and} \quad \mathcal{J}(U_1) = (y).$$

Next we will show that

$$\Gamma(\mathbb{P}^1_k, \mathcal{O}_{\mathbb{P}^1_k}) = \mathcal{O}_{\mathbb{P}^1_k}(\mathbb{P}^1_k) = k.$$

For $f \in \mathcal{O}_{\mathbb{P}^1_k}(\mathbb{P}^1_k)$, let $F = \rho_{U_0, \mathbb{P}^1_k}(f)$ and $G = \rho_{U_1, \mathbb{P}^1_k}(f)$. Then $f \in k[x]$ and $G \in k[y]$. We also get $U_{01} = U_0 \cap U_1 = \operatorname{Spec} k[x, \frac{1}{x}]$ and $\rho_{U_{01}, U_0}(F) = \rho_{U_{01} U_1}(G)$, where in U_{01}, y (more precisely $\rho_{U_{01}, U_1}(y)$) equals $1/x$. Hence $F(x) = G(1/x)$. Since F and G are polynomials

in x and y, respectively, F and G are the same constant. Namely, we get $f \in k$. Unless $f = 0$, $f \in k$ does not become 0 at $\mathfrak{p}_0$ and $\mathfrak{p}_\infty$. Therefore,
$$\mathcal{J}(\mathbb{P}^1_k) = 0.$$

Under the above preparation, we will study the presheaf $\mathcal{C}$ determined by the homomorphism $\iota : \mathcal{J} \to \mathcal{O}_{\mathbb{P}^1_k}$. Since for each open set U of $\mathbb{P}^1_k$ we have
$$\mathcal{C}(U) = \operatorname{Coker} \iota_U = \mathcal{O}_{\mathbb{P}^1_k}(U)/\mathcal{J}(U),$$
in particular we get
$$\mathcal{C}(U_0) = k[x]/(x) \overset{\sim}{\to} k \quad \text{and} \quad \mathcal{C}(U_1) = k[y]/(y) \overset{\sim}{\to} k.$$
That is, we regard $\mathcal{C}(U_0) = k$ and $\mathcal{C}(U_1) = k$. Since $\mathfrak{p}_0 \notin U_{01}$ and $\mathfrak{p}_\infty \notin U_{01}$, we have $\mathcal{J}(U_{01}) = \mathcal{O}_{\mathbb{P}^1_k}(U_{01})$. Consequently,
$$\mathcal{C}(U_{01}) = 0.$$
For the open covering $\{U_0, U_1\}$ of $\mathbb{P}^1_k$, let a and b be arbitrary elements in $k = \mathcal{C}(U_0)$ and $k = \mathcal{C}(U_1)$, respectively. Then
$$\rho_{U_{01}, U_0}(a) = 0 \quad \text{and} \quad \rho_{U_{01}, U_1}(b) = 0.$$
In particular, for the case where $a \neq b$, since
$$\mathcal{C}(\mathbb{P}^1_k) = \mathcal{O}_{\mathbb{P}^1_k}(\mathbb{P}^1_k)/\mathcal{J}(\mathbb{P}^1_k) = k,$$
one cannot find $f \in \mathcal{C}(\mathbb{P}^1_k)$ to satisfy $\rho_{U_0, \mathbb{P}^1_k}(f) = a$ and $\rho_{U_1, \mathbb{P}^1_k}(f) = b$. Therefore, the presheaf $\mathcal{C}$ does not satisfy (F2), i.e., $\mathcal{C}$ is not a sheaf. □

PROBLEM 4. For an n-dimensional projective space
$$\mathbb{P}^n_k = \operatorname{Proj} k[x_0, x_1, \ldots, x_n]$$
(see Example 2.32), show that
$$\Gamma(\mathbb{P}^n_k, \mathcal{O}_{\mathbb{P}^n_k}) = k.$$

Thus, for a sheaf homomorphism $\varphi : \mathcal{G} \to \mathcal{H}$, we have the associated sheaf $\operatorname{Coker} \varphi$, i.e., the sheafification of the presheaf $\mathcal{C}$ defined as in (4.8). The sheaf $\operatorname{Coker} \varphi$ is said to be the *cokernel* of φ. As was explained, the image and cokernel of a sheaf homomorphism need to be sheafified. However, at stalks those associated sheaves are as simple as additive groups.

THEOREM 4.5. *Let $\varphi_x : \mathcal{G}_x \to \mathcal{H}_x$ be the induced homomorphism of stalks at x by a homomorphism $\varphi : \mathcal{G} \to \mathcal{H}$ of sheaves of additive groups. Then the stalks $(\operatorname{Ker}\varphi)_x, (\operatorname{Im}\varphi)_x, (\operatorname{Coker}\varphi)_x$ of $\operatorname{Ker}\varphi, \operatorname{Im}\varphi, \operatorname{Coker}\varphi$, respectively, at x coincide with the kernel, image, and cokernel of an additive group homomorphism φ_x. Namely, we have*

$$(\operatorname{Ker}\varphi)_x = \operatorname{Ker}\varphi_x,$$
$$(\operatorname{Im}\varphi)_x = \operatorname{Im}\varphi_x = \varphi_x(\mathcal{G}_x),$$
$$(\operatorname{Coker}\varphi)_x = \operatorname{Coker}\varphi_x = \mathcal{H}_x/\varphi_x(\mathcal{G}_x).$$

PROOF. For an open set U of X, let

$$\mathcal{F}(U) = \operatorname{Ker}\{\varphi_U : \mathcal{G}(U) \to \mathcal{H}(U)\},$$
$$\mathcal{I}(U) = \varphi_V(\mathcal{G}(U)),$$
$$\mathcal{C}(U) = \mathcal{H}(U)/\mathcal{I}(U).$$

Then we have the following exact sequences of additive groups:

$$0 \to \mathcal{F}(U) \to \mathcal{G}(U) \to \mathcal{I}(U) \to 0,$$
$$0 \to \mathcal{I}(U) \to \mathcal{H}(U) \to \mathcal{C}(U) \to 0.$$

An inductive limit preserves exactness (see Problem 5, below). Hence, we get

$$0 \to \varinjlim_{x \in U} \mathcal{F}(U) \to \varinjlim_{x \in U} \mathcal{G}(U) \to \varinjlim_{x \in U} \mathcal{I}(U) \to 0,$$
$$0 \to \varinjlim_{x \in U} \mathcal{I}(U) \to \varinjlim_{x \in U} \mathcal{H}(U) \to \varinjlim_{x \in U} \mathcal{C}(U) \to 0.$$

Then Example 4.2 implies that the above exact sequences become

$$0 \to (\operatorname{Ker}\varphi)_x \to \mathcal{G}_x \to (\operatorname{Im}\varphi)_x \to 0,$$
$$0 \to (\operatorname{Im}\varphi)_x \to \mathcal{H}_x \to (\operatorname{Coker}\varphi)_x \to 0.$$

Since φ_x is precisely the map $\mathcal{G}_x \to (\operatorname{Im}\varphi)_x \to \mathcal{H}_x$, which is obtained from the above exact sequences, the proof is completed. □

PROBLEM 5. Let Λ be a preordered set and let $\{L_\lambda, f_{\mu\lambda}\}$, $\{M_\lambda, g_{\mu\lambda}\}$ and $\{N_\lambda, h_{\mu\lambda}\}$ be inductive systems of additive groups indexed by Λ. For an exact sequence

$$0 \to L_\lambda \to M_\lambda \to N_\lambda \to 0, \quad \lambda \in \Lambda,$$

satisfying the commutativity of the diagram

$$\begin{array}{ccccccccc} 0 & \longrightarrow & L_\lambda & \longrightarrow & M_\lambda & \longrightarrow & N_\lambda & \longrightarrow & 0 \\ & & \downarrow f_{\mu\lambda} & & \downarrow g_{\mu\lambda} & & \downarrow h_{\mu\lambda} & & \\ 0 & \longrightarrow & L_\mu & \longrightarrow & M_\mu & \longrightarrow & N_\mu & \longrightarrow & 0, \end{array}$$

prove that the sequence

$$0 \to \varinjlim_{\lambda \in \Lambda} L_\lambda \to \varinjlim_{\lambda \in \Lambda} M_\lambda \to \varinjlim_{\lambda \in \Lambda} N_\lambda \to 0$$

is exact.

Let $\mathcal{F}$ be a subsheaf of a sheaf $\mathcal{G}$. For the natural map $\iota : \mathcal{F} \to \mathcal{G}$, denote its cokernel by $\mathcal{G}/\mathcal{F}$. The sheaf $\mathcal{G}/\mathcal{F}$ is said to be the *quotient sheaf* of $\mathcal{G}$ by the subsheaf $\mathcal{F}$. The quotient sheaf $\mathcal{G}/\mathcal{F}$ is precisely the sheaf associated to the presheaf $\mathcal{G}(U)/\mathcal{F}(U)$ of additive groups.

EXAMPLE 4.6. Recall the subsheaf $\mathcal{J}$ of the structure sheaf $\mathcal{O}^1_{\mathbb{P}^k}$ of a one-dimensional projective space $\mathbb{P}^1_k$ over a field k (see Example 4.4). Since for $x \neq \mathfrak{p}_0, \mathfrak{p}_\infty$ we have $\mathcal{J}_x = \mathcal{O}_{\mathbb{P}^1_{k,x}}$, the associated sheaf $\mathcal{O}_{\mathbb{P}^1_k}/\mathcal{J}$ of the presheaf $\mathcal{C}$ in Example 4.4 has a zero stalk (i.e., trivial additive group) at x. Furthermore, since $\mathcal{J}_{\mathfrak{p}_0}$ is the ideal of $\mathcal{O}_{\mathbb{P}^1_k,\mathfrak{p}_0}$ generated at x, we obtain $(\mathcal{O}_{\mathbb{P}^1_k}/\mathcal{J})_{\mathfrak{p}_0} = \mathcal{O}_{\mathbb{P}^1_k,\mathfrak{p}_0}/\mathcal{J}_{\mathfrak{p}_0} \xrightarrow{\sim} k$. Therefore, the stalks of $\mathcal{O}_{\mathbb{P}^1_k}/\mathcal{J}$ at $\mathfrak{p}_0$ and $\mathfrak{p}_\infty$ are k, and 0 otherwise. This result gives an impression that skyscrapers stand only at $\mathfrak{p}_0$ and $\mathfrak{p}_\infty$. Hence, the sheaf $\mathcal{O}_{\mathbb{P}^1_k}/\mathcal{J}$ is sometimes called a *skyscraper sheaf*. We have $\Gamma(\mathbb{P}^1_k, \mathcal{O}_{\mathbb{P}^1_k}/\mathcal{J}) \xrightarrow{\sim} k \oplus k$. □

(c) **Exact Sequences**

Let

(4.9) $$\cdots \to \mathcal{F}_{i-1} \xrightarrow{\varphi_{i-1}} \mathcal{F}_i \xrightarrow{\varphi_i} \mathcal{F}_{i+1} \xrightarrow{\varphi_{i+1}} \cdots$$

be a sequence of sheaves $\mathcal{F}_i$ of additive groups over a topological space X, where each φ_i is an additive group homomorphism. When $\operatorname{Im}\varphi_{i-1} = \operatorname{Ker}\varphi_i$ for each i, the sequence (4.9) is said to be *exact*. Let 0 be the sheaf which assigns a trivial additive group for each open set of X.

For a sheaf homomorphism $\varphi : \mathcal{F} \to \mathcal{G}$, as in the case of additive groups, φ is said to be *injective* when $\operatorname{Ker}\varphi = 0$, and *surjective* when $\operatorname{Im}\varphi = \mathcal{G}$. In terms of exact sequences, $\varphi : \mathcal{F} \to \mathcal{G}$ is injective if the sequence

$$0 \to \mathcal{F} \to \mathcal{G}$$

is exact and φ is surjective if the sequence
$$\mathcal{F} \to \mathcal{G} \to 0$$
is exact. Furthermore, the sequence
$$(4.10) \qquad 0 \to \mathcal{F} \xrightarrow{\varphi} \mathcal{G} \xrightarrow{\psi} \mathcal{H} \to 0$$
is exact when φ is injective, ψ is surjective, and $\operatorname{Im}\varphi = \operatorname{Ker}\psi$. When (4.10) is exact, it is called a *short exact sequence*. Short exact sequences will appear often in this book.

In particular, if $\mathcal{F}$ is a subsheaf of $\mathcal{G}$, the inclusion $\mathcal{F}(U) \subset \mathcal{G}(U)$ induces the injective natural map $\iota : \mathcal{F} \to \mathcal{G}$. Then we have the exact sequence
$$0 \to \mathcal{F} \xrightarrow{\iota} \mathcal{G} \to \mathcal{G}/\mathcal{F} \to 0.$$
This and the exact sequence (4.10) imply that $\mathcal{H}$ is isomorphic to $\mathcal{G}/\mathcal{F}$.

PROBLEM 6. Prove that an injective and surjective homomorphism $\varphi : \mathcal{F} \to \mathcal{G}$ is an isomorphism, i.e., for each open set U, $\varphi_U : \mathcal{F}(U) \to \mathcal{G}(U)$ is an isomorphism.

From Theorem 4.5, we clearly obtain the following proposition.

PROPOSITION 4.7. *For sheaves of additive groups over a topological space X, the sequence (4.9) is exact if and only if the sequence of stalks at each x in X*
$$\cdots \to \mathcal{F}_{i-1,x} \xrightarrow{\varphi_{i-1,x}} \mathcal{F}_{i,x} \xrightarrow{\varphi_{i,x}} \mathcal{F}_{i+1,x} \xrightarrow{\varphi_{i+1,x}} \cdots$$
is an exact sequence of additive group homomorphisms. $\square$

PROPOSITION 4.8. *For an exact sequence*
$$0 \to \mathcal{F} \xrightarrow{\varphi} \mathcal{G} \xrightarrow{\psi} \mathcal{H}$$
of sheaves of additive groups over a topological space X and for each open set U of X, we have the following exact sequence of additive groups:
$$0 \to \Gamma(U, \mathcal{F}) \xrightarrow{\varphi_U} \Gamma(U, \mathcal{G}) \xrightarrow{\psi_U} \Gamma(U, \mathcal{H}).$$
However, ψ_U need not be surjective even if ψ is surjective.

PROOF. For the induced sequence of additive groups
$$0 \to \Gamma(U, \mathcal{F}) = \mathcal{F}(U) \xrightarrow{\gamma_U} \Gamma(U, \mathcal{G}) = \mathcal{G}(U) \xrightarrow{\psi_U} \Gamma(U, \mathcal{H}) = \mathcal{H}(U),$$

since φ is injective, the sheaf property (F1) implies the injectivity of φ_U. Consider $t \in \mathcal{G}(U)$ such that $\psi_U(t) = 0$. Since $\operatorname{Ker}\psi_x = \operatorname{Im}\varphi_x$, there is $s_x \in \mathcal{F}_x$ satisfying $\varphi_x(s_x) = t_x$. Since φ_x is injective, this s_x is uniquely determined. Choose an open neighborhood V of x and $s_V \in \mathcal{F}(V)$ so that the germ of s_V at x is s_x. Then there exists a small enough open neighborhood $W \subset V$ such that $\rho_{W,V}(\varphi_V(s_V)) = \rho_{W,V}(t)$. Thus, we get an open covering $\{U_j\}_{j \in J}$ of U and $s_j \mathcal{F}(U_j)$ so that $\varphi_{U_j}(s_j) = \rho_{U_j,U}(t)$. Since φ_{U_j} is injective, s_j is uniquely determined. On $U_{jk} = U_j \cap U_k \neq \varnothing$, $\varphi_{U_{jk}}$ is also injective. Hence we have $\rho_{U_{jk},U_j}(s_j) = \rho_{U_{jk},U_k}(s_k)$. Therefore, there exists $s \in \mathcal{F}(U)$ such that $\phi_{U_j,U}(s) = s_j$, $j \in J$. Namely, $\operatorname{Im}\varphi_U = \operatorname{Ker}\psi_U$.

From Example 4.4, we have the exact sequence
$$0 \to \mathcal{J} \to \mathcal{O}_{\mathbb{P}_k^1} \to \mathcal{O}_{\mathbb{P}_k^1}/\mathcal{J} \to 0.$$
As shown in Example 4.4, we have $\Gamma(\mathbb{P}_k^1, \mathcal{O}_{\mathbb{P}_k^1}) = k$. From Example 4.6, we have $\Gamma(\mathbb{P}_k^1, \mathcal{O}_{\mathbb{P}_k^1}/\mathcal{J}) = k \oplus k$. Hence the homomorphism
$$\Gamma(\mathbb{P}_k^1, \mathcal{O}_{\mathbb{P}_k^1}) \to \Gamma(\mathbb{P}_k^1, \mathcal{O}_{\mathbb{P}_k^1}/\mathcal{J})$$
cannot be surjective. Therefore, for a surjective ψ, ψ_U need not be surjective. $\square$

Hence, for a short exact sequence of sheaves
$$0 \to \mathcal{F} \to \mathcal{G} \to \mathcal{H} \to 0,$$
the sequence of sections over an open set U is guaranteed to be exact only as
$$0 \to \Gamma(U, \mathcal{F}) \to \Gamma(U, \mathcal{G}) \to \Gamma(U, \mathcal{H}).$$
This lack of exactness on the right necessitates the cohomology of sheaves, which makes algebraic geometry difficult and more interesting.

Note that, as explained above, for a surjective sheaf homomorphism $\psi : \mathcal{G} \to \mathcal{H}$ and for a section $t \in \mathcal{H}(U)$ over an open set U, at each point x in U there exists an open neighborhood $V \subset U$ of x satisfying $\psi_V(s_V) = \rho_{V,U}(t)$ for some $s_V \in \mathcal{G}(V)$. In general $V \neq U$, i.e., ψ_U need not be surjective. One can find an open covering $\{U_j\}_{j \in J}$ of U and $s_j \in \mathcal{G}(U_j)$ so that $\psi_{U_j}(s_j) = \rho_{U_j,U}(t)$. Cohomology's role is to describe when one can choose $\{s_j\}$ to get $s \in \mathcal{G}(U)$ and $\psi_U(s) = t$.

EXAMPLE 4.9. Let $\mathcal{O}_X$ be the sheaf of holomorphic functions over the complex plane $X = \mathbb{C}$ (i.e., regular functions). See Example 2.18(3). Let $\mathcal{M}_X$ be the sheaf of meromorphic functions associated to the set of all meromorphic functions on an open set U of X (locally the quotient f/g of holomorphic, i.e., regular functions). By the natural inclusion $\mathcal{O}_X(U) \subset \mathcal{M}_X$, $\mathcal{O}_X$ may be considered as a subsheaf of $\mathcal{M}_X$. We have the exact sequence

(4.11) $$0 \to \mathcal{O}_X \to \mathcal{M}_X \to \mathcal{M}_X/\mathcal{O}_X \to 0.$$

For an open set U, we will study an element of $\Gamma(U, \mathcal{M}_X/\mathcal{O}_X)$. By the sheafification (4.1) of a presheaf, $t \in \Gamma(U, \mathcal{M}_X/\mathcal{O}_X)$ may be considered such that the $t_j \in \mathcal{M}_X(U_j)/\mathcal{O}(U_j)$, $j \in J$, where $\{U_j\}_{j \in J}$ is a properly chosen open covering of U, satisfy the following. Namely, the restrictions of t_j and t_k on $U_{jk} = U_j \cap U_k \neq \varnothing$ coincide. Let $\widetilde{t_j}$ be a meromorphic function on U_j satisfying $t_j \equiv \widetilde{t_j} \bmod \mathcal{O}_X(U_j)$. Then we get $\widetilde{t_j} - \widetilde{t_k} \in \mathcal{O}(U_{jk})$. On the other hand, a meromorphic function on U_j can have only isolated poles. If necessary, take a smaller U_j so that in U_j, $\widetilde{t_j}$ may have finitely many poles $a_1^{(j)}, \ldots, a_{n_j}^{(j)}$ whose principle part (the negative exponent terms) of the Laurent expansion is

$$p_i^{(j)} = \frac{\alpha_{k_j^{(i)}}^{(j)}}{(z - a_i^{(j)})^{k_j^{(i)}}} + \frac{\alpha_{k_j^{(i)}-1}^{(j)}}{(z - a_i^{(j)})^{k_j^{(i)}-1}} + \cdots + \frac{\alpha_{-1}^{(j)}}{z - a_i^{(j)}}.$$

Then $\widetilde{t_j} - \sum_{i=1}^{n_j} p_i^{(j)}$ is holomorphic in U_j. Therefore,

$$t_j \in \mathcal{M}_X(U_j)/\mathcal{O}_X(U_j),$$

which is determined by $\widetilde{t_j}$, associates points $a_i^{(j)}$, $1 \leq i \leq n_j$, and the principal part $p_i^{(j)}$ of the Laurent expansion. On $U_{jk} \neq \varnothing$, $\widetilde{t_j} - \widetilde{t_k}$ being holomorphic means that the principal parts of the Laurent expansions of $\widetilde{t_j}$ and $\widetilde{t_k}$ at the poles in U_{jk} coincide. Hence, to give $t \in \Gamma(U, \mathcal{M}_X/\mathcal{O}_X)$ is to give a sequence $\{a_\lambda\}$, without an accumulating point in U, and the principal part

(4.12) $$\frac{\alpha_{k_\lambda}^{(\lambda)}}{(z - a_\lambda)^{k_\lambda}} + \frac{\alpha_{k_\lambda - 1}^{(\lambda)}}{(z - a_\lambda)^{k_\lambda - 1}} + \cdots + \frac{\alpha_{-1}^{(\lambda)}}{z - a_\lambda}$$

of the pole at a_λ. In the induced exact sequence

$$0 \to \Gamma(U, \mathcal{O}_X) \to \Gamma(U, \mathcal{M}_X) \xrightarrow{f} \Gamma(U, \mathcal{M}_X/\mathcal{O}_X)$$

from the exact sequence (4.11), in order for $t \in \Gamma(U, \mathcal{M}_X/\mathcal{O}_X)$ to be the image of the above homomorphism f, t must have (4.12) as the principal part of the Laurent expansion at a_j. There also exists a meromorphic function that is holomorphic in $U \setminus \{a_\lambda\}$. The Mittag-Leffler theorem in complex analysis implies that f is indeed surjective. □

We can extend the above example to the case of a domain D in $\mathbb{C}^n$. For $n \geq 2$, the poles of a meromorphic function are not isolated. Hence, the situation is more complicated. Then an element of $\Gamma(D, \mathcal{M}_D/\mathcal{O}_D)$ is said to be a *Cousin distribution*. The Cousin problem is to determine whether the Cousin distribution is the image of a meromorphic function or not. A Cousin problem is one of the intriguing problems for the development of the theory of holomorphic functions in several complex variables.

The above sheaf $\mathcal{M}_X$ corresponds to the *sheaf field of fractions* $\mathcal{K}_X$ for a scheme X.

In general, for a commutative ring R, the totality S of non-zero-divisors is multiplicatively closed. Then $S^{-1}R$ is said to be the *ring of total quotients*, denoted by $Q(R)$. If R is an integral domain, then $S = R \setminus \{0\}$. Then the ring of total quotients is exactly the *quotient field*. When R possesses a zero divisor, the ring of total quotients $Q(R)$ is not a field.

For an affine open set U of a scheme X, the ring of total quotients $Q(\Gamma(U, \mathcal{O}_X))$ of $\Gamma(U, \mathcal{O}_X)$ defines a presheaf. Let $\mathcal{K}_X$ be the associated sheaf, i.e., the sheafification of the presheaf $Q(\Gamma(U,, \mathcal{O}_X))$. Note also that for affine open sets $V \subset U$, the restriction map $\Gamma(U, \mathcal{O}_X) \to \Gamma(V, \mathcal{O}_X)$ induces a homomorphism $Q(\Gamma(U, \mathcal{O}_X)) \to Q(\Gamma(V, \mathcal{O}_X))$ of rings of total quotients.

EXERCISE 4.10. For an affine open set U of a Noetherian scheme X, we have
$$\Gamma(U, \mathcal{K}_X) = Q(\Gamma(U, \mathcal{O}_X)).$$
Furthermore, for each $x \in X$, we also have $\mathcal{K}_{X,x} = Q(\mathcal{O}_{X,x})$.

PROOF. Let $U = \operatorname{Spec} R$ be an affine open set of X, where R is a Noetherian ring. Choose $f_1, f_2, \ldots, f_n \in R$ so that $\{U_i = D(f_i)\}$, $i = 1, 2, \ldots, n$, is an open covering of U. Namely, $1 \in (f_1, f_2, \ldots, f_n)$.

We will prove the following two assertions.
 (1) If the image in $Q(R_{f_i})$ of $\alpha \in Q(R)$ is 0 for $1 \leq i \leq n$, then $\alpha = 0$.
 (2) For arbitrary i and j, if the image in $Q(R_{f_i f_j})$ of $\alpha_i \in Q(R_{f_i})$ coincides with the image in $Q(R_{f_i f_j})$ of $\alpha_j \in Q(R_{f_j})$, then there exists $\alpha \in Q(R)$ whose image in $Q(R_{f_i})$ is α_i.

The above assertions imply that for an affine open set U of X, the presheaf $Q(\Gamma(U, \mathcal{O}_X))$ is a sheaf, i.e., it satisfies (F1) and (F2). Therefore, for an affine open set, we have $\Gamma(U, \mathcal{K}_X) = Q(\Gamma(U, \mathcal{O}_X))$.

PROOF OF (1). Let $\alpha = \frac{b}{a}$, $a, b \in R$, where a not a zero divisor. From the hypothesis, there exists a positive integer m_i so that $f_i^{m_i} b = 0$. Let $m = \max_{1 \leq i \leq n} m_i$. Then, for all i, we get $f_i^m b = 0$. Since $1 \in (f_1, \ldots, f_n)$, we have $1 = \sum_{i=1}^{n} a_i f_i$, $a_i \in R$. The nm-th power of both sides gives $1 = \sum_{i=1}^{n} c_i f_i^m$ for some $c_i \in R$. Then, $b = 1 \cdot b = \sum c_i(f_i^m b) = 0$, i.e., $\alpha = 0$. □

PROOF OF (2). Let $\alpha_i = \frac{b_i}{a_i}$, $a_i, b_i \in R_{f_i}$. If necessary, by replacing a_i and b_i by $f_i^l a_i$ and $f_i^l b_i$, respectively, one may assume $a_i, b_i \in R$. When $\frac{b_i}{a_i} = \frac{b_j}{a_j}$ over $U_i \cap U_j = D(f_i) \cap D(f_j) = D(f_i f_j)$, we can find N satisfying $(f_i f_j)^N (a_i b_j - a_j b_i) = 0$. Then, by multiplying a power of f_i to a_i and b_i, we can assume that $a_i b_j - a_j b_i = 0$ for all i and j. Let

$$I = \{r \in R \mid \text{ for all } i, \ rb_i \text{ belongs to the ideal } (a_i) \text{ of } R_{f_i}\}.$$

Then I is an ideal. Since $a_j b_i = a_i b_j \in (a_i)$, we have $a_1, a_2, \ldots, a_n \in I$. The Noetherianness of R implies that we can write $I = (c_1, \ldots, c_s)$. If $cc_j = 0$, $1 \leq j \leq n$, then since $a_i \in I$, we get $ca_i = 0$, $1 \leq i \leq s$. Since a_i is not a zero divisor in R_{f_i}, c is 0 in R_{f_i}. Namely, there exists a positive integer M so that $f_i^M c = 0$ for all i. As before, we conclude that $c = 0$. Thus I contains a non-zero-divisor α. The definition of I implies that there is $\alpha_i \in R_{f_i}$ to satisfy $\alpha b_i = \alpha_i a_i$. That is, $\frac{\alpha b_i}{a_i} \in \Gamma(U_i, \mathcal{O}_X)$. Since $\frac{b_i}{a_i} = \frac{b_j}{a_j}$ over $U_i \cap U_j$, we get $\frac{\alpha b_i}{a_i} = \frac{\alpha b_j}{a_j}$. Then $\frac{\alpha b_i}{a_i}$, $1 \leq i \leq n$, define an element β in $\Gamma(U, \mathcal{O}_X)$. Therefore, the image of $\frac{\beta}{\alpha} \in Q(\Gamma(U, \mathcal{O}_X))$ in $\Gamma(U_i, \mathcal{O}_X)$ is $\frac{b_i}{a_i}$, i.e., $\frac{\beta}{\alpha}$ is the element that we seek. The last claim is obvious from $\Gamma(U, \mathcal{K}_X) = Q(\Gamma(U, \mathcal{O}_X))$. □

Since for an affine open set U, $Q(\Gamma(U, \mathcal{O}_X))$ is a $\Gamma(U, \mathcal{O}_X)$-module, $\mathcal{K}_X$ is an $\mathcal{O}_X$-module. However, $\mathcal{K}_X$ is not a quasicoherent $\mathcal{O}_X$-module in general (see the next section for quasicoherent sheaves).

4.2. Quasicoherent Sheaves and Coherent Sheaves

All sheaves that we have seen so far are sheaves of additive groups. In this section, we will study sheaves of $\mathcal{O}_X$-modules. Then we will focus on quasicoherent sheaves and coherent sheaves, which play an important role in algebraic geometry. Even though the theory can be built on ringed spaces, we will develop it over schemes.

(a) $\mathcal{O}_X$-Modules

For an affine scheme, we have already considered $\mathcal{O}_X$-modules in §2.3(a). We will begin with the definition of an $\mathcal{O}_X$-module over a general scheme.

A sheaf $\mathcal{F}$ over a scheme $(X, \mathcal{O}_X)$ is said to be an $\mathcal{O}_X$-*module* when the following condition is satisfied: to each open set U, $\mathcal{F}(U)$ is an $\mathcal{O}_X(U)$-module such that for open sets $V \subset U$ the diagram

$$\begin{array}{ccc} \mathcal{O}_X(U) \times \mathcal{F}(U) & \longrightarrow & \mathcal{F}(U) \\ \downarrow & & \downarrow \\ \mathcal{O}_X(V) \times \mathcal{F}(V) & \longrightarrow & \mathcal{F}(V) \end{array}$$

commutes, where the horizontal maps indicate $\mathcal{O}_X(U)$ and $\mathcal{O}_X(V)$ module structures on $\mathcal{F}(U)$ and $\mathcal{F}(V)$, respectively. Note that the stalk $\mathcal{F}_x$ at x of $\mathcal{F}$ is an $\mathcal{O}_{X,x}$-module. Namely, let $s \in \mathcal{O}_X(U)$ and $t \in \mathcal{F}(V)$ induce $a \in \mathcal{O}_{X,x}$ and $f \in \mathcal{F}_x$ at x. Then choose W containing x so that $W \subset V \cap U$. Define $\hat{s} = \rho_{W,U}(s)$ and $\hat{t} = \rho^{\mathcal{F}}_{W,V}(t)$. Then the germ determined by $\hat{s} \cdot \hat{t} \in \mathcal{F}(W)$ is precisely af.

For $\mathcal{O}_X$-modules $\mathcal{F}$ and $\mathcal{G}$, let $\varphi : \mathcal{F} \to \mathcal{G}$ be a homomorphism of sheaves of additive groups. When φ is compatible with the $\mathcal{O}_X$-module structures of $\mathcal{F}$ and $\mathcal{G}$, namely, for an open set U, the diagram

$$\begin{array}{ccc} \mathcal{O}_X(U) \times \mathcal{F}(U) & \xrightarrow{(\mathrm{id}_{\mathcal{O}_X(U)}, \varphi_U)} & \mathcal{O}_X(U) \times \mathcal{G}(U) \\ \downarrow & & \downarrow \\ \mathcal{F}(U) & \xrightarrow{\varphi_U} & \mathcal{G}(U) \end{array}$$

commutes, then φ is said to be a *homomorphism of $\mathcal{O}_X$-modules*, or an $\mathcal{O}_X$-*module homomorphism*. Notice that at stalks, $\varphi_x : \mathcal{F}_x \to \mathcal{G}_x$ is an $\mathcal{O}_{X,x}$-module homomorphism.

For $\mathcal{O}_X$-modules $\mathcal{F}$ and $\mathcal{G}$, the totality of all the $\mathcal{O}_X$-module homomorphisms from $\mathcal{F}$ to $\mathcal{G}$ is denoted by $\mathrm{Hom}_{\mathcal{O}_X}(\mathcal{F}, \mathcal{G})$.

PROBLEM 7. Show that $\mathrm{Hom}_{\mathcal{O}_X}(\mathcal{F}, \mathcal{G})$ is a $\Gamma(X, \mathcal{O}_X)$-module.

PROBLEM 8. For an $\mathcal{O}_X$-module $\mathcal{F}$ over X and for an open set U,

$$\mathrm{Hom}_{\mathcal{O}_X(U)}(\mathcal{O}_X|U, \mathcal{F}|U) \xrightarrow{\sim} \mathcal{F}(U)$$

is an isomorphism.

Various $\mathcal{O}_X$-modules will be defined in this section. The next lemma will play a crucial role for those $\mathcal{O}_X$-modules.

LEMMA 4.11. *Let $\mathcal{G}$ be a presheaf of additive groups over a scheme $(X, \mathcal{O}_X)$. A presheaf of $\mathcal{O}_X$-modules can be defined just as we defined a sheaf of $\mathcal{O}_X$-modules. Then the sheafification $^a\mathcal{G}$, i.e., the associated sheaf to $\mathcal{G}$, is a sheaf of $\mathcal{O}_X$-modules.*

PROOF. We will show that $^a\mathcal{G}(U)$ in (4.1) is an $\mathcal{O}_X(U)$-module. Let $b \in \mathcal{O}_X(U)$ and let $\{s(x)\}_{x \in U} \in {}^a\mathcal{G}(U)$. Then define $b \cdot \{s(x)\}$ by $b_x\{s(x)\}$, where b_x is the germ of b at x. For an arbitrary point x in U, choose an open set $V \subset U$ and $t \in \mathcal{G}(U)$ to satisfy $t_y = s(y)$, $y \in V$. Then we have $\tilde{t} = \rho_{V,U}(b)t \in \mathcal{G}(V)$, and we have $\tilde{t}_y = b_y s(y)$ for all $y \in V$. Hence, we get $b \cdot \{s(x)\} \in {}^a\mathcal{G}(U)$. It is clear that with this operation $^a\mathcal{G}(U)$ becomes an $\mathcal{O}_X(U)$-module. The compatibility with the restriction map can be easily shown. □

COROLLARY 4.12. *The kernel, image, and cokernel of an $\mathcal{O}_X$-module homomorphism $\varphi : \mathcal{G} \to \mathcal{H}$ are $\mathcal{O}_X$-modules.*

PROOF. For an open set U we get $\mathcal{O}_X(U)$-modules $\operatorname{Ker} \varphi_U$, $\operatorname{Im} \varphi_U$ and $\operatorname{Coker} \varphi_U$, which are compatible with the restriction map. Therefore $\operatorname{Ker} \varphi_U$ is an $\mathcal{O}_X$-module, and from Lemma 4.11, $\operatorname{Im} \varphi_U$ and $\operatorname{Coker} \varphi_U$ are also $\mathcal{O}_X$-modules. □

Let $\mathcal{F}$ and $\mathcal{G}$ be sheaves of additive groups. For each open set U, the direct sum $\mathcal{F}(U) \oplus \mathcal{G}(U)$ of additive groups is a sheaf. We shall denote this sheaf by $\mathcal{F} \oplus \mathcal{G}$, and call it the direct sum of sheaves $\mathcal{F}$ and $\mathcal{G}$. When $\mathcal{F}$ and $\mathcal{G}$ are sheaves of $\mathcal{O}_X$-modules, $\mathcal{F} \oplus \mathcal{G}$ is also an $\mathcal{O}_X$-module. We often write $\mathcal{O}_X^{\oplus n}$ for $\underbrace{\mathcal{O}_X \oplus \cdots \oplus \mathcal{O}_X}_{n}$, or even more simply $\mathcal{O}_X^n$. If an $\mathcal{O}_X$-module $\mathcal{F}$ is isomorphic to $\mathcal{O}_X^n$ as $\mathcal{O}_X$-modules, $\mathcal{F}$ is said to be a *free module of rank n*. An $\mathcal{O}_X$-module $\mathcal{F}$ is said to be a *locally free $\mathcal{O}_X$-module of rank n* if there exists an open covering $\{U_j\}_{j \in J}$ of X such that the restriction $\mathcal{F}|U_j$ of $\mathcal{F}$ to U_j is a free module of rank n over $\mathcal{O}_{U_j} = \mathcal{O}_X|U_j$. A locally free $\mathcal{O}_X$-module of rank n is also called a *locally free sheaf* of rank n. In particular, when $n = 1$, a locally free sheaf of rank one is said to be an *invertible sheaf* over X. The notion of an invertible sheaf is very important in algebraic geometry.

EXAMPLE 4.13. A morphism of schemes $f : W \to X$ is said to be a *vector bundle* when the following conditions (V1) and (V2) are satisfied:

4.2. QUASICOHERENT SHEAVES AND COHERENT SHEAVES

(V1) For an open covering $\{U_i\}_{i \in I}$ and for each $i \in I$, there exists a scheme isomorphism over U_i

$$\varphi_i : f^{-1}(U_i) \xrightarrow{\sim} \mathbb{A}^n_{U_i} = \mathbb{A}^n_{\mathbb{Z}} \times_{\operatorname{Spec} \mathbb{Z}} U_i.$$

(V2) When $U_i \cap U_j \neq \varnothing$, for an arbitrary affine scheme $V = \operatorname{Spec} R \subset U_i \cap U_j$, the isomorphism $\varphi_{ij} = \varphi_i \circ \varphi_j^{-1} | \mathbb{A}^n_V : \mathbb{A}^n_V \to \mathbb{A}^n_V$ is the scheme isomorphism induced by a linear automorphism

$$\theta_{ij} : R[x_1, x_2, \ldots, x_n] \to R[x_1, x_2, \ldots, x_n],$$

$$\theta_{ij}(x_k) = \sum_{l=1}^{n} a_{kl} x_l, \quad a_{kl} \in R.$$

A vector bundle of rank 1 is called a *line bundle*, and they are especially important in algebraic geometry.

Let $f : W \to X$ be a vector bundle of rank n. For an arbitrary open set U of X, define

$$\mathcal{F}(U) = \{s : U \to W | s \text{ is a scheme morphism satisfying } f \circ s = \operatorname{id}_U\}.$$

An element of $\mathcal{F}(U)$ is called a *section* of $f : W \to X$ over U. For an affine open set $U \subset U_i$, where $U = \operatorname{Spec} A$, $f^{-1}(U)$ can be identified with $\mathbb{A}^n_U = \operatorname{Spec} A[x_1, x_2, \ldots, x_n]$ as schemes over U. Since $s(U) \subset f^{-1}(U)$, this identification implies that a section $s : U \to W \in \mathcal{F}(U)$ corresponds to an A-homomorphism $\sigma : A[x_1, x_2, \ldots, x_n] \to A$. On the other hand, when σ is given, we are given $a_i = \sigma(x_i) \in A$, $i = 1, 2, \ldots, n$. This correspondence is one-to-one. Therefore, as sets, we get $\mathcal{F}(Y) \xrightarrow{\sim} A^{\oplus n}$. This isomorphism defines a free A-module structure on $\mathcal{F}(U)$.

When $U \subset U_i \cap U_j$, two free A-module structures are induced on $\mathcal{F}(U)$. However, (V2) implies that they are isomorphic.

For a general U, choose an affine open covering $\{V_\lambda\}_{\lambda \in \Lambda}$ of U that has the above property. Then, for $s, t \in \mathcal{F}(U)$, define $s + t$ to be $\rho_{V_\lambda, U}(s) + \rho_{V_\lambda, U}(t)$ over V_λ. Thus, we can define an $\mathcal{O}_X(U)$-module structure on $\mathcal{F}(U)$. Namely, $\mathcal{F}$ becomes an $\mathcal{O}_X$-module. The above argument and condition (V1) imply that $\mathcal{F}$ is a locally free sheaf of rank n. This sheaf is sometimes said to be the *sheaf of local sections* of a vector bundle $f : W \to X$.

Conversely, for a given locally free sheaf $\mathcal{F}$ of rank n over X, one can show that there exists a sheaf of local sections over X isomorphic to $\mathcal{F}$ as $\mathcal{O}_X$-modules. $\square$

EXERCISE 4.14. Let $\mathcal{F}$ and $\mathcal{G}$ be sheaves of $\mathcal{O}_X$-modules. Then, for an open set U, the presheaf $\operatorname{Hom}_{\mathcal{O}_X | U}(\mathcal{F}|U, \mathcal{G}|U)$ is a sheaf. This sheaf is denoted as $\underline{\mathrm{Hom}}_{\mathcal{O}_X}(\mathcal{F},\mathcal{G})$. If $\mathcal{F}$ is a free $\mathcal{O}_X$-module of rank n, then $\underline{\mathrm{Hom}}_{\mathcal{O}_X}(\mathcal{F},\mathcal{G})$ is isomorphic to $\mathcal{G}^{\oplus n}$ as $\mathcal{O}_X$-modules. Furthermore, for an exact sequence of $\mathcal{O}_X$-modules

$$0 \to \mathcal{H}_1 \to \mathcal{H}_2 \to \mathcal{H}_3 \to 0,$$

we have exact sequences

$$0 \to \underline{\mathrm{Hom}}_{\mathcal{O}_X}(\mathcal{H}_3,\mathcal{F}) \to \underline{\mathrm{Hom}}_{\mathcal{O}_X}(\mathcal{H}_2,\mathcal{F}) \to \underline{\mathrm{Hom}}_{\mathcal{O}_X}(\mathcal{H}_1,\mathcal{F}),$$
$$0 \to \underline{\mathrm{Hom}}_{\mathcal{O}_X}(\mathcal{F},\mathcal{H}_1) \to \underline{\mathrm{Hom}}_{\mathcal{O}_X}(\mathcal{F},\mathcal{H}_2) \to \underline{\mathrm{Hom}}_{\mathcal{O}_X}(\mathcal{F},\mathcal{H}_3).$$

PROOF. For simplicity's sake, set $\mathcal{H}(U) = \mathrm{Hom}_{\mathcal{O}_X|U}(\mathcal{F}|U,\mathcal{G}|U)$. Let $\{U_j\}_{j\in J}$ be an open covering of U and let $\varphi \in \mathcal{H}(U)$. Then the restriction map $\rho_{U_j,U}$ is the natural restriction of a homomorphism of sheaves. Notice that $\mathcal{H}$ is a presheaf of $\mathcal{O}_X$-modules. Suppose $\varphi_j = \rho_{U_j,U}(\varphi) = 0$, $j \in J$. For an arbitrary open set $V \subset U$ and an arbitrary section $s \in \mathcal{F}(V)$, let $V_j = U_j \cap V$. Then

$$\varphi_{jV_j}(\rho_{V_j,V}(s)) = 0, \quad j \in J,$$

i.e., $\varphi_V(s) = 0$. Consequently, $\varphi = 0$. Next, for $\varphi_j \in \mathcal{H}(U_j)$, $j \in J$, suppose $\rho_{U_{ij},U_i}(\varphi_i) = \rho_{U_{ij},U_j}(\varphi_j)$. For an arbitrary open set $V \subset U$ and an arbitrary section $s \in \mathcal{F}(V)$, let $t_j = \varphi_{jV_j}(\rho_{V_j,V}(s))$, $j \in J$. Then we have $\rho^{\mathcal{G}}_{V_{ij},V_i}(t_i) = \rho^{\mathcal{G}}_{V_{ij},V_j}(t_j)$ by the assumption. Therefore, there exists $t \in \mathcal{F}(U)$ satisfying $\rho_{V_j,V}(t) = t_j$, $j \in J$. This t is uniquely determined. Then define $\varphi_V(s) = t$, i.e., obtaining $\varphi_V \in \mathcal{H}(V)$. Since V is an arbitrary open set of U, we have just proved the existence of $\varphi \in \mathcal{H}(U)$ to satisfy $\varphi_j = \rho_{U_j,U}(\varphi)$, $j \in J$.

For $\mathcal{F} \xrightarrow{\sim} \mathcal{O}_X^{\oplus n}$, we have

$$\mathcal{H}(U) \xrightarrow{\sim} \mathrm{Hom}_{\mathcal{O}_X|U}((\mathcal{O}_X|U)^{\oplus n}, \mathcal{G}|U) \xrightarrow{\sim} \mathcal{G}(U)^{\oplus n},$$

i.e., $\mathcal{H} \xrightarrow{\sim} \mathcal{G}^{\oplus n}$.

It is easy to see that

$$0 \to \underline{\mathrm{Hom}}_{\mathcal{O}_X}(\mathcal{H}_3,\mathcal{F}) \xrightarrow{g^*} \underline{\mathrm{Hom}}_{\mathcal{O}_X}(\mathcal{H}_2,\mathcal{F}) \xrightarrow{f^*} \underline{\mathrm{Hom}}_{\mathcal{O}_X}(\mathcal{H}_1,\mathcal{F})$$

is an exact sequence of $\mathcal{O}_X$-modules. For an open set U of X and for $\varphi \in \mathrm{Hom}_{\mathcal{O}_U}(\mathcal{H}_3|U,\mathcal{F}|U)$, let us assume $\varphi \circ g|U = 0$. For an open set V and $t \in \mathcal{H}_3(V)$, one can choose an open covering $\{W_j\}_{j\in J}$ and $s_j \in \mathcal{H}_2(W_j)$ satisfying $g_{W_j}(s_j) = \rho_{W_j,V}(t)$. Since we have $\varphi_{W_j}(\rho_{W_j,V}(t)) = \varphi_{W_j}(g_{W_j}(s_j)) = 0$, we get $\varphi_V(t) = 0$. That is, $\varphi = 0$. Hence g^* is injective.

Next, suppose that $\psi \in \mathrm{Hom}_{\mathcal{O}_U}(\mathcal{H}_2|U,\mathcal{F}|U)$ satisfies $\psi \circ f|U = 0$. Then ψ is identically a zero map on $\mathrm{Im}\, f|U$. Hence, we can consider

$\psi \in \mathrm{Hom}_{\mathcal{O}_U}((\mathcal{H}_2/\mathrm{Im}\,f)|U, \mathcal{F}|U)$. There is $\varphi \in \mathrm{Hom}_{\mathcal{O}_U}(\mathcal{H}_3|U, \mathcal{F}|U)$ to satisfy $\varphi = \psi \circ g|U$. Consequently, we obtain $\mathrm{Im}\,g^* = \mathrm{Ker}\,f^*$.

The exactness of the last sequence can be proved in a similar way. □

PROBLEM 9. For $\mathcal{O}_X$-modules $\mathcal{F}, \mathcal{G}$ and $\mathcal{H}$, prove the following isomorphisms of $\mathcal{O}_X$-modules:

$$\underline{\mathrm{Hom}}_{\mathcal{O}_X}(\mathcal{F} \oplus \mathcal{G}, \mathcal{H}) \xrightarrow{\sim} \underline{\mathrm{Hom}}_{\mathcal{O}_X}(\mathcal{F}, \mathcal{H}) \oplus \underline{\mathrm{Hom}}_{\mathcal{O}_X}(\mathcal{G}, \mathcal{H}),$$
$$\underline{\mathrm{Hom}}_{\mathcal{O}_X}(\mathcal{F}, \mathcal{G} \oplus \mathcal{H}) \xrightarrow{\sim} \underline{\mathrm{Hom}}_{\mathcal{O}_X}(\mathcal{F}, \mathcal{G}) \oplus \underline{\mathrm{Hom}}_{\mathcal{O}_X}(\mathcal{F}, \mathcal{H}).$$

Let $\mathcal{F}$ and $\mathcal{G}$ be $\mathcal{O}_X$-modules and let U be an open set in X. Then the assignment of an $\mathcal{O}_X(U)$-module

$$\mathcal{F}(U) \otimes_{\mathcal{O}_X(U)} \mathcal{G}(U)$$

is a presheaf of $\mathcal{O}_X$-modules. The sheaf associated to this presheaf is denoted as $\mathcal{F} \otimes_{\mathcal{O}_X} \mathcal{G}$, and is called the *tensor product* of the $\mathcal{O}_X$-modules $\mathcal{F}$ and $\mathcal{G}$. Note that $\mathcal{F} \otimes_{\mathcal{O}_X} \mathcal{G}$ is also an $\mathcal{O}_X$-module.

EXERCISE 4.15. (1) The stalk $(\mathcal{F} \otimes_{\mathcal{O}_X} \mathcal{G})_x$ at x of the tensor product $\mathcal{F} \otimes_{\mathcal{O}_X} \mathcal{G}$ of $\mathcal{O}_X$-modules $\mathcal{F}$ and $\mathcal{G}$ is isomorphic to $\mathcal{F}_x \otimes_{\mathcal{O}_{X,x}} \mathcal{G}_x$ as $\mathcal{O}_{X,x}$-modules.

(2) For an exact sequence of $\mathcal{O}_X$-modules

$$\mathcal{F}_1 \to \mathcal{F}_2 \to \mathcal{F}_3 \to 0,$$

the sequence obtained by tensoring with an $\mathcal{O}_X$-module $\mathcal{G}$ over $\mathcal{O}_X$,

$$\mathcal{F}_1 \otimes_{\mathcal{O}_X} \mathcal{G} \to \mathcal{F}_2 \otimes_{\mathcal{O}_X} \mathcal{G} \to \mathcal{F}_3 \otimes_{\mathcal{O}_X} \mathcal{G} \to 0,$$

is exact.

PROOF. (1) By the definition of sheafification, we have

$$(\mathcal{F} \otimes_{\mathcal{O}_X} \mathcal{G}) = \varinjlim_{x \in U} \mathcal{F}(U) \otimes_{\mathcal{O}_X(U)} \mathcal{G}(U).$$

For an open set U containing x, the $\mathcal{O}_X(U)$-module homomorphism

$$\mathcal{F}(U) \otimes_{\mathcal{O}_X(U)} \mathcal{G}(U) \to \mathcal{F}_x \oplus_{\mathcal{O}_{X,x}} \mathcal{G}_x$$

induces an $\mathcal{O}_{X,x}$-module homomorphism

$$\varphi : (\mathcal{F} \otimes_{\mathcal{O}_X} \mathcal{G})_x \to \mathcal{F}_x \otimes_{\mathcal{O}_{X,x}} \mathcal{G}_x.$$

Note that a bilinear homomorphism of $\mathcal{O}_{X,x}$-modules

$$\Psi : \mathcal{F}_x \times \mathcal{G}_x \to (\mathcal{F} \otimes_{\mathcal{O}_X} \mathcal{G})_x$$

can be defined. For $f_x \in \mathcal{F}_x$ and $g_x \in \mathcal{G}_x$, choose an open set U containing x, and choose $f \in \mathcal{F}(U)$ and $g \in \mathcal{G}(U)$ so that the germs of f and g at x are f_x and g_x, respectively. Let $\Psi(f_x, g_x)$ be the element of $(\mathcal{F} \otimes_{\mathcal{O}_X} \mathcal{G})_x$ determined by $f \otimes g \in \mathcal{F}(U) \otimes_{\mathcal{O}_X(U)} \mathcal{G}(U)$. It is clear that $\Psi(f_x, g_x)$ does not depend upon the choices of U and f and g. One can also show that, for $a_x, b_x \in \mathcal{O}_{X,x}, f_x, f_{1x}, f_{2x} \in \mathcal{F}_x$ and $g_x, g_{1x}, g_{2x} \in \mathcal{G}_x$,

$$\Psi(a_x f_{1x} + b_x f_{2x}, g_x) = a_x \Psi(f_{1x}, g_x) + b_x \Psi(f_{2x}, g_x),$$
$$\Psi(f_x, a_x g_{1x} + b_x g_{2x}) = a_x \Psi(f_x, g_{1x}) + b_x \Psi(f_x, g_{2x}).$$

Therefore, by the universal mapping property of tensor products (see §2.3), we obtain the following commutative diagram:

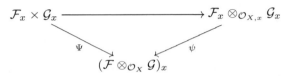

where ψ is a uniquely determined $\mathcal{O}_{X,x}$-homomorphism. Since $\varphi(\Psi(f_x, g_x)) = f_x \otimes g_x$, we have $\varphi \circ \psi = \mathrm{id}$. The definition of φ implies that for $f \otimes g \in \mathcal{F}(U) \otimes_{\mathcal{O}_X(U)} \mathcal{G}(U)$ we have

$$\varphi((f \otimes g)_x) = f_x \otimes g_x \in \mathcal{F}_x \otimes_{\mathcal{O}_{X,x}} \mathcal{G}_x,$$

and from the definition of ψ, we also have $\psi(f_x \otimes g_x) = (f \otimes g)_x$. Namely, $\psi \circ \varphi = \mathrm{id}$. Therefore, φ and ψ are isomorphisms of $\mathcal{O}_{X,x}$-modules.

(2) For an exact sequence of modules over a commutative ring R

$$M_1 \to M_2 \to M_3 \to 0,$$

and for an R-module N, we have an exact sequence

$$M_1 \otimes_R N \to M_2 \otimes_R N \to M_3 \otimes_R N \to 0,$$

by tensoring with N over R. Since $(\mathcal{F}_j \otimes_{\mathcal{O}_X} \mathcal{G})_x = \mathcal{F}_{j,x} \otimes_{\mathcal{O}_{X,x}} \mathcal{G}_x$, we get the exact sequence in (2). □

EXAMPLE 4.16. For invertible sheaves $\mathcal{L}$ and $\mathcal{M}$ over a scheme X, $\mathcal{L} \otimes_{\mathcal{O}_X} \mathcal{M}$ is also an invertible sheaf. Put $\mathcal{L}^{-1} = \underline{\mathrm{Hom}}_{\mathcal{O}_X}(\mathcal{L}, \mathcal{O}_X)$. Then one can define a natural $\mathcal{O}_X$-homomorphism

$$\varphi : \mathcal{L} \otimes_{\mathcal{O}_X} \mathcal{L}^{-1} \to \mathcal{O}_X,$$
$$a \otimes f \mapsto f(a).$$

For an affine open U satisfying $\mathcal{L}|U \xrightarrow{\sim} \mathcal{O}_U$, we get

$$\mathcal{L}^{-1}|U \xrightarrow{\sim} \underline{\mathrm{Hom}}_{\mathcal{O}_U}(\mathcal{O}_U, \mathcal{O}_U) \xrightarrow{\sim} \mathcal{O}_U.$$

Hence, over U, φ is an $\mathcal{O}_U$-isomorphism. Namely, we have $\mathcal{L} \otimes_{\mathcal{O}_X} \mathcal{L}^{-1} \xrightarrow{\sim} \mathcal{O}_X$. Therefore, under the tensor product, isomorphic classes of invertible sheaves as $\mathcal{O}_X$-modules form a group. This group is called the *Picard group* of X, and is denoted as $\operatorname{Pic} X$. Define $\mathcal{L}^{\otimes n}$ (or simply $\mathcal{L}^n$) as $\underbrace{\mathcal{L} \otimes \cdots \otimes \mathcal{L}}_{n}$, and for $n = -m$, $m \geq 1$, define $\mathcal{L}^{\otimes n} = (\mathcal{L}^{-1})^{\otimes m}$. Note that we define $\mathcal{L}^0 = \mathcal{O}_X$.

When, for an arbitrary exact sequence of R-modules

$$0 \to M_1 \to M_2 \to M_3 \to 0,$$

we have an exact sequence

$$0 \to M_1 \otimes_R N \to M_2 \otimes_R N \to M_3 \otimes_R N \to 0,$$

then the R-module N is said to be an *R-flat module*. Similarly, for an arbitrary exact sequence of $\mathcal{O}_X$-modules

$$0 \to \mathcal{F}_1 \to \mathcal{F}_2 \to \mathcal{F}_3 \to 0,$$

if the induced sequence

$$0 \to \mathcal{F}_1 \otimes_{\mathcal{O}_X} \mathcal{G} \to \mathcal{F}_2 \otimes_{\mathcal{O}_X} \mathcal{G} \to \mathcal{F}_3 \otimes_{\mathcal{O}_X} \mathcal{G} \to 0$$

is also exact, then the $\mathcal{O}_X$-module $\mathcal{G}$ is said to be an $\mathcal{O}_X$-*flat sheaf*. Since $\mathcal{F} \otimes_{\mathcal{O}_X} \mathcal{O}_X \xrightarrow{\sim} \mathcal{F}$, $\mathcal{O}_X$ is an $\mathcal{O}_X$-flat sheaf. Since we have $\mathcal{F} \otimes_{\mathcal{O}_X} \mathcal{O}_X^{\oplus n} \xrightarrow{\sim} \mathcal{F}^{\oplus n}$, an $\mathcal{O}_X$-free sheaf is also an $\mathcal{O}_X$-flat sheaf. We can generalize the above result as follows.

LEMMA 4.17. (i) *A locally free $\mathcal{O}_X$-module (i.e., locally $\mathcal{O}_X$-free sheaf) is an $\mathcal{O}_X$-flat sheaf.*

(ii) *An $\mathcal{O}_X$-module $\mathcal{G}$ is an $\mathcal{O}_X$-flat sheaf if and only if $\mathcal{G}_x$ is an $\mathcal{O}_{X,x}$-flat sheaf at each point $x \in X$.*

PROOF. The stalk $\mathcal{G}_x$ of a locally $\mathcal{O}_X$-free sheaf $\mathcal{G}$ at x is an $\mathcal{O}_{X,x}$-flat module. Hence, (ii) implies (i). Since $(\mathcal{F} \otimes_{\mathcal{O}_X} \mathcal{G})_x \xrightarrow{\sim} \mathcal{F}_{X,x} \otimes_{\mathcal{O}_{X,x}} \mathcal{G}_x$, we get (ii). □

(b) **Quasicoherent Sheaves**

Among modules over a commutative ring R, *finitely generated R-modules* and *finitely presented R-modules* are important (a finitely presented R-module is a module isomorphic to the cokernel of a homomorphism $\varphi : R^{\oplus m} \to R^{\oplus n}$). As for $\mathcal{O}_X$-modules, coherent sheaves play an important role. We shall begin with a definition. We will denote the sheaf $\mathcal{O}_X | U$ restricted to an open set U simply by $\mathcal{O}_U$.

4. COHERENT SHEAVES

DEFINITION 4.18. Let $\mathcal{F}$ be an $\mathcal{O}_X$-module.

(i) If for each point x in X there is an open neighborhood U of x so that the sequence of $\mathcal{O}_X$-modules

$$\mathcal{O}_U^{\oplus I} \to \mathcal{O}_U^{\oplus J} \to \mathcal{F}|U \to 0$$

is exact, then $\mathcal{F}$ is said to be *quasicoherent*, or a *quasicoherent sheaf*. Note that I and J need not be finite sets, and also that the cardinalities of I and J may vary for various points.

(ii) If for each x in X, there exists an open set U containing x so that the sequence of $\mathcal{O}_U$-modules

$$\mathcal{O}_U^{\oplus n} \to \mathcal{F}|U \to 0$$

is exact, then $\mathcal{F}$ is said to be a *finitely generated* $\mathcal{O}_X$-module. Note that n may vary at different points x. (Such a sheaf $\mathcal{F}$ might better be called a locally finitely generated $\mathcal{O}_X$-module. But we follow the customary phrase.)

EXAMPLE 4.19. (1) The sheaf $\mathcal{O}_X$ is finitely generated and quasicoherent as an $\mathcal{O}_X$-module.

(2) Let $X = \operatorname{Spec} R$ be an affine scheme. For R-modules M and N, we showed in Example 2.25 that an $\mathcal{O}_X$-module $\widetilde{M}$ is induced. We also showed that for an R-module homomorphism $\varphi : M \to N$ there is induced an $\mathcal{O}_X$-module homomorphism $\tilde{\varphi} : \widetilde{M} \to \widetilde{N}$. For a point $\mathfrak{p} \in \operatorname{Spec} R$, a homomorphism between the stalks is precisely the localized homomorphism at $\mathfrak{p}$, i.e., $\varphi_\mathfrak{p} : M_\mathfrak{p} \to N_\mathfrak{p}$. The localization of R-modules preserves exactness (see Problem 10 below). For an exact sequence of R-modules

$$M_1 \to M_2 \to M_3 \to M_4 \to \cdots,$$

the induced sequence of $\mathcal{O}_X$-modules

$$\widetilde{M_1} \to \widetilde{M_2} \to \widetilde{M_3} \to \widetilde{M_4} \to \cdots$$

is exact. From this fact, the $\mathcal{O}_X$-module $\widetilde{M}$ determined by an R-module M is quasicoherent. This is because every R-module M can satisfy an exact sequence of the form

$$R^{\oplus I} \to R^{\oplus J} \to M \to 0.$$

4.2. QUASICOHERENT SHEAVES AND COHERENT SHEAVES

For a finitely generated R-module M, let $s_1, s_2, \ldots, s_n$ be a set of generators for M as an R-module. Then we get a surjective homomorphism

$$R^{\oplus n} \to M \to 0,$$

$$(a_1, \ldots, a_n) \mapsto \sum_{j=1}^{n} a_j s_j,$$

inducing the surjective $\mathcal{O}_X$-homomorphism

$$\mathcal{O}_X^{\oplus n} \to \widetilde{M} \to 0.$$

Namely, $\widetilde{M}$ is a finitely generated $\mathcal{O}_X$-module. (For an affine scheme such as the above case, one can take $U = X$ in Definition 4.18(ii). As we will show later, we may not take $U = X$ if X is a general scheme.) □

PROBLEM 10. For an exact sequence of R-modules

$$\cdots \to M^{(1)} \to M^{(2)} \to M^{(3)} \to \cdots,$$

show that the corresponding sequence

$$\cdots \to M_{\mathfrak{p}}^{(1)} \to M_{\mathfrak{p}}^{(2)} \to M_{\mathfrak{p}}^{(3)} \to \cdots$$

obtained by the localization at a prime ideal $\mathfrak{p}$ of R is also exact.

The above example of an affine scheme is essential. We have more precise results for an affine scheme, as follows.

PROPOSITION 4.20. *Let $(X, \mathcal{O}_X)$ be the affine scheme determined by a commutative ring R.*

(i) *The $\mathcal{O}_X$-module $\widetilde{M}$ determined by an R-module M is quasicoherent, and for an open set $D(f)$ of X*

(4.13) $$\Gamma(D(f), \widetilde{M}) = M_f,$$

and in particular

$$\Gamma(X, \widetilde{M}) = M.$$

(ii) *For an R-module homomorphism $\varphi : M \to N$, the map*

$$\Phi : \mathrm{Hom}_R(M, N) \to \underline{\mathrm{Hom}}_{\mathcal{O}_X}(\widetilde{M}, \widetilde{N})$$

assigning an $\mathcal{O}_X$-module homomorphism $\tilde{\varphi}$ is an isomorphism of R-modules.

(iii) *For R-modules M and N we have isomorphisms of $\mathcal{O}_X$-modules $\widetilde{M} \oplus \widetilde{N} \stackrel{\sim}{\to} (M \oplus N)^\sim$ and $\widetilde{M} \otimes_{\mathcal{O}_X} \widetilde{N} \stackrel{\sim}{\to} (M \otimes_R N)^\sim$. Furthermore, if M is a finitely presented R-module, we have an isomorphism*

$$\underline{\mathrm{Hom}}_{\mathcal{O}_X}(\widetilde{M}, \widetilde{N}) \stackrel{\sim}{\to} (\mathrm{Hom}_R(M, N))^\sim.$$

PROOF. (i) The coherency of the sheaf $\widetilde{M}$ has been shown in Example 4.19, and (4.13) follows from the construction of $\widetilde{M}$ in Example 2.25.

(ii) For $f \in \underline{\mathrm{Hom}}_{\mathcal{O}_X}(\widetilde{M}, \widetilde{N})$, assign $\varphi = f_X : \Gamma(X, \widetilde{M}) = M \to \Gamma(X, \widetilde{N}) = N$. We denote this map by Ψ. Then we have $\Psi \circ \Phi = \mathrm{id}$. On the other hand, for $\mathfrak{p} \in X = \mathrm{Spec}\, R$, f induces an $\mathcal{O}_{X,\mathfrak{p}} = R_\mathfrak{p}$-homomorphism $f_\mathfrak{p} : M_\mathfrak{p} \to N_\mathfrak{p}$ between the stalks at $\mathfrak{p}$. (Notice that from Example 2.25 we have $\widetilde{M}_\mathfrak{p} = M_\mathfrak{p}$.) The $\mathcal{O}_X$-module homomorphism $\tilde{\varphi} : \widetilde{M} \to \widetilde{N}$ induced by $\Psi(f) = \varphi = f_X : M \to N$ determines the $R_\mathfrak{p}$-homomorphism $\tilde{\varphi}_\mathfrak{p} : M_\mathfrak{p} \to N_\mathfrak{p}$ on the stalks at $\mathfrak{p}$. Notice that for $\frac{m}{s} \in M_\mathfrak{p}$, $m \in M$, and $s \in R \setminus \mathfrak{p}$ we have

$$\tilde{\varphi}_\mathfrak{p}\left(\frac{m}{s}\right) = \frac{\varphi(m)}{s} = \frac{f_X(m)}{s}.$$

Since $m \in M$ determines an element $\frac{m}{1}$ in $M_\mathfrak{p}$, the $R_\mathfrak{p}$-homomorphism $f_\mathfrak{p}$ satisfies

$$f_\mathfrak{p}\left(\frac{m}{s}\right) = f_\mathfrak{p}\left(\frac{1}{s} \cdot \frac{m}{1}\right) = \frac{1}{s} f_\mathfrak{p}\left(\frac{m}{1}\right) = \frac{1}{s} f_X(m),$$

i.e., $\tilde{\varphi}_\mathfrak{p} = f_\mathfrak{p}$. Hence we obtain $\Phi \circ \Psi = \mathrm{id}$. Since Φ is a homomorphism of R-modules, Φ is an R-module isomorphism.

(iii) As in Example 2.25, for an open set $D(f)$, the sheaf $\widetilde{M} \oplus \widetilde{N}$ is associated with $M_f \oplus N_f$. The first isomorphism in (iii) follows from $M_f \oplus N_f = (M \oplus N)_f$. We get the sheaf $\widetilde{M} \otimes_{\mathcal{O}_X} \widetilde{N}$ by sheafifying the presheaf $\widetilde{M}(U) \otimes_{\mathcal{O}_X(U)} \widetilde{N}(U)$ for an open set U. In particular, for $U = D(f)$ the corresponding additive group is $M_f \otimes_{R_f} N_f$. On the other hand, as R_f-modules, $M_f \otimes_{R_f} N_f$ is isomorphic to $(M \otimes_R N)_f$ (see Problem 11, below). Hence, as $\mathcal{O}_X$-modules, $\widetilde{M} \otimes_{\mathcal{O}_X} \widetilde{N}$ is isomorphic to $(M \otimes_{\mathcal{O}_X} N)^\sim$. By the definition, $\underline{\mathrm{Hom}}_{\mathcal{O}_X}(\widetilde{M}, \widetilde{N})(D(f)) = \mathrm{Hom}_{R_f}(M_f, N_f)$. Since M is a finitely presented R-module, we have an exact sequence

$$R^{\oplus A} \xrightarrow{\varphi} R^{\oplus b} \xrightarrow{\psi} M \to 0,$$

where a and b are positive integers. This exact sequence implies that
$$R_f^{\oplus a} \xrightarrow{\varphi_f} R_f^{\oplus b} \xrightarrow{\psi_f} M_f \to 0$$
is exact. Therefore, we obtain the following exact sequence:
$$0 \to \operatorname{Hom}_{R_f}(M_f, N_f) \xrightarrow{\psi_f^*} \operatorname{Hom}_{R_f}(R_f^{\oplus b}, N_f) \xrightarrow{\varphi_f^*} \operatorname{Hom}_{R_f}(R_f^{\oplus a}, N_f).$$
Since we have isomorphisms
$$\operatorname{Hom}_{R_f}(R_f^{\oplus b}, N_f) \xrightarrow{\sim} N_f^{\oplus b} \xrightarrow{\sim} (\operatorname{Hom}_R(R^{\oplus b}, N))_f,$$
$$\operatorname{Hom}_{R_f}(R_f^{\oplus a}, N_f) \xrightarrow{\sim} N_f^{\oplus a} \xrightarrow{\sim} (\operatorname{Hom}_R(R^{\oplus a}, N))_f,$$
and also φ_f^* is the localized $\varphi^* : \operatorname{Hom}_R(R^{\oplus b}, N) \to \operatorname{Hom}_R(R^{\oplus a}, N)$ at f, we conclude that $\operatorname{Ker} \varphi_f^*$ is the localized $\operatorname{Ker} \varphi^*$ at f. Since $\operatorname{Im}\{\psi^* : \operatorname{Hom}_R(M, N) \to \operatorname{Hom}_R(R^{\oplus a}, N)\} = \operatorname{Ker} \varphi^*$ and $\operatorname{Im} \psi_f^* = \operatorname{Ker} \varphi_f^*$, there exists an R_f-isomorphism
$$\operatorname{Hom}_{R_f}(M_f, N_f) \xrightarrow{\sim} (\operatorname{Hom}_R(M, N))_f.$$
Consequently, as $\mathcal{O}_X$-modules we get an isomorphism
$$\underline{\operatorname{Hom}}_{\mathcal{O}_X}(\widetilde{M}, \widetilde{N}) \xrightarrow{\sim} (\operatorname{Hom}_R(M, N))^\sim. \quad \square$$

PROBLEM 11. Prove that for R-modules M and N there is an isomorphism
$$M_f \otimes_{R_f} N_f \xrightarrow{\sim} (M \otimes_R N)_f$$
of R_f-modules.

The structure of a quasicoherent sheaf over an affine scheme can be explained as follows.

THEOREM 4.21. *An $\mathcal{O}_X$-module $\mathcal{F}$ over an affine scheme $X = \operatorname{Spec} R$ is quasicoherent if and only if $\mathcal{F}$ is isomorphic to the associated sheaf $\widetilde{M}$ with an R-module M. Then $\Gamma(X, \mathcal{F})$ is isomorphic to M as R-modules.*

PROOF. We have already shown in Example 4.19 that for an R-module M, the sheaf $\widetilde{M}$ is a quasicoherent $\mathcal{O}_X$-module. Conversely, for a quasicoherent $\mathcal{O}_X$-module $\mathcal{F}$, let $M = \Gamma(X, \mathcal{F})$. We will show that $\widetilde{M}$ is isomorphic to $\mathcal{F}$ as $\mathcal{O}_X$-modules. Since $\mathcal{F}$ is quasicoherent, for an arbitrary point $\mathfrak{p} \in \operatorname{Spec} R$ there exists an open neighborhood $D(f)$ of $\mathfrak{p}$ such that
$$(\mathcal{O}_X|D(f))^{\oplus I} \xrightarrow{\varphi} (\mathcal{O}_X|D(f))^{\oplus J} \xrightarrow{\psi} \mathcal{F}|D(f) \to 0$$

is exact. Then define

$$M_{D(f)} = \operatorname{Coker}\{\varphi|D(f) : R_f^{\oplus I} \to R_f^{\oplus J}\}.$$

From 4.19(2), $\mathcal{F}|D(f)$ is isomorphic to $\widetilde{M}_{D(f)}$ as $\mathcal{O}_{D(f)}$-modules. In particular, as R_f-modules, $\Gamma(D(f), \mathcal{F})$ is isomorphic to $M_{D(f)}$. For the sake of simplicity, $M_{D(f)} = \Gamma(D(f), \mathcal{F})$. For each point $\mathfrak{p} \in \operatorname{Spec} R$, choose such an open set $D(f)$ as above. Then we get an open covering of X. Since X is quasicompact (Corollary 2.9), one can choose finitely many $D(f_i)$, $i \in I$, covering $X : X = \bigcup_{i \in I} D(f_i)$. We will write M_i for $M_{D(f_i)}$. Let $M = \Gamma(X, \mathcal{F})$. The restriction map $\rho_i = \rho_{D(f_i), X} : M = \Gamma(X, \mathcal{F}) \to \Gamma(D(f_i), \mathcal{F}) = M_i$ induces an R_{f_i}-module homomorphism $\tilde{\rho}_i : M_{f_i} \to M_i$. We will show that $\hat{\rho}_i$ is an isomorphism. First we will show the injectivity of $\hat{\rho}_i$. If, for $s \in M$, $\hat{\rho}_i(\frac{s}{f_i^m}) = 0$, we get $\hat{\rho}_i(\frac{s}{f_i^m}) = \frac{1}{f_i^m} \hat{\rho}_i(\frac{s}{1}) = 0$, namely $\hat{\rho}_i(\frac{s}{1}) = 0$. Since $\hat{\rho}_i$ is the localization of ρ_i at f_i, we obtain $\rho_i(s) = 0$. On the other hand, since $D(f_i) \cap D(f_j) = D(f_i f_j)$, we can regard $\mathcal{F}|D(f_i f_j) = (\mathcal{F}|D(f_j))|D(f_i f_j) = \widetilde{M}_j|D(f_i f_j) = ((M_j)_{(f_i f_j)})^\sim$. Since one has $\rho_{D(f_i f_j), X}(s) = \rho_{D(f_i f_j), D(f_i)}(\rho_i(s)) = 0$ and also $\rho_{D(f_i f_j), X}(s) = \rho_{D(f_i f_j), D(f_j)}(\rho_j(s)) = 0$, one can find a positive integer n_j to satisfy $(f_i f_j)^{n_j} \rho_j(s) = 0$ in $(M_j)_{(f_i f_j)}$. Since $\frac{1}{f_j} \in \mathcal{O}_X(D(f_j)) = R_{f_j}$, we get $f_i^{n_j} \rho_j(s) = \rho_j(f_i^{n_j} s) = 0$. Since I is a finite set, for a positive integer $n \geq \max_{j \in I} n_j$ we obtain $\rho_j(f_i^n s) = 0$ for all $j \in I$. Recall that $\rho_j = \rho_{D(f_j), X}$, to conclude that sheaf condition (F1) implies $f_i^n s = 0$ in $M = \Gamma(X, \mathcal{F})$. Therefore, $\frac{s}{f_i^m} = 0$ in M, i.e., $\hat{\rho}_i$ is injective.

We will show that $\hat{\rho}_i$ is surjective. As before, one regards

$$\rho_{D(f_i f_j), D(f_i)}(s_i) \in \mathcal{F}(D(f_i f_j)) = (M_j)_{(f_i f_j)}.$$

Namely, one can choose a positive integer n_j to satisfy $(f_i f_j)^{n_j} s_i \in M_j$. Since M_j is an R_{f_j}-module, we have $f_i^{n_j} s_i \in M_j$. Again take $n \geq \max_{j \in I} n_j$. Then for all $j \in I$, we have $f_i^n s_i \in M_j$. Then sheaf condition (F2) implies that $s \in M = \mathcal{F}(X)$ is determined by $f_i^n s_i$, namely, $s_i = \frac{s}{f_i^n}$. Hence $\hat{\rho}_i$ is surjective.

Consequently, for all $i \in I$, we have $\widetilde{M}|D(f_i) \xrightarrow{\sim} \mathcal{F}|D(f_i)$. That is, $\widetilde{M}$ is isomorphic to $\mathcal{F}$. □

By combining Theorem 4.21 with Example 4.19(2), we obtain the following important result.

4.2. QUASICOHERENT SHEAVES AND COHERENT SHEAVES

COROLLARY 4.22. *For an exact sequence of quasicoherent $\mathcal{O}_X$-modules*

$$0 \to \mathcal{F}_1 \to \mathcal{F}_2 \to \mathcal{F}_3 \to 0$$

over an affine scheme $X = \operatorname{Spec} R$, we obtain the following exact sequence of R-modules:

$$0 \to \Gamma(X, \mathcal{F}_1) \to \Gamma(X, \mathcal{F}_2) \to \Gamma(X, \mathcal{F}_3) \to 0.$$

PROOF. Let $M_j = \Gamma(X, \mathcal{F}_j)$. From Proposition 4.8, we have the exact sequence of R-modules

$$0 \to M_1 \to M_2 \to M_3.$$

By Theorem 4.21, $\mathcal{F}_j \xrightarrow{\sim} \widetilde{M}_j$. Our assumption is that $\tilde{\varphi}_2 : \widetilde{M}_2 \to \widetilde{M}_3$ is surjective. Then put $N = \operatorname{Coker} \varphi_2$. The exact sequence

$$M_2 \xrightarrow{\varphi_2} M_3 \to N \to 0$$

induces the exact sequence

$$\widetilde{M}_2 \xrightarrow{\tilde{\varphi}_2} \widetilde{M}_3 \to \widetilde{N} \to 0$$

of $\mathcal{O}_X$-modules. (Note that the above exactness comes from Problem 10 and Example 4.19(2).) Since $\tilde{\varphi}_2$ is surjective, we have $\widetilde{N} = 0$. Then Theorem 4.21 implies $N = \Gamma(X, \widetilde{N}) = 0$, i.e., φ_3 is surjective. □

Corollary 4.22 is important. As we showed in the proof of Theorem 4.21, when X is an affine scheme, a surjective map $\varphi : \mathcal{F} \to \mathcal{G}$ of quasicoherent sheaves induces surjectivity on $\Gamma(X, \mathcal{F}) \to \Gamma(X, \mathcal{G})$. This result tells us that the higher cohomology groups of an affine scheme with coefficient in a quasicoherent sheaf vanish. We will return to this topic in Chapter 6.

PROBLEM 12. Prove that for an exact sequence of quasicoherent sheaves over an affine scheme $X = \operatorname{Spec} R$,

$$\cdots \to \mathcal{F}_1 \to \mathcal{F}_2 \to \mathcal{F}_3 \to \mathcal{F}_4 \to \cdots,$$

we have an exact sequence of R-modules of sections over X

$$\cdots \to \Gamma(X, \mathcal{F}_1) \to \Gamma(X, \mathcal{F}_2) \to \Gamma(X, \mathcal{F}_3) \to \Gamma(X, \mathcal{F}_4) \to \cdots.$$

Note that the totality of quasicoherent sheaves over an affine scheme $X = \operatorname{Spec} R$ as objects and their sheaf homomorphisms as morphisms form a category. We denote the set of morphisms from $\mathcal{F}$ to $\mathcal{G}$ by $\operatorname{Hom}_{\mathcal{O}_X}(\mathcal{F}, \mathcal{G})$. We write $(\mathcal{O}_X\text{-q.c.Mod})$ for this category of quasicoherent sheaves. The assignment sending an R-module M

to an $\mathcal{O}_X$-module $\widetilde{M}$ is a functor from the category (R-mod) of R-modules to the category ($\mathcal{O}_X$-q.c.Mod). The following corollary gives a stronger result.

COROLLARY 4.23. *For an affine scheme $X = \operatorname{Spec} R$, the functor*
$$\Phi : (R\text{-mod}) \to (\mathcal{O}_X\text{-q.c. Mod}),$$
$$M \mapsto \widetilde{M},$$
establishes an equivalence of categories.

PROOF. For a quasicoherent sheaf $\mathcal{F}$, assigning an R-module $\Gamma(X, \mathcal{F})$ defines a functor
$$\Psi : (\mathcal{O}_X\text{-q.c.Mod}) \to (R\text{-mod}),$$
$$\mathcal{F} \mapsto \Gamma(X, \mathcal{F}).$$
Then Theorem 4.21 implies
$$\Psi \circ \Phi \xrightarrow{\sim} \operatorname{id} \quad \text{and} \quad \Phi \circ \Psi \xrightarrow{\sim} \operatorname{id}. \quad \square$$

COROLLARY 4.24. *Let $\mathcal{F}$ and $\mathcal{G}$ be quasicoherent sheaves over a scheme $(X, \mathcal{O}_X)$. Then $\operatorname{Ker} \varphi, \operatorname{Im} \varphi$, and $\operatorname{Coker} \varphi$ of an $\mathcal{O}_X$-homomorphism $\varphi : \mathcal{F} \to \mathcal{G}$ are quasicoherent.*

PROOF. Since X can be covered by affine schemes, the coherency of those sheaves may be verified over an affine scheme. Hence, we may assume that X is an affine scheme. For $X = \operatorname{Spec} R$, let $M = \Gamma(X, \mathcal{F})$ and $N = \Gamma(X, \mathcal{G})$. Then we have $\mathcal{F} = \widetilde{M}$ and $\mathcal{G} = \widetilde{N}$. Hence we get $\operatorname{Ker} \varphi = \widetilde{\operatorname{Ker} \varphi_X}, \operatorname{Im} \varphi = \widetilde{\operatorname{Im} \varphi_X}$, and $\operatorname{Coker} \varphi = \widetilde{\operatorname{Coker} \varphi_X}$. That is, these three sheaves are quasicoherent. $\square$

(c) **Coherent Sheaves**

We observed that quasicoherent sheaves are analogous to R-modules. As the analogous sheaf to a finitely generated R-module, we will consider a coherent sheaf.

DEFINITION 4.25. An $\mathcal{O}_X$-module $\mathcal{F}$ over a scheme X is said to be *coherent*, or to be a *coherent sheaf*, if conditions (i) and (ii) are satisfied.

(i) $\mathcal{F}$ is a finitely generated $\mathcal{O}_X$-module.

(ii) For an arbitrary open set U of X and for an arbitrary $\mathcal{O}_U$-module homomorphism
$$\varphi : \mathcal{O}_U^{\oplus n} \to \mathcal{F}|U,$$
$\operatorname{Ker} \varphi$ is a finitely generated $\mathcal{O}_X$-module. $\square$

4.2. QUASICOHERENT SHEAVES AND COHERENT SHEAVES

First we will show that a coherent sheaf is quasicoherent. Since a coherent sheaf is a finitely generated $\mathcal{O}_X$-module, for each x in X there exists an open set U containing x so that the sequence of $\mathcal{O}_U$-modules

$$\psi : \mathcal{O}_U^{\oplus n} \to \mathcal{F}|U \to 0$$

is exact. Then $\operatorname{Ker}\psi$ is finitely generated as an $\mathcal{O}_U$-module. Hence, there exists an open set V containing x so that the sequence

$$\varphi : \mathcal{O}_V^{\oplus m} \to \operatorname{Ker}\psi|V \to 0$$

is exact. Take such a V contained in U to obtain the following exact sequence:

(4.14) $$\mathcal{O}_V^{\oplus m} \xrightarrow{\varphi} \mathcal{O}_V^{\oplus n} \xrightarrow{\psi|V} \mathcal{F}|V \to 0.$$

That is, $\mathcal{F}$ is quasicoherent.

However, the next example shows that a finitely generated $\mathcal{O}_X$-module can be quasicoherent but not coherent.

EXAMPLE 4.26. Let I be an ideal of the ring $k[x_1, x_2, \ldots, x_n, \ldots]$ of polynomials in infinitely many variables over a field k defined by $I = (x_1 x_2, \ldots, x_1 x_n, \ldots, x_i x_j, \ldots)$, and let $R = k[x_1, x_2, \ldots]/I$ be the residue ring. Let J be the ideal generated by the images in R of $x_2, x_3, \ldots, x_n, \ldots$. Then we have an exact sequence of R-modules

$$0 \to J \to R \xrightarrow{\times \overline{x}_1} R,$$

where $\overline{x}_1$ indicates the image of x_1 in R. Over the affine scheme $X = \operatorname{Spec} R$, there corresponds the exact sequence of $\mathcal{O}_X$-quasicoherent sheaves

$$0 \to \widetilde{J} \to \mathcal{O}_X \xrightarrow{\varphi} \mathcal{O}_X.$$

Then $\widetilde{J} = \operatorname{Ker}\varphi$ is not finitely generated as an $\mathcal{O}_X$-module. Hence, $\mathcal{O}_X$ is not a coherent sheaf as an $\mathcal{O}_X$-module. □

In this example, R is not a Noetherian ring, and J is not a finitely generated ideal. For a Noetherian ring we have the following fact.

PROPOSITION 4.27. *Let $\mathcal{F}$ be a quasicoherent $\mathcal{O}_X$-module over the affine scheme $X = \operatorname{Spec} R$ determined by a Noetherian ring R. Then $\mathcal{F}$ is coherent if and only if $\Gamma(X, \mathcal{F})$ is a finitely generated R-module. In particular, $\mathcal{O}_X$ is a coherent sheaf. Conversely, a finitely generated $\mathcal{O}_X$-module is coherent.*

PROOF. If $\mathcal{F}$ is a coherent sheaf, an open affine covering $X = \bigcup_{i \in I} D(f_i)$ can be chosen so that for each $i \in I$ the sequence
$$\mathcal{O}_{D(f_i)}^{\oplus n_i} \to \mathcal{F}|D(f_i) \to 0$$
is exact. Since $D(f_i)$ has an affine scheme structure, the above exact sequence implies the following exact sequence of R_{f_i}-modules:
$$R_{f_i}^{\oplus n_i} \xrightarrow{\varphi_i} M_i = \Gamma(D(f_i), \mathcal{F}) \to 0.$$
Let $M = \Gamma(X, \mathcal{F})$. Then $M_i = M_i$. As an R_{f_i}-module, we can choose generators $t_{ij} = \varphi((0, \ldots, 0, \overset{j}{1}, 0, \ldots, 0)) \in M_i$ for M_i. Since $M_i = M_{f_i}$, there exists a positive integer m_{ij} to satisfy $\tilde{t}_{ij} = f_i^{m_{ij}} t_{ij} \in M$. For an arbitrary element s of M, regard $\frac{s}{1} \in M_{f_i}$. We can write
$$\frac{s}{1} = \sum_{j=1}^{n_i} a_j t_{ij}, \qquad a_j \in R_{f_i}.$$
Therefore, for a sufficiently large positive integer r_i, we can have
$$f_i^{r_i} s = \sum_{j=1}^{n_i} b_j \tilde{t}_{ij}, \qquad b_j \in R.$$
Recall that X is quasicompact (Corollary 2.9). Hence, I can be taken as a finite set. Let $m = \max_{i,j}\{r_i, m_{ij}\}$. Then $f_i^m s \in M$. Notice that $X = \bigcup_{i \in I} D(f_i) = \bigcup_{i \in I} D(f_i^r)$. Proposition 2.8 implies that $c_i \in R$ and $i \in I$ can be chosen to satisfy $\sum_{i \in I} c_i f_i^r = 1$. Therefore,
$$s = \sum_{i \in I} \sum_{j=1}^{n_i} (c_i b_j) \tilde{t}_{ij},$$
namely, M is generated by finitely many $\{\tilde{t}_{ij}\}$ as an R-module.

Conversely, assume that $M = \Gamma(X, \mathcal{F})$ is a finitely generated R-module. Then $\mathcal{F} = \widetilde{M}$. Since for $f \in R$, M_f is a finitely generated R_f-module, we have a surjective homomorphism
$$R_f^{\oplus n} \to M_f \to 0.$$
This homomorphism induces the surjective homomorphism of $\mathcal{O}_{D(f)}$-modules
$$\mathcal{O}_{D(f)}^{\oplus n} \to \widetilde{M_f} \to 0.$$
By regarding $M_f = \Gamma(D(f), \mathcal{F})$, we see that $\widetilde{M_f} = \mathcal{F}|D(f)$. That is, $\mathcal{F}$ is a finitely generated $\mathcal{O}_X$-module.

4.2. QUASICOHERENT SHEAVES AND COHERENT SHEAVES 33

Let U be an open set of X and let
$$\varphi : \mathcal{O}_U^{\oplus m} \to \mathcal{F}|U$$
be an $\mathcal{O}_U$-homomorphism. For $x \in U$, choose an affine open set $D(f) \subset U$ containing x. Then the restriction of φ to $D(f)$ is determined by the homomorphism of R_f-modules
$$\psi : R_f^{\oplus m} \to M_f.$$
Then we have $\operatorname{Ker}\varphi|D(f) = \widetilde{\operatorname{Ker}\psi}$. Note that R_f is Noetherian since R is Noetherian. Since $\operatorname{Ker}\psi$ is an R_f-submodule of the Noetherian R_f-module $R_f^{\oplus m}$, $\operatorname{Ker}\psi$ is a finitely generated R_f-module. Hence $\widetilde{\operatorname{Ker}\psi}$ is a finitely generated $\mathcal{O}_{D(f)}$-module. On the other hand, $\widetilde{\operatorname{Ker}\psi} = \operatorname{Ker}\varphi|D(f)$ implies that $\operatorname{Ker}\varphi$ is a finitely generated $\mathcal{O}_U$-module. Consequently, $\mathcal{F} = \widetilde{M}$ is a coherent sheaf.

Let $\mathcal{F}$ be a finitely generated $\mathcal{O}_X$-module. Choose an open covering $\{U_i\}_{i \in I}$ of X to satisfy the exactness of
$$\mathcal{O}_{U_i}^{\oplus n} \to \mathcal{F}|U_i \to 0.$$
Since X is a Noetherian space, there exists a finite covering $\{U_i\}_{i \in I}$, $U_i = D(f_i)$, $f_i \in R$. Then
$$\varphi_i((0,\ldots,0,\overset{j}{\check{1}},0,\ldots,0)) = g_j^{(i)} \in \Gamma(D(f_i), \mathcal{F}), \quad j = 1, 2, \ldots, n_i,$$
are generators of $\mathcal{F}|U_i$ as an $\mathcal{O}_{U_i}$-module. Take l to be a positive integer large enough so that $f_i^l g_j^{(i)} = g_{ij} \in M = \Gamma(X, \mathcal{F})$. Since $g_{ij}|U_i$, $j = 1, 2, \ldots, n_i$, generate $\mathcal{F}|U_i$ as an $\mathcal{O}_{U_i}$-module, the f_{ij}, $1 \le i \le m$, $1 \le j \le n_i$, generate $\mathcal{F}$ as an $\mathcal{O}_X$-module. Set $n = \sum_{i=1}^m n_i$. Then the sequence of $\mathcal{O}_X$-modules
$$\mathcal{O}_X^{\oplus n} \to \mathcal{F} \to 0,$$
$$(a_{ij}) \mapsto \sum_{i,j} a_{ij} f_{ij},$$
is exact. Therefore, M is a finitely generated R-module, and $\mathcal{F} = \widetilde{M}$ is a coherent sheaf. □

Thus, coherency and finiteness are related in the sense of the above proposition. Coherent sheaves will play a significant role in what will follow. The next theorem, whose proof is left as an exercise (see Exercise **4.4**), is important.

THEOREM 4.28. *Let*

$$0 \to \mathcal{F} \to \mathcal{G} \to \mathcal{H} \to 0$$

be an exact sequence of $\mathcal{O}_X$-modules over a scheme X. If any two sheaves among $\mathcal{F}, \mathcal{G}$ and $\mathcal{H}$ are coherent, the third is also coherent.

COROLLARY 4.29. *Let $\varphi : \mathcal{F} \to \mathcal{G}$ be a homomorphism of coherent $\mathcal{O}_X$-modules $\mathcal{F}$ and $\mathcal{G}$ over a scheme X. Then $\operatorname{Ker}\varphi, \operatorname{Im}\varphi$, and $\operatorname{Coker}\varphi$ are all coherent.*

PROOF. For $x \in X$, there exists an open set U so that the map

$$\psi : \mathcal{O}_U^{\oplus m} \to \mathcal{F}|U \to 0$$

is a surjective homomorphism. Compose ψ with the surjective homomorphism $\mathcal{F}|U \to \operatorname{Im}\varphi|U \to 0$ to obtain the surjective homomorphism

$$\mathcal{O}_U^{\oplus m} \to \operatorname{Im}\varphi|U \to 0$$

of $\mathcal{O}_U$-modules. Therefore, $\operatorname{Im}\varphi$ is a finitely generated $\mathcal{O}_X$-module. Furthermore, for an open set V and an $\mathcal{O}_V$-homomorphism

$$\eta : \mathcal{O}_V^{\oplus n} \to \operatorname{Im}\varphi|V,$$

compose η with the canonical injective $\mathcal{O}_X$-homomorphism $\iota : \operatorname{Im}\varphi \to \mathcal{G}$. Then for $\iota \circ \eta : \mathcal{O}_V^{\oplus n} \to \mathcal{G}|V$, we have $\operatorname{Ker}\eta = \operatorname{Ker}\iota \circ \eta$. Since $\mathcal{G}$ is coherent, $\operatorname{Ker}\eta = \operatorname{Ker}\iota \circ \eta$ is a finitely generated $\mathcal{O}_V$-module. Therefore, $\operatorname{Im}\varphi$ is a coherent sheaf. From the exact sequences of $\mathcal{O}_X$-modules

$$0 \to \operatorname{Ker}\varphi \to \mathcal{F} \to \operatorname{Im}\varphi \to 0,$$
$$0 \to \operatorname{Im}\varphi \to \mathcal{G} \to \operatorname{Coker}\varphi \to 0,$$

we conclude from Theorem 4.28 that $\operatorname{Ker}\varphi$ and $\operatorname{Coker}\varphi$ are coherent sheaves. □

PROBLEM 13. For coherent $\mathcal{O}_X$-modules $\mathcal{F}$ and $\mathcal{G}$, prove that $\mathcal{F} \oplus \mathcal{G}$ is also a coherent sheaf.

Moreover, from the above corollary we will prove that the notion of coherence is closed under the tensor product and under Hom.

COROLLARY 4.30. *For coherent $\mathcal{O}_X$-modules $\mathcal{F}$ and $\mathcal{G}$ over a scheme X, $\mathcal{F} \otimes_{\mathcal{O}_X} \mathcal{G}$ and $\underline{\operatorname{Hom}}_{\mathcal{O}_X}(\mathcal{F}, \mathcal{G})$ are coherent.*

4.2. QUASICOHERENT SHEAVES AND COHERENT SHEAVES

PROOF. From (4.14), for an arbitrary point $x \in X$ there exists an open neighborhood U of x so that we have the exact sequence

$$\mathcal{O}_U^{\oplus m} \xrightarrow{\varphi} \mathcal{O}_U^{\oplus n} \to \mathcal{F}|U \to 0$$

of $\mathcal{O}_U$-modules. Then, tensor this exact sequence with $\mathcal{G}|U$ to get the exact sequence

$$\mathcal{G}^{\oplus m}|U \xrightarrow{\tilde{\varphi}} \mathcal{G}^{\oplus n}|U \to \mathcal{F}|U \otimes \mathcal{G}|U \to 0$$

of $\mathcal{O}_U$-modules. Since $\mathcal{F}|U \otimes \mathcal{G}|U$ and $\mathcal{F} \otimes \mathcal{G}|U$ are the sheafifications of the presheaf $\mathcal{F}(V) \otimes_{\mathcal{O}_X(V)} \mathcal{G}(V)$ for open sets $V \subset U$, we have $\mathcal{F}|U \otimes \mathcal{G}|U = (\mathcal{F} \otimes \mathcal{G})|U$. Hence $(\mathcal{F} \otimes \mathcal{G})|U = \text{Coker}\,\tilde{\varphi}$. For a coherent sheaf $\mathcal{G}$, $\mathcal{G}|U$ is also coherent, and therefore $\mathcal{G}^{\oplus m}|U$ and $\mathcal{G}^{\oplus n}|U$ are coherent. Corollary 4.29 implies that $\text{Coker}\,\tilde{\varphi}$ is coherent. In general, for an open covering $\{U_j\}_{j \in J}$ of X, if $\mathcal{F}|U_j$ is a coherent $\mathcal{O}_X$-module for each $j \in J$, then $\mathcal{F}$ is a coherent sheaf. (See Problem 14, below.) Hence $\mathcal{F}$ is a coherent sheaf. $\square$

PROBLEM 14. Prove that if, for an $\mathcal{O}_X$-module $\mathcal{F}$, $\mathcal{F}|U_j$ is coherent as an $\mathcal{O}_{U_j}$-module for each $j \in J$, where $\{U_j\}_{j \in J}$ is an open covering of a scheme X, then $\mathcal{F}$ is coherent.

EXERCISE 4.31. Let $\mathcal{F}$ and $\mathcal{G}$ be $\mathcal{O}_X$-modules over a scheme X. If $\mathcal{F}$ is a coherent sheaf, for an arbitrary point $x \in X$ there is an isomorphism of $\mathcal{O}_{X,x}$-modules

$$\underline{\text{Hom}}_{\mathcal{O}_X}(\mathcal{F}, \mathcal{G})_x \xrightarrow{\sim} \text{Hom}_{\mathcal{O}_{X,x}}(\mathcal{F}_x, \mathcal{G}_x).$$

PROOF. For a coherent sheaf $\mathcal{F}$, as in (4.14), there is an open neighborhood U of x such that we have an exact sequence of $\mathcal{O}_U$-modules

(4.15) $$\mathcal{O}_U^{\oplus m} \xrightarrow{\varphi} \mathcal{O}_U^{\oplus n} \to \mathcal{F}|U \to 0.$$

By Example 4.14, $\underline{\text{Hom}}_{\mathcal{O}_X|U}(\mathcal{F}|U, \mathcal{G}|U) = \underline{\text{Hom}}_{\mathcal{O}_X}(\mathcal{F}, \mathcal{G})|U$. Therefore, (4.15) induces the exact sequence

$$0 \to \underline{\text{Hom}}_{\mathcal{O}_X}(\mathcal{F}, \mathcal{G})|U \to \underline{\text{Hom}}_{\mathcal{O}_X}(\mathcal{O}_X^{\oplus n}, \mathcal{G})|U \xrightarrow{\hat{\varphi}} \underline{\text{Hom}}_{\mathcal{O}_X}(\mathcal{O}_X^{\oplus m}, \mathcal{G})|U$$

of $\mathcal{O}_U$-modules. From this exact sequence, the $\mathcal{O}_{X,x}$-module exact sequence

$$0 \to \underline{\text{Hom}}_{\mathcal{O}_X}(\mathcal{F}, \mathcal{G})_x \to \underline{\text{Hom}}_{\mathcal{O}_X}(\mathcal{O}_X^{\oplus n}, \mathcal{G})_x \xrightarrow{\hat{\varphi}_x} \underline{\text{Hom}}_{\mathcal{O}_X}(\mathcal{O}_X^{\oplus m}, \mathcal{G})_x$$

is obtained.

In general, for $\mathcal{O}_X$-modules $\mathcal{A}$ and $\mathcal{B}$ and an open neighborhood V of x, Exercise 4.14 implies
$$\underline{\mathrm{Hom}}_{\mathcal{O}_X}(\mathcal{A}, \mathcal{B})(V) = \mathrm{Hom}_{\mathcal{O}_{X|V}}(\mathcal{A}|V, \mathcal{B}|V),$$
inducing a natural $\mathcal{O}_{X,x}$-homomorphism
$$\underline{\mathrm{Hom}}_{\mathcal{O}_X}(\mathcal{A}, \mathcal{B})_x \to \mathrm{Hom}_{\mathcal{O}_{X,x}}(\mathcal{A}_x, \mathcal{B}_x).$$
From (4.15) we get the exact sequence of $\mathcal{O}_{X,x}$-modules
$$\mathcal{O}_{X,x}^{\oplus m} \to \mathcal{O}_{X,x}^{\oplus n} \to \mathcal{F}_x \to 0.$$
Hence we have the exact sequence
$$0 \to \mathrm{Hom}_{\mathcal{O}_{X,x}}(\mathcal{F}_x, \mathcal{G}_x) \to \mathrm{Hom}_{\mathcal{O}_{X,x}}(\mathcal{O}_{X,x}^{\oplus n}, \mathcal{G}_x) \to \mathrm{Hom}_{\mathcal{O}_{X,x}}(\mathcal{O}_{X,x}^{\oplus m}, \mathcal{G}_x).$$
Consequently, we obtain the commutative diagram of $\mathcal{O}_{X,x}$-modules

$$\begin{array}{ccccccc}
0 & \to & \underline{\mathrm{Hom}}_{\mathcal{O}_X}(\mathcal{F}, \mathcal{G})_x & \to & \underline{\mathrm{Hom}}_{\mathcal{O}_X}(\mathcal{O}_X^{\oplus n}, \mathcal{G})_x & \xrightarrow{\hat{\varphi}_x} & \underline{\mathrm{Hom}}_{\mathcal{O}_X}(\mathcal{O}_X^{\oplus m}, \mathcal{G})_x \\
& & \eta_1 \downarrow & & \eta_2 \downarrow & & \eta_3 \downarrow \\
0 & \to & \mathrm{Hom}_{\mathcal{O}_{X,x}}(\mathcal{F}_x, \mathcal{G}_x) & \to & \mathrm{Hom}_{\mathcal{O}_{X,x}}(\mathcal{O}_{X,x}^{\oplus n}, \mathcal{G}_x) & \xrightarrow{\varphi_x^*} & \mathrm{Hom}_{\mathcal{O}_{X,x}}(\mathcal{O}_{X,x}^{\oplus m}, \mathcal{G}_x).
\end{array}$$

Since $\underline{\mathrm{Hom}}_{\mathcal{O}_X}(\mathcal{O}_X, \mathcal{G}) = \mathcal{G}$, η_2 and η_3 are isomorphisms. Hence η_3 is also an isomorphism. □

4.3. Direct Image and Inverse Image

(a) Direct Image and Inverse Image of a Sheaf under a Continuous Map

Let $f : X \to Y$ be a continuous map of topological spaces. For a sheaf $\mathcal{F}$ of additive groups over X, we get the sheaf $f_*\mathcal{F}$ over Y associated with the presheaf $\mathcal{F}(f^{-1}(U))$. This $f_*\mathcal{F}$ is called the direct image of $\mathcal{F}$ (see Exercise 2.26). Then for a sheaf $\mathcal{G}$ over Y, is it possible to construct a sheaf over X? For an open set V in X, define an order among open sets U of Y containing $f(V)$ as follows: $U_1 > U_2$ if $U_1 \subset U_2$. We can consider an inductive limit over this inductively ordered set. Hence we get

(4.16) $$\varinjlim_{f(V) \subset U} \mathcal{G}(U).$$

For open sets V and restriction maps, we get the presheaf

$$\varinjlim_{f(V) \subset U} \mathcal{G}(U)$$

over X. Then let $f^{-1}\mathcal{G}$ be the sheafification of this presheaf. This

4.3. DIRECT IMAGE AND INVERSE IMAGE

sheaf $f^{-1}\mathcal{G}$ is said to be the *inverse image* of $\mathcal{G}$ under f. By the definition of sheafification, the stalk of $f^{-1}\mathcal{G}$ at $x \in X$ is given as

(4.17) $$(f^{-1}\mathcal{G})_x = \varinjlim_{f(x) \in U} \mathcal{G}(U) = \mathcal{G}_{f(x)}.$$

An important relation exists between the direct image and the inverse image, as follows.

LEMMA 4.32. *Let $f : X \to Y$ be a continuous map between topological spaces X and Y. Then for a sheaf $\mathcal{F}$ over X and a sheaf $\mathcal{G}$ over Y, there exists an isomorphism of additive groups*

$$\operatorname{Hom}_X(f^{-1}\mathcal{G}, \mathcal{F}) \xrightarrow{\sim} \operatorname{Hom}_Y(\mathcal{G}, f_*\mathcal{F}),$$

where $\operatorname{Hom}_X(\cdot, \cdot)$ and $\operatorname{Hom}_Y(\cdot, \cdot)$ denote the totalities of sheaf homomorphisms over X and Y, respectively.

PROOF. For the sake of simplicity, let $f^\bullet \mathcal{G}$ be the presheaf over X defined as in (4.16). We first show that

(4.18) $$\operatorname{Hom}_X(f^{-1}\mathcal{G}, \mathcal{F}) \xrightarrow{\sim} \operatorname{Hom}_{\text{presheaf}}(f^\bullet \mathcal{F}, \mathcal{G}).$$

For an open set U of X, one can choose an open covering $\{U_\alpha\}_{\alpha \in A}$ of U so that for $s \in f^{-1}\mathcal{G}(U_\alpha)$ we have $s_\alpha \in f^\bullet \mathcal{G}(U_\alpha)$ for $\alpha \in A$; for $U_\alpha \cap U_\beta \neq \varnothing$, the germs of s_α and s_β coincide at an arbitrary point $x \in U_\alpha \cap U_\beta$. Namely, $s_{\alpha,x} = s_{\beta,x}$ (see (4.1)).

Let $\varphi \in \operatorname{Hom}_X(f^{-1}\mathcal{G}, \mathcal{F})$. For the above s, let $\varphi_U(s) = t$, $t_\alpha = \varphi_{U_\alpha}(s_\alpha)$, $\alpha \in A$. Then we get $\varphi_x(s_{\alpha,x}) = \varphi_x(s_x) = t_x$. Hence $t_\alpha = \varphi_{U_\alpha,U}(t)$. Thus, for a given $\hat{s} \in f^\bullet \mathcal{G}(U)$, the element $s \in f^{-1}\mathcal{G}(U)$ determined by $\hat{s}$ induces a unique element $t = \varphi_U(s) \in \mathcal{F}(U)$. Then let $\hat{\varphi}_U(\hat{s}) = t$. Observe that $\{\hat{\varphi}_U\}$ determines $\hat{\varphi} \in \operatorname{Hom}_{\text{presheaf}}(f^\bullet \mathcal{G}, \mathcal{F})$.

Conversely, for a given $\hat{\psi} \in \operatorname{Hom}_{\text{presheaf}}(f^\bullet \mathcal{G}, \mathcal{F})$, as in the above, choose $s_\alpha \in f^\bullet \mathcal{G}(U_\alpha)$, $\alpha \in A$ (where $\{U_\alpha\}_{\alpha \in A}$ is an open covering of U). Then $t_\alpha = \hat{\psi}_{U_\alpha}(s_\alpha)$, $\alpha \in A$, determines $t \in \mathcal{F}(U)$. By defining $\psi_U(s) = t$, $\psi = \{\psi_U\}$ is a sheaf homomorphism from $f^{-1}\mathcal{G}$ to $\mathcal{F}$. Note that the maps $\varphi \mapsto \hat{\varphi}$ and $\hat{\psi} \mapsto \psi$ are inverse to each other. Hence (4.18) is proved.

Next, we will prove that

(4.19) $$\operatorname{Hom}_{\text{presheaf}}(f^\bullet \mathcal{G}, \mathcal{F}) \xrightarrow{\sim} \operatorname{Hom}_Y(\mathcal{G}, f_*\mathcal{F}).$$

When $\hat{\psi} \in \mathrm{Hom}_{\mathrm{presheaf}}(f^{\bullet}\mathcal{G}, \mathcal{F})$ is given, for an open set V of Y, from (4.19) we get homomorphisms of additive groups
$$\tilde{\psi}_{f^{-1}(V)} : \mathcal{G}(V) \to f^{\bullet}\mathcal{G}(f^{-1}(V)) \xrightarrow{\hat{\psi}_{f^{-1}(V)}} \mathcal{F}(f^{-1}(V)) = f_{*}\mathcal{F}(V).$$
Let $\psi_V = \tilde{\psi}_{f^{-1}(V)}$. Then $\{\psi_V\}$ are sheaf homomorphisms of additive groups from $\mathcal{G}$ to $f_*\mathcal{F}$. Conversely, for a given $\psi \in \mathrm{Hom}_Y(\mathcal{G}, f_*\mathcal{F})$, we get homomorphisms of additive groups
$$\varphi_V : \mathcal{G}(V) \to f_*\mathcal{F}(V) = \mathcal{F}(f^{-1}(V)),$$
where V is an open set of Y. For an open set U of X, take an open set V satisfying $f(U) \subset V$. From φ_V we obtain homomorphisms of additive groups
$$\tilde{\varphi}_{U,V} : \mathcal{G}(V) \xrightarrow{\varphi_V} \mathcal{F}(f^{-1}(V)) \xrightarrow{\rho_{U,f^{-1}(V)}} \mathcal{F}(U).$$
Since $\{\varphi_V\}$ are homomorphisms of sheaves, for $f(U) \subset V \subset V'$ we get the commutative diagram

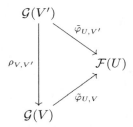

Hence, the homomorphism of additive groups
$$\hat{\varphi}_U : f^{\bullet}\mathcal{G}(U) = \varinjlim_{f(U) \subset V} \mathcal{G}(V) \to \mathcal{F}(U)$$
is obtained. Moreover, for open sets $W \subset U$ one can show as demonstrated in the above that the diagram
$$\begin{array}{ccc} f^{\bullet}\mathcal{G}' & \xrightarrow{\hat{\varphi}_U} & \mathcal{F}(U) \\ \rho_{W,U} \downarrow & & \downarrow \rho_{W,U} \\ f^{\bullet}\mathcal{G}(W) & \xrightarrow{\hat{\varphi}_W} & \mathcal{F}(W) \end{array}$$
is commutative. Therefore, $\{\hat{\varphi}_U\}$ are presheaf homomorphisms from $f^{\bullet}\mathcal{G}$ to $\mathcal{F}$. One can easily show that $\hat{\psi} \mapsto \psi$ and $\varphi \mapsto \hat{\varphi}$ are inverse to each other, proving (4.19). Our lemma now follows from (4.18) and (4.19). $\square$

PROBLEM 15. Prove that, for an exact sequence of sheaves of additive groups
$$0 \to \mathcal{F}_1 \to \mathcal{F}_2 \to \mathcal{F}_3 \to 0$$
over a topological space Y, a continuous map $f : X \to Y$ induces the following exact sequence of inverse images over X:
$$0 \to f^{-1}\mathcal{F}_1 \to f^{-1}\mathcal{F}_2 \to f^{-1}\mathcal{F}_3 \to 0.$$

(b) **Direct Image and Inverse Image under a Scheme Morphism**

For a scheme morphism $f : X \to Y$ and an $\mathcal{O}_X$-module $\mathcal{F}$, we have the direct image $f_*\mathcal{F}$. More precisely, the scheme morphism f should be written as $\Phi = (f, \theta) : (X, \mathcal{O}_X) \to (Y, \mathcal{O}_Y)$, where f is a continuous map and where $\theta : \mathcal{O}_Y \to f_*\mathcal{O}_X$ is a homomorphism of sheaves of commutative rings (see §2.3(b)). For an open set U in Y,

(4.20) $\qquad \mathcal{O}_X(f^{-1}(U)) \times \mathcal{F}(f^{-1}(U)) \to \mathcal{F}(f^{-1}(U))$

induces an $f_*\mathcal{O}_X$-module structure on $f_*\mathcal{F}$. Moreover, a homomorphism of commutative rings

$$\theta_U : \mathcal{O}_Y(U) \to \mathcal{O}_X(f^{-1}(U))$$

combined with (4.20) gives an action of $\mathcal{O}_Y(U)$ on $\mathcal{F}(f^{-1}(U)) = (f_*\mathcal{F})(U)$, i.e.,

$$\mathcal{O}_Y(U) \times (f_*\mathcal{F})(U) \to (f_*\mathcal{F})(U).$$

Namely, $f_*\mathcal{F}$ is an $\mathcal{O}_Y$-module. For a morphism of affine schemes, the direct image of a coherent sheaf can be interpreted as follows.

EXERCISE 4.33. Let $(\varphi^a, \varphi^\#) : (X, \mathcal{O}_X) = (\operatorname{Spec} S, \mathcal{O}_{\operatorname{Spec} S}) \to (Y, \mathcal{O}_Y) = (\operatorname{Spec} R, \mathcal{O}_{\operatorname{Spec} R})$ be an affine scheme morphism induced by a homomorphism of commutative rings $\varphi : R \to S$. Let $\mathcal{F} = \widetilde{M}$ be the $\mathcal{O}_X$-module determined by an S-module M. Then the direct image $\varphi^a_*\mathcal{F}$ coincides with the $\mathcal{O}_Y$-module determined by the R-module structure on M via φ. Consequently, the direct image of a quasicoherent sheaf is a quasicoherent sheaf.

PROOF. For $f \in R$, let $g = \varphi(f) \in S$. Then the inverse image of an open set $U = D(f)$ of Y under φ^a can be written as $(\varphi^a)^{-1}(D(f)) = D(g)$. The homomorphism

$$\varphi^\#_U : \mathcal{O}_Y(U) \to (\varphi^a_*\mathcal{O}_X)(U) = \mathcal{O}_X((\varphi^a)^{-1}(U))$$

is precisely the homomorphism $\varphi_f : R_f \to S_g$ induced by φ. Let $M_{[\varphi]}$ denote the S-module M regarded as an R-module via φ. Then, by φ_f,

M_g may be considered as an R_g-module which we denote by $(M_g)_{[\varphi_f]}$. Then $(M_g)_{[\varphi_f]}$ coincides with $(M_{[\varphi]})_f$. Hence we get $\varphi_*^a \mathcal{F} = \widetilde{M}_{[\varphi]}$. □

Next, for a scheme morphism $f : X \to Y$, we will consider the inverse image $f^{-1}\mathcal{G}$ of an $\mathcal{O}_Y$-module $\mathcal{G}$. In general, for a given sheaf $\mathcal{A}$ of commutative rings over a topological space X, a sheaf $\mathcal{B}$ of $\mathcal{A}$-modules can be defined in the same way as an $\mathcal{O}_X$-module. That is, for any open set U of X, $\mathcal{B}(U)$ is an $\mathcal{A}(U)$-module. For open sets $V \subset U$, we have the commutative diagram

$$\begin{array}{ccc} \mathcal{A}(U) \times \mathcal{B}(U) & \longrightarrow & \mathcal{B}(U) \\ \rho^{\mathcal{A}}_{V,U} \times \rho^{\mathcal{B}}_{V,U} \downarrow & & \downarrow \rho^{\mathcal{B}}_{V,U} \\ \mathcal{A}(V) \times \mathcal{B}(V) & \longrightarrow & \mathcal{B}(V) \end{array}$$

Then $\mathcal{B}$ is said to be an $\mathcal{A}$-module. Using this terminology, the inverse image $f^{-1}\mathcal{G}$ of an $\mathcal{O}_Y$-module $\mathcal{G}$ under a scheme morphism $f : X \to Y$ can be considered as an $f^{-1}\mathcal{O}_Y$-module. The scheme morphism $f : X \to Y$ determines a homomorphism of sheaves of commutative rings $\theta : \mathcal{O}_Y \to f_*\mathcal{O}_X$. Then, by Lemma 4.32, this homomorphism determines a homomorphism of sheaves of commutative rings

$$\hat{\theta} : f^{-1}\mathcal{O}_Y \to \mathcal{O}_X.$$

By this $\hat{\theta}$, $\mathcal{O}_X$ can be regarded as an $f^{-1}\mathcal{O}_Y$-algebra. Then, for the $f^{-1}\mathcal{O}_Y$-module $f^{-1}\mathcal{G}$ and $\mathcal{O}_X$, denote the tensor product over $f^{-1}\mathcal{O}_Y$ as

(4.21) $$f^*\mathcal{G} = f^{-1}\mathcal{G} \otimes_{f^{-1}\mathcal{O}_Y} \mathcal{O}_X.$$

The $\mathcal{O}_X$-module $f^*\mathcal{G}$ is said to be the *inverse image of $\mathcal{G}$ under a scheme morphism f*. More formally speaking, for $\Phi = (f, \theta) : (X, \mathcal{O}_X) \to (Y, \mathcal{O}_Y)$, the inverse image of $\mathcal{G}$ should be written as $\Phi^*\mathcal{G}$ instead of $f^*\mathcal{G}$. As we noted, $f^*\mathcal{G}$ is an $\mathcal{O}_X$-module. From (4.21), the stalk $(f^*\mathcal{G})_x$ at $x \in X$ is given by

(4.22) $$(f^*\mathcal{G})_x = \mathcal{G}_{f(x)} \otimes_{\mathcal{O}_{Y,f(x)}} \mathcal{O}_{X,x}.$$

The structure of the inverse image of a quasicoherent sheaf under an affine scheme morphism will be given in the following lemma. Let $\tilde{\varphi} : X = \operatorname{Spec} S \to Y = \operatorname{Spec} R$ be the morphism of affine schemes determined by a homomorphism $\varphi : R \to S$ of commutative rings. For an R-module M, let $\mathcal{F} = \widetilde{M}$ be the induced $\mathcal{O}_Y$-quasicoherent sheaf.

4.3. DIRECT IMAGE AND INVERSE IMAGE

LEMMA 4.34. *In the above notation, the inverse image $\tilde{\varphi}^*\mathcal{F}$ of $\mathcal{F}$ under the morphism $\tilde{\varphi}$ is the $\mathcal{O}_X$-module determined by the S-module $M \otimes_R S$.*

PROOF. By $\varphi : R \to S$, the S-module $M \otimes_R S$ can be considered as an R-module which we denote by $M_{[\varphi]} \otimes_R S$. For $m \in M$, assign $m \otimes 1 \in M \otimes_R S$. Then we get an R-module homomorphism $h : M \to M_{[\varphi]} \otimes_R S$. Then h induces an $\mathcal{O}_Y$-module homomorphism

$$\tilde{h} : \widetilde{M} \to \widetilde{M_{[\varphi]} \otimes_R S}.$$

From Example 4.33, this $\mathcal{O}_Y$-module homomorphism $\tilde{h}$ can be written as

$$\tilde{h} : \mathcal{F} \to \tilde{\varphi}_*(\widetilde{M \otimes_R S}).$$

Lemma 4.32 implies that there is a sheaf homomorphism

$$\hat{h} : \tilde{\varphi}^{-1}\mathcal{F} \to \widetilde{M \otimes_R S}.$$

Notice that $\hat{h}$ is a homomorphism of $\tilde{\varphi}^{-1}\mathcal{O}_Y$-modules. Since $\widetilde{M \otimes_R S}$ is an $\mathcal{O}_X$-module, $\hat{h}$ induces an $\mathcal{O}_X$-module homomorphism

$$\overline{h} : \tilde{\varphi}^*\mathcal{F} = \tilde{\varphi}^{-1}\mathcal{F} \otimes_{f^{-1}\mathcal{O}_Y} \mathcal{O}_X \to \widetilde{M \otimes_R S}.$$

On the other hand, for a prime ideal $\mathfrak{p}$ of S, set $\mathfrak{q} = \varphi^{-1}(\mathfrak{p})$. Then $\varphi^a(\mathfrak{p}) = \mathfrak{q}$, and from (4.22) we get

$$(\tilde{\varphi}^*\mathcal{F})_\mathfrak{p} = (\tilde{\varphi}^{-1}\mathcal{F})_\mathfrak{p} \otimes_{\mathcal{O}_{Y,\mathfrak{q}}} \mathcal{O}_{X,\mathfrak{p}} = M_\mathfrak{q} \otimes_{R_\mathfrak{q}} S_\mathfrak{p} = (M \otimes_R S)_\mathfrak{p}.$$

Note also that

$$(\widetilde{M \otimes_R S})_\mathfrak{p} = (M \otimes_R S)_\mathfrak{p}.$$

Hence $\overline{h}$ is an isomorphism between the stalks at each point in X. Namely, $\overline{h}$ is an isomorphism. □

From Lemma 4.34 we obtain the following proposition.

PROPOSITION 4.35. *Let $f : X \to Y$ be a morphism of schemes.*
(i) For a quasicoherent $\mathcal{O}_Y$-module $\mathcal{G}$, the inverse image $f^*\mathcal{G}$ under f is a quasicoherent $\mathcal{O}_X$-module.
(ii) When X and Y are Noetherian schemes, the inverse image $f^*\mathcal{G}$ of a coherent $\mathcal{O}_Y$-module $\mathcal{G}$ is a coherent $\mathcal{O}_X$-module.

PROOF. For $x \in X$ and $y = f(x) \in Y$, choose affine open neighborhoods $U = \operatorname{Spec} S$ and $V = \operatorname{Spec} R$ of x and y, respectively, so that $f(U) \subset V$. Then $f|U$ is induced from a ring homomorphism $\varphi : R \to S$. Since $\mathcal{G}$ is quasicoherent, Theorem 4.21 implies that

$\mathcal{G}|V$ is isomorphic to the $\mathcal{O}_X$-module determined by the R-module $M = \Gamma(V, \mathcal{G})$. Hence, by Lemma 4.34, we get $f^*\mathcal{G}|U = \widetilde{M \otimes_R S}$. Namely, $f^*\mathcal{G}|U$ is a quasicoherent $\mathcal{O}_U$-module. Since x is an arbitrary point in X, we conclude that $f^*\mathcal{G}$ is a quasicoherent sheaf.

For Noetherian schemes X and Y, we can choose Noetherian rings R and S. By Proposition 4.27, for a coherent sheaf $\mathcal{G}$, the R-module M is finitely generated. Hence, $M \otimes_R S$ is a finitely generated S-module. That is, $f^*\mathcal{G}|U$ is a coherent sheaf. Since x is an arbitrary point in X, $f^*\mathcal{G}$ is coherent. □

In order to prove the analogous statements for the direct image, we need a new notion. For a given scheme morphism $f : X \to Y$, if there is an open covering $\{V_j\}_{j \in J}$ of Y so that $f^{-1}(V_j)$ may be quasicompact, then f is said to be a *quasicompact morphism*. Since an affine scheme is quasicompact, a morphism of affine schemes $f : X \to Y$ is a quasicompact morphism.

PROPOSITION 4.36. *If either a scheme morphism $f : X \to Y$ is quasicompact and separated, or X is a Noetherian scheme, then the direct image $f_*\mathcal{F}$ of a quasicoherent sheaf $\mathcal{F}$ is quasicoherent.*

PROOF. It is sufficient to prove the quasi-coherency of $f_*\mathcal{F}$ in a neighborhood of a point in Y. Hence, Y may be assumed to be an affine scheme. Furthermore, since f is a quasicompact morphism, we may assume X to be quasicompact as well. Note that a Noetherian scheme is quasicompact. Therefore, there exists a finite affine covering $\{U_i\}_{i \in I}$ of X, i.e., $X = \bigcup_{i \in I} U_i$. For a separated morphism f, $U_{ij} = U_i \cap U_j$ is an affine open set. (See Exercise 4.39(3), below.) If X is a Noetherian scheme, $U_{ij} = U_i \cap U_j$ can be covered by finitely many affine open sets $\{U_{ijk}\}$. A section $s \in \Gamma(f^{-1}(V), \mathcal{F})$ coincides with the system of sections $\{s_i\}_{i \in I}$ over $f^{-1}(V) \cap U_i$ whose restrictions to each U_{ijk} coincide. For simplicity, denote the restrictions to U_i and U_{ijk} by f. Namely, we have the following exact sequence of sheaves:

$$0 \to f_*\mathcal{F} \xrightarrow{\varphi} \bigoplus_{i \in I} f_*(\mathcal{F}|U_i) \xrightarrow{\psi} \bigoplus_{i,j,k} f_*(\mathcal{F}|U_{ijk}).$$

Note that φ is determined by the restriction map $\rho_{U_i \cap f^{-1}(V), f^{-1}(V)}$, and ψ is induced by the map on $f_*(\mathcal{F}|U_i)$ to $f_*(\mathcal{F}|U_{ijk})$ which is determined by the restriction map $\rho_{U_{ijk} \cap f^{-1}(V), U_i \cap f^{-1}(V)}$ and to $f_*(\mathcal{F}|U_{jik})$ which is determined by $-\rho_{U_{jik} \cap f^{-1}(V), U_i \cap f^{-1}(V)}$ (notice the minus sign). From Example 4.33, $f_*(\mathcal{F}|U_i)$ and $f_*(\mathcal{F}|U_{ijk})$ are quasicoherent. Hence, Corollary 2.24 implies that $f_*\mathcal{F}$ is a quasicoherent sheaf. □

For a morphism $f : X \to Y$ of Noetherian schemes, the direct image of a coherent sheaf $\mathcal{F}$ of $\mathcal{O}_X$-modules need not be coherent. A k-homomorphism $\varphi : k \to k[x]$ from a field k to the polynomial ring $k[x]$ over k induces an affine scheme morphism $\varphi^a : X = \operatorname{Spec} k[x] \to Y = \operatorname{Spec} k$. From Proposition 4.27, $\mathcal{O}_X$ is a coherent sheaf. Then $\varphi^a_* \mathcal{O}_X$ is a sheaf over $Y = \operatorname{Spec} k$ which is determined by the k-module $k[x]$. Notice that $k[x]$ is not a finitely generated k-module. (As a ring, $k[x]$ is finitely generated over k, but as a k-vector space, the dimension of $k[x]$ is infinite.) Therefore, by Proposition 4.27, $\varphi^a_* \mathcal{O}_X$ is not coherent.

As we will witness in the following, coherent sheaves are extremely important in algebraic geometry, and scheme morphisms that give the coherent direct images are important. Projective and proper morphisms will be discussed in Chapter 5.

We have the result corresponding to Lemma 4.32 for a direct image and an inverse image.

LEMMA 4.37. *Let* $f : X \to Y$ *be a scheme morphism. For an* $\mathcal{O}_X$-*module* $\mathcal{F}$ *and an* $\mathcal{O}_Y$-*module* $\mathcal{G}$, *there exist an isomorphism of additive groups*

$$\operatorname{Hom}_{\mathcal{O}_X}(f^*\mathcal{G}, \mathcal{F}) \xrightarrow{\sim} \operatorname{Hom}_{\mathcal{O}_Y}(\mathcal{G}, f_*\mathcal{F}),$$

where for a scheme Z, $\operatorname{Hom}_{\mathcal{O}_Z}(\cdot, \cdot)$ *indicates the totality of* $\mathcal{O}_Z$-*module homomorphisms.*

PROOF. We will use the same notation as in the proof for Lemma 4.32. Regard $f^{\bullet}\mathcal{G}$ as a presheaf of $f^{\bullet}\mathcal{O}_Y$-modules. Consider the tensor product of presheaves $f^{\bullet}\mathcal{G} \otimes_{f^{\bullet}\mathcal{O}_Y} \mathcal{O}_X$. One can easily show that the sheafification of this presheaf is $f^*\mathcal{G}$. As in the proof for Lemma 4.32, one can also prove the isomorphisms as follows:

$$\operatorname{Hom}_{\mathcal{O}_X}(f^*\mathcal{G}, \mathcal{F}) \xrightarrow{\sim} \operatorname{Hom}_{\mathcal{O}_X\text{-presheaf}}(f^{\bullet}\mathcal{G} \otimes_{f^{\bullet}\mathcal{O}_Y} \mathcal{O}_X, \mathcal{F}),$$
$$\operatorname{Hom}_{\mathcal{O}_X\text{-presheaf}}(f^{\bullet}\mathcal{G} \otimes_{f^{\bullet}\mathcal{O}_Y} \mathcal{O}_X, \mathcal{F}) \xrightarrow{\sim} \operatorname{Hom}_{\mathcal{O}_Y}(\mathcal{G}, f_*\mathcal{F}). \quad \square$$

PROBLEM 16. Let $f : X \to Y$ be a scheme morphism and let $\mathcal{F}$ be a quasicoherent sheaf over X. Prove that there exists a canonical $\mathcal{O}_X$-module homomorphism $\eta : f^*f_*\mathcal{F} \to \mathcal{F}$.

PROBLEM 17. For a scheme morphism $f : X \to Y$, let
$$0 \to \mathcal{F}_1 \to \mathcal{F}_2 \to \mathcal{F}_3 \to 0$$
be an exact sequence of $\mathcal{O}_X$-modules, and also let
$$0 \to \mathcal{G}_1 \to \mathcal{G}_2 \to \mathcal{G}_3 \to 0$$
be an exact sequence of $\mathcal{O}_Y$-modules. Prove that the direct image induces the exact sequence of $\mathcal{O}_Y$-modules
$$0 \to f_*\mathcal{F}_1 \to f_*\mathcal{F}_2 \to f_*\mathcal{F}_3,$$
and that the inverse image induces the exact sequence of $\mathcal{O}_Y$-modules
$$f^*\mathcal{G}_1 \to f^*\mathcal{G}_2 \to f^*\mathcal{G}_3 \to 0.$$

4.4. Schemes and Quasicoherent Sheaves

We will give more examples of schemes which can be defined in terms of quasicoherent sheaves.

(a) **Closed Subschemes and Ideal Sheaves**

Let $\mathcal{J}$ be a subsheaf of the structure sheaf $\mathcal{O}_X$ which is also an $\mathcal{O}_X$-submodule. Namely, for an arbitrary open set U of X, $\mathcal{J}(U) \subset \mathcal{O}_X(U)$ is an $\mathcal{O}_X(U)$-submodule (i.e., $\mathcal{J}(U)$ is an ideal of $\mathcal{O}_X(U)$). Then $\mathcal{J}$ is said to be an *ideal sheaf* of $\mathcal{O}_X$. In particular, when $\mathcal{J}$ is a quasicoherent sheaf, $\mathcal{J}$ is said to be a *quasicoherent ideal sheaf*. In what will follow, we will focus on quasicoherent ideal sheaves.

For a sheaf $\mathcal{F}$ on X, let

(4.23) $$\operatorname{supp}(\mathcal{F}) = \{x \in X | \mathcal{F}_x \neq 0\},$$

which is called the *support* of $\mathcal{F}$. Let us consider the support of the quotient sheaf $\mathcal{O}_X/\mathcal{J}$ of a quasicoherent ideal sheaf $\mathcal{J}$ over a scheme X. For an affine covering $\{U_j\}_{j \in J}$, $U_j = \operatorname{Spec} A_j$, of X, let $I_j = \Gamma(U_j, \mathcal{J})$. Then I_j is an ideal of the commutative ring A_j, and $(\mathcal{O}_X/\mathcal{J})|U_j$ is a sheaf of commutative rings determined by A_j/I_j. We have

(4.24) $$\operatorname{supp}(\mathcal{O}_X/\mathcal{J}) \cap U_j = V(I_j) \subset \operatorname{Spec} A_j.$$

This is because for $\mathfrak{p} \supset I_j$ we have $(A_j/I_j)_\mathfrak{p} \neq 0$ and for $\mathfrak{p} \not\supset I_j$ we have $A_{j,\mathfrak{p}} = I_{j,\mathfrak{p}}$, i.e., $(A_j/I_j)_\mathfrak{p} = 0$. Therefore, since $\operatorname{supp}(\mathcal{O}_X/\mathcal{J}) \cap U_j$ is a closed set in U_j and $\{U_j\}$ is an open covering of X, the support of $\mathcal{O}_X/\mathcal{J}$ is a closed set in X. That is, we have obtained the following lemma.

LEMMA 4.38. *The support* $Y = \operatorname{supp}(\mathcal{O}_X/\mathcal{J})$ *of the quotient sheaf* $\mathcal{O}_X/\mathcal{J}$ *of a quasicoherent ideal sheaf* $\mathcal{J}$ *of a scheme* X *is a closed set in* X. *Then* $(Y, \mathcal{O}_X/\mathcal{J})$ *is a closed subscheme of* X. □

Note that the restriction $(\mathcal{O}_X/\mathcal{J})|Y$ of the sheaf $\mathcal{O}_X/\mathcal{J}$ to Y (namely writing $\mathcal{O}_Y$ for $\iota^{-1}\mathcal{O}_X/\mathcal{J}$ of the natural injection $\iota : Y \to X$) could better be written as $(Y, \mathcal{O}_Y)$. However, since the stalks of $\mathcal{O}_X/\mathcal{J}$ outside Y are all zero, we use the abbreviated notation as in the above lemma. Notice that $\iota_*\mathcal{O}_Y = \mathcal{O}_X/\mathcal{J}$.

A definition of a closed subscheme was given in §2.4(c). That definition is not quite complete. We are ready to give the full-fledged definition of a closed subscheme as follows. For a quasicoherent ideal sheaf $\mathcal{J}$ of a scheme X, let $Y = \operatorname{supp}(\mathcal{O}_X/\mathcal{J})$. Then $(Y, \mathcal{O}_X/\mathcal{J})$ is said to be the *closed subscheme* determined by the ideal sheaf $\mathcal{J}$. In §2.4(c), we made the following definition. For a given scheme morphism $\iota = (\iota, \iota^\#) : (Y, \mathcal{O}_Y) \to (X, \mathcal{O}_X)$, if Y is a closed subset of X, $\iota : Y \to X$ is a natural injection, and the kernel of the surjection $\iota^\# : \mathcal{O}_X \to \iota_*\mathcal{O}_Y$ is a quasicoherent ideal sheaf of $\mathcal{O}_X$, then $(Y, \mathcal{O}_Y)$ or (Y, ι) is said to be a *closed subscheme* of X. Notice that the quasicoherency of $\operatorname{Ker} \iota^\#$ is a part of the conditions. Hence, a closed immersion gets associated with a stronger condition. Namely, a scheme morphism $f : W \to X$ is said to be a *closed immersion* if for a closed subscheme (Y, ι) there exists an isomorphism of schemes $\theta : W \xrightarrow{\sim} Y$ satisfying $f = \iota \circ \theta$. All the statements about a closed scheme and a closed immersion in Chapter 2 are valid with this new definition.

EXERCISE 4.39. (1) An affine scheme $f : \operatorname{Spec} B \to \operatorname{Spec} A$ is a closed immersion if and only if the ring homomorphism $\varphi : A \to B$ is surjective. Then B is isomorphic to the quotient ring A/I of the ideal $I = \operatorname{Ker} \varphi$.

(2) For an affine open set U in Y, the inverse image $j^{-1}(U)$ of a closed immersion $j : X \to Y$ is an affine open set of X unless $j^{-1}(U) = \varnothing$.

(3) When $f : X \to Y = \operatorname{Spec} R$ is a separated morphism, for affine open sets U and V the intersection $U \cap V$ is an affine open set in X unless $U \cap V = \varnothing$.

PROOF. (1) By Theorem 4.21 a quasicoherent ideal sheaf $\mathcal{J}$ of $Y = \operatorname{Spec} A$ coincides with the sheaf $\widetilde{I}$ induced by the ideal $I = \Gamma(Y, \mathcal{J})$. Hence, the closed subscheme $(Z, \mathcal{O}_Y/\mathcal{J})$ determined by the

ideal sheaf $\mathcal{J}$, where $Z = \mathrm{supp}(\mathcal{O}_Y/\mathcal{J})$, is isomorphic to
$$(\mathrm{Spec}(A/I), \mathcal{O}_{\mathrm{Spec}\,A/I}).$$
On the other hand, the closed immersion f induces an isomorphism between the closed subscheme $(Z, \mathcal{O}_Z)$ in $\mathrm{Spec}\,A$ and $\mathrm{Spec}\,B$. Therefore, there is an isomorphism of rings $A/I \xrightarrow{\sim} B$. Namely, f is determined by the homomorphisms $A \to A/I \xrightarrow{\sim} B$.

Conversely, for a given surjective homomorphism $\varphi : A \to B$ of commutative rings, as we saw in the above, $f = \varphi^a : \mathrm{Spec}\,B \to \mathrm{Spec}\,A$ is the composition of the isomorphism from $\mathrm{Spec}\,B$ to the closed subscheme $\mathrm{Spec}\,A/I$ and the natural morphism $\mathrm{Spec}\,A/I \to \mathrm{Spec}\,A$. That is, f is a closed immersion.

(2) The kernel $\mathcal{J} = \mathrm{Ker}\,j^\#$ of the surjective homomorphism $j^\# : \mathcal{O}_Y \to j_*\mathcal{O}_X$ is a quasicoherent ideal sheaf of $\mathcal{O}_Y$. Then $j_*\mathcal{O}_X|U$ is isomorphic to $(\mathcal{O}_Y/\mathcal{J})|U$. Let $A = \Gamma(U, \mathcal{O}_Y)$ and let $J = \Gamma(U, \mathcal{J})$. Then $(\mathcal{O}_Y/\mathcal{J})|U$ coincides with $\widetilde{A/J}$. That is, $j^{-1}(U)$ is isomorphic to $\mathrm{Spec}\,A/J$. Hence $j^{-1}(U)$ is an affine open set in X.

(3) Since f is a separated morphism, the diagonal morphism $\Delta_{X/Y} : X \to X \times_Y X$ is a closed immersion. Let $U = \mathrm{Spec}\,A$ and $V = \mathrm{Spec}\,B$. Then the morphism f is an affine scheme morphism determined by the homomorphisms $g : R \to A$ and $f : R \to B$ over U and V, respectively. Therefore, we obtain $U \times_Y V = \mathrm{Spec}\,A \otimes_R B$, an affine open set of $X \times_Y X$. Since $U \cap V = \Delta_{X/Y}^{-1}(U \times_Y V)$, from (2), $U \cap V$ is an affine open set. □

For a subscheme W in a scheme Y, the direct image $\iota_*\mathcal{F}$ of a sheaf $\mathcal{F}$ over Y under the natural scheme morphism $\iota : W \to Y$ satisfies $(\iota_*\mathcal{F})_y = 0$ for $y \in Y \setminus W$. Then $\iota_*\mathcal{F}$ is sometimes said to be the *sheaf obtained by extending $\mathcal{F}$ by zero outside W*. If $\mathcal{F}$ is an $\mathcal{O}_W$-module, $\iota_*\mathcal{F}$ may be regarded as an $\mathcal{O}_Y$-module by the surjective homomorphism $\iota^\# : \mathcal{O}_Y \to \iota_*\mathcal{O}_W$.

(b) **Affine Morphisms and Quasicoherent $\mathcal{O}_Y$-Algebras**

A sufficient condition for the direct image of a quasicoherent sheaf to be quasicoherent is given in Proposition 4.36. As a typical morphism satisfying the sufficient condition, an affine morphism will be studied.

Let $f : X \to Y$ be a morphism of schemes. If there exists an affine covering $\{U_j\}_{j \in J}$ of Y such that $f^{-1}(U_j)$ is an affine open set in X, then f is said to be an *affine morphism*. When $f : X \to Y$ is an affine morphism, X is said to be an affine scheme over Y. Since an affine scheme is quasicompact, an affine morphism is a quasicompact morphism. See Corollary 2.9. Furthermore, since $f^{-1}(U_j) \to U_j$ is an affine scheme morphism, by Lemma 3.22, $f^{-1}(U_j) \to U_j$ is separated. Proposition 3.23 implies f is separated. Therefore, from Proposition 4.36, the direct image $f_*\mathcal{O}_X$ of the structure sheaf over X is quasicoherent. For an open set U of Y, $(f_*\mathcal{O}_X)(U) = \mathcal{O}_X(f^{-1}(U))$ becomes an $\mathcal{O}_Y(U)$-algebra through the homomorphism $\theta : \mathcal{O}_Y \to f_*\mathcal{O}_X$ induced by f. This $\mathcal{O}_Y$-algebra structure is compatible with the restriction map. Then $f_*\mathcal{O}_X$ is an $\mathcal{O}_Y$-*commutative algebra*, or simply, an $\mathcal{O}_Y$-*algebra*.

Next, we will show that an affine morphism $f : X \to Y$ can be recovered by the $\mathcal{O}_Y$-algebra $\mathcal{A}_X = f_*\mathcal{O}_X$. First choose an affine open covering $\{U_j\}_{j \in J}$ of Y so that $V_j = f^{-1}(U_j)$ is an affine open set in X. Set $A_j = \Gamma(f^{-1}(U_j), \mathcal{O}_X)$ and $B_j = \Gamma(U_j, \mathcal{O}_Y)$. Then, $\theta_{U_j} : B_j \to A_j$ is a commutative ring homomorphism, and A_j may be considered as a B_j-algebra. By the assumption, $f^{-1}(U_j) = \operatorname{Spec} A_j$. On the other hand, $A_j = \Gamma(U_j, \mathcal{A}_X)$. Hence the affine scheme $f^{-1}(U_j) = \operatorname{Spec} A_j$ is determined by $\mathcal{A}_X$. Moreover, since

$$\Gamma(f^{-1}(U_i) \cap f^{-1}(U_j), \mathcal{O}_X) = \Gamma(U_i \cap U_j, \mathcal{A}_X),$$

it follows that $\operatorname{Spec} A_i \cap f^{-1}(U_i)$ and $\operatorname{Spec} A_j \cap f^{-1}(U_j)$ are isomorphic as local ringed spaces. Therefore, the scheme X can be recovered from the $\mathcal{O}_Y$-algebra $\mathcal{A}_X = f_*\mathcal{O}_X$. Note also that the $\mathcal{O}_Y$-algebra structure is determined by the homomorphism $\theta_{U_j} : B_j \to A_j$, which induces the affine scheme morphism $\theta_{U_j}^a : \operatorname{Spec} A_j \to \operatorname{Spec} B_j$. Then $\theta_{U_j}^a$ is the restriction of f to $f^{-1}(U_j)$. That is, the $\mathcal{O}_Y$-algebra structure of $\mathcal{A}_X$ gives an $\mathcal{O}_Y$-algebra homomorphism

$$\eta : \mathcal{O}_Y \to \mathcal{A}_X,$$
$$a \mapsto a \cdot 1,$$

satisfying $\eta_{U_j} = \theta_{U_j}$. Consequently, the $\mathcal{O}_Y$-algebra structure of $\mathcal{A}_X$ determines the scheme morphism $f : X \to Y$. Namely, an affine morphism $f : X \to Y$ is induced by the $\mathcal{O}_Y$-algebra $\mathcal{A}_X = f_*\mathcal{O}_X$.

PROBLEM 18. Let $f : X \to Y$ be an affine morphism and let U be an affine open set in Y. Prove that the restricted morphism of f to $f^{-1}(U)$ is also an affine morphism $f^{-1}(U) \to U$.

Conversely, let us consider the case when a quasicoherent $\mathcal{O}_Y$-algebra $\mathcal{A}$ over a scheme Y is given. For an affine open set U of Y, let $A_U = \Gamma(U, \mathcal{A})$. Then $\mathcal{A}|U = \widetilde{A_U}$, and A_U is an $\mathcal{O}_Y(U)$-algebra. Therefore, as in the above, we obtain an affine morphism

$f_U : \operatorname{Spec} A_U \to \operatorname{Spec} \mathcal{O}_Y(U) = U$. Let V be another affine open set in Y. Similarly, we get an affine scheme morphism $f_V : \operatorname{Spec} A_V \to V$, where $A_V = \Gamma(V, \mathcal{A})$. If $U \cap V \neq \varnothing$, then, as we will show, $f_U^{-1}(U \cap V)$ and $f_V^{-1}(U \cap V)$ are isomorphic schemes. Notice that $f_{U*}\mathcal{O}_{\operatorname{Spec} A_U} = \mathcal{A}|U$ and $f_{V*}\mathcal{O}_{\operatorname{Spec} A_V} = \mathcal{A}|V$. For an affine open set $W \subset U \cap V$, we have $f_U^{-1}(W) = \operatorname{Spec} A_U \times_U W$ and $f_V^{-1}(W) = \operatorname{Spec} A_V \times_V W$. Hence $f_U^{-1}(W)$ and $f_V^{-1}(W)$ are affine open sets of $\operatorname{Spec} A_U$ and $\operatorname{Spec} A_V$, respectively. Then we have

$$\Gamma(f_U^{-1}(W), \mathcal{O}_{\operatorname{Spec} A_U}) = \Gamma(U, \mathcal{A}) \otimes_{\mathcal{O}_Y(U)} \mathcal{O}_Y(W) \overset{\sim}{\to} \Gamma(W, \mathcal{A})$$

and also

$$\Gamma(f_V^{-1}(W), \mathcal{O}_{\operatorname{Spec} A_V}) \overset{\sim}{\to} \Gamma(W, \mathcal{A}).$$

Namely, as stated, $f_U^{-1}(W)$ and $f_V^{-1}(W)$ are isomorphic as affine schemes. Therefore, $f_U^{-1}(U \cap V)$ and $f_V^{-1}(U \cap V)$ are isomorphic schemes. Let $\{U_i\}_{i \in I}$ be an open covering of Y and let $A_i = \Gamma(U_i, \mathcal{A})$. For $U_i \cap U_j \neq \varnothing$, glue $f_i : \operatorname{Spec} A_i \to U_i$ as above to obtain a scheme morphism $f : X \to Y$, $X = \bigcup_{i \in I} \operatorname{Spec} A_i$. By the construction, f is an affine morphism. Then, let $X = \operatorname{Spec} \mathcal{A}$, which is said to be the affine scheme over Y determined by the quasicoherent $\mathcal{O}_Y$-commutative algebra $\mathcal{A}$. The map $f : \operatorname{Spec} \mathcal{A} \to Y$ is called the structure morphism.

Thus, for an affine morphism $f : X \to Y$, the affine scheme $\operatorname{Spec} \mathcal{A}_X$ over Y induced by the quasicoherent $\mathcal{O}_Y$-algebra $\mathcal{A}_X = f_*\mathcal{O}_X$ and the structure morphism $\operatorname{Spec} \mathcal{A}_X \to Y$ are nothing but X and $f : X \to Y$. Furthermore, we have the following.

THEOREM 4.40. *For an affine scheme over Y, i.e., $f : X \to Y$, there is induced a quasicoherent $\mathcal{O}_Y$-algebra $\mathcal{A}_X = f_*\mathcal{O}_X$. This correspondence establishes a contravariant equivalence functor between the category (Affine)/Y of affine schemes over Y and the category (q.c.$\mathcal{O}_Y$-alg) of quasicoherent $\mathcal{O}_Y$-algebras over Y.*

PROOF. A morphism from an affine scheme over Y, i.e., $f : X \to Y$, to another $g : Z \to Y$ in (Affine)/Y is a morphism $h : X \to Z$ so that the diagram

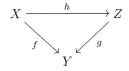

is commutative. Then, $h : X \to Z$ induces a sheaf homomorphism of commutative rings $\theta : \mathcal{O}_Z \to h_*\mathcal{O}_X$. Moreover, θ induces a homomorphism of sheaves of $\mathcal{O}_Y$-algebras $\tilde{\theta} : g_*\mathcal{O}_Z \to g_*(h_*\mathcal{O}_X)$. Since $f = g \circ h$, we have

$$g_*(h_*\mathcal{O}_X) = f_*\mathcal{O}_X = \mathcal{A}_X,$$

i.e., $\tilde{\theta} : \mathcal{A}_Z \to \mathcal{A}_X$. Namely, we have a contravariant functor

$$F : (\text{Affine})/Y \to (\text{q.c.}\mathcal{O}_Y\text{-Alg}),$$
$$f : X \to Y \mapsto \mathcal{A}_X = f_*\mathcal{O}_X.$$

Conversely, for $\mathcal{A} \in \text{Ob}(\text{q.c.}\mathcal{O}_Y\text{-Alg})$, $f : \text{Spec}\,\mathcal{A} \to Y$ is obtained. Then for an $\mathcal{O}_Y$-algebra homomorphism $\varphi : \mathcal{A} \to \mathcal{B}$, we get a scheme morphism $\varphi^a : \text{Spec}\,\mathcal{B} \to \text{Spec}\,\mathcal{A}$ in $(\text{Affine})/Y$. That is, we have a contravariant functor

$$G : (\text{q.c.}\mathcal{O}_Y\text{-Alg}) \to (\text{Affine})/Y,$$
$$\mathcal{A} \mapsto f : \text{Spec}\,\mathcal{A} \to Y.$$

From the argument before this theorem, we get $F \circ G \xrightarrow{\sim} \text{id}$ and $G \circ F \xrightarrow{\sim} \text{id}$. □

PROBLEM 19. Let $g^*\mathcal{A}$ be the pull-back of a quasicoherent $\mathcal{O}_Y$-algebra $\mathcal{A}$ under a morphism $g : Z \to Y$. Prove that $\text{Spec}\,g^*\mathcal{A} \to Z$ is isomorphic to the base change of $\text{Spec}\,\mathcal{A} \to Y$ by the morphism g. Therefore, the fibre $f^{-1}(y)$ at y of $f : X \to Y$ is an affine scheme.

Summary

4.1. The definitions of the sheafification of a presheaf, the image $\text{Im}\,\varphi$, the kernel $\text{Ker}\,\varphi$ and the cokernel $\text{Coker}\,\varphi$ of a sheaf homomorphism $\varphi : \mathcal{F} \to \mathcal{G}$.

4.2. The definition of an exact sequence of sheaves.

4.3. The definitions of a quasicoherent $\mathcal{O}_X$-module and a coherent $\mathcal{O}_X$-module over a scheme X.

4.4. A quasicoherent $\mathcal{O}_X$-module $\mathcal{F}$ over an affine scheme $X = \text{Spec}\,R$ is isomorphic to the sheaf $\widetilde{M}$ determined by the R-module $M = \Gamma(X, \mathcal{F})$.

4.5. An exact sequence of quasicoherent sheaves of $\mathcal{O}_X$-modules $0 \to \mathcal{F}_1 \to \mathcal{F}_2 \to \mathcal{F}_3 \to 0$ over an affine scheme $X = \text{Spec}\,R$ induces an exact sequence of R-modules $0 \to \Gamma(X, \mathcal{F}_1) \to \Gamma(X, \mathcal{F}_2) \to \Gamma(X, \mathcal{F}_3) \to 0$.

4.6. Definitions of the tensor product $\mathcal{F} \otimes_{\mathcal{O}_X} \mathcal{G}$ of $\mathcal{O}_X$-modules $\mathcal{F}$ and $\mathcal{G}$ and the $\mathcal{O}_X$-module $\underline{\mathrm{Hom}}_{\mathcal{O}_X}(\mathcal{F}, \mathcal{G})$.

4.7. The definition of an $\mathcal{O}_X$-flat sheaf.

4.8. The definitions of the inverse image $f^*\mathcal{F}$ of an $\mathcal{O}_Y$-module $\mathcal{F}$ under a scheme morphism $f : X \to Y$.

4.9. The definition of an affine morphism and the definition of the affine scheme $\pi : \mathrm{Spec}\,\mathcal{A} \to X$ associated with a quasicoherent sheaf $\mathcal{A}$ of commutative $\mathcal{O}_X$-algebras, and its fundamental properties.

Exercises

4.1. Prove that the support of a finitely generated $\mathcal{O}_X$-module $\mathcal{F}$ over a scheme X, i.e.,

$$\mathrm{supp}\,\mathcal{F} = \{x \in X | \mathcal{F}_x \neq 0\}$$

is a closed set of X.

4.2. Let $\mathcal{F}$ be a quasicoherent $\mathcal{O}_X$-module $\mathcal{F}$ over an affine scheme $X = \mathrm{Spec}\,R$. For a section t of $\mathcal{F}$ over $D(f)$, $f \in R$, i.e., $t \in \Gamma(D(f), \mathcal{F})$, prove that $f^n t \in \Gamma(X, \mathcal{F})$ for some positive integer n.

4.3. (1) For an open set U in a scheme, let $\widetilde{\mathcal{N}}(U)$ be the nilpotent radical of $\Gamma(U, \mathcal{O}_X)$. Prove that $\widetilde{\mathcal{N}}(U)$ forms a presheaf. Then the sheafification $\mathcal{N}$ of $\widetilde{\mathcal{N}}$ is said to be the nilpotent ideal sheaf. Prove that $\mathcal{N}$ is a quasicoherent sheaf.

(2) Denote $(X, \mathcal{O}_X/\mathcal{N})$ as $(X_{\mathrm{red}}, \mathcal{O}_{X_{\mathrm{red}}})$. Show that $(X_{\mathrm{red}}, \mathcal{O}_{X_{\mathrm{red}}})$ is a closed subscheme of X and that the underlying space coincides with X. Prove that each stalk $\mathcal{O}_{X_{\mathrm{red}},x}$ does not have nilpotent elements. (A scheme is said to be a reduced scheme if each stalk of the structure sheaf has no nilpotent elements.)

(3) For a scheme morphism $f : X \to Y$, prove that one obtains a scheme morphism $f_{\mathrm{red}} : X_{\mathrm{red}} \to Y_{\mathrm{red}}$. Prove also that for a separated morphism f, the morphism f_{red} is separated.

4.4. Let $\mathcal{A}$ be a sheaf of commutative rings over a topological space X. Then we can define a sheaf of $\mathcal{A}$-modules as in §4.2(a). As we showed in Definition 4.18(ii) and Definition 4.25, the notions of a finitely generated $\mathcal{A}$-module and a coherent $\mathcal{A}$-module can be defined. For a short exact sequence of $\mathcal{A}$-module sheaves

$$0 \to \mathcal{F} \xrightarrow{\varphi} \mathcal{G} \xrightarrow{\psi} \mathcal{H} \to 0,$$

if any two sheaves are coherent, so is the third. Prove this fact as you prove the following assertions.

(1) For an epimorphism of $\mathcal{A}$-module sheaves $f : \mathcal{B} \to \mathcal{E}$, if $\mathcal{B}$ is a finitely generated $\mathcal{A}$-module, so is $\mathcal{E}$.

(2) For a homomorphism of $\mathcal{A}$-modules $g : \mathcal{B} \to \mathcal{E}$, if $\mathcal{B}$ is a finitely generated $\mathcal{A}$-module and $\mathcal{E}$ is coherent, then $\operatorname{Im} g$ is coherent.

(3) In the above short exact sequence, if $\mathcal{G}$ and $\mathcal{H}$ are coherent, so is $\mathcal{F}$.

(4) If $\mathcal{F}$ and $\mathcal{G}$ are coherent, so is $\mathcal{H}$.

(5) If $\mathcal{F}$ and $\mathcal{H}$ are coherent, then $\mathcal{G}$ is a finitely generated $\mathcal{A}$-module.

(6) If $\mathcal{F}$ and $\mathcal{H}$ are coherent, so is $\mathcal{G}$.

4.5. Define a symmetric algebra $\mathbb{S}(E)$ of a module E over a commutative ring R as follows. Denote the tensor product $\underbrace{E \otimes_R \cdots \otimes_R E}_{n}$ by $T^n(E)$. In particular, define $T^0(E) = E$. Let

$$\mathbb{T}(E) = \bigoplus_{n=0}^{\infty} T^n(E),$$

which is called *the tensor algebra over* E, where the product in $\mathbb{T}(E)$ is defined as $(a_1 \otimes \cdots \otimes a_m) \cdot (b_1 \otimes \cdots \otimes b_n) = a_1 \otimes \cdots \otimes a_m \otimes b_1 \otimes \cdots \otimes b_n$. With this product, $\mathbb{T}(E)$ is an R-algebra. Let I be the two-sided ideal of $\mathbb{T}^n(E)$ generated by $x \otimes y - y \otimes x$, $x, y \in E$. The quotient ring $\mathbb{T}(E)/I$ is called *the symmetric algebra over* E, and is denoted by $\mathbb{S}(E)$ or $\mathbb{S}_R(E)$.

Furthermore, for R-modules E and F, we have

$$\mathbb{S}(E \oplus F) = \mathbb{S}(E) \otimes_R \mathbb{S}(F),$$

and $\mathbb{S}(R^{\oplus n})$ is isomorphic to the ring $R[z_1, z_2, \ldots, z_n]$ of polynomials in n variables. For a multiplicatively closed set S in R,

$$\mathbb{S}(S^{-1}E) = S^{-1}\mathbb{S}(E).$$

Thus, one can define the sheaf $\mathbb{S}(\mathcal{E})$ of symmetric algebras for a quasicoherent $\mathcal{O}_S$-module $\mathcal{E}$ over a scheme S. That is, for an open set U in S, assign a presheaf $\mathbb{S}(\mathcal{E}(U))$ of symmetric algebras. Then $\mathbb{S}(\mathcal{E})$ is the sheafification of the presheaf $\mathbb{S}(\mathcal{E}(U))$. Note that $\mathbb{S}(\mathcal{E})$ is a commutative $\mathcal{O}_S$-algebra. In particular, for an affine scheme $S = \operatorname{Spec} R$ and for $\mathcal{E} = \widetilde{E}$, we have $\mathbb{S}(E_f) = \mathbb{S}(E)_f$, $f \in R$. Namely, $\mathbb{S}(\mathcal{E}) = \widetilde{\mathbb{S}(E)}$.

For a quasicoherent $\mathcal{O}_S$-module $\mathcal{E}$ over a scheme S, let $F_\mathcal{E}$ be the functor from the category $(\mathrm{Sch})/S$ of schemes over S to the category (Mod) of modules defined by

$$F_\mathcal{E} : (\mathrm{Sch})/S \to (\mathrm{Mod}),$$
$$g : T \to S \mapsto \mathrm{Hom}_{\mathcal{O}_T}(\mathcal{E}_{(T)}, \mathcal{O}_T).$$

Notice that $\mathcal{E}_{(T)}$ is defined as the inverse image $g^*\mathcal{E}$ of the morphism $g : T \to S$. Prove the following.

(1) Since $\mathbb{T}^1(E) = E$, there is a natural map $\sigma : E \to \mathbb{S}(E)$. Then the symmetric algebra $\mathbb{S}(E)$ is characterized by the following universal mapping property:

Any R-linear map φ from E to a commutative R-algebra A, $\varphi : E \to A$, is uniquely factored through

$$E \xrightarrow{\sigma} \mathbb{S}(E) \xrightarrow{f} A,$$

where f is a homomorphism of commutative R-algebras. (That is, there is an isomorphism $\mathrm{Hom}_R(E, A) \xrightarrow{\sim} \mathrm{Hom}_{R\text{-Alg}}(\mathbb{S}(E), A)$.)

(2) $F_\mathcal{E} : (\mathrm{Sch})/S \to (\mathrm{Mod})$ is a contravariant functor.

(3) The functor $F_\mathcal{E}$ is represented by the $\mathcal{O}_{\mathbb{V}(\mathcal{E})}$-homomorphism $\mathcal{E}_{\mathbb{V}(\mathcal{E})} = f^*\mathcal{E} \to \mathcal{O}_{\mathbb{V}(\mathcal{E})}$, which is induced by

$$f : \mathbb{V}(\mathcal{E}) = \mathrm{Spec}\,\mathbb{S}(\mathcal{E}) \to S$$

and the natural homomorphism

$$\mathcal{E} \otimes_{\mathcal{O}_S} \mathbb{S}(\mathcal{E}) \to \mathbb{S}(\mathcal{E}),$$
$$a \otimes [b_1 \otimes \cdots \otimes b_n] \mapsto [a \otimes b_1 \otimes \cdots \otimes b_n].$$

Then $\mathbb{V}(\mathcal{E})$ is said to be the *vector fiber space* associated with the $\mathcal{O}_S$-module $\mathcal{E}$.

(4) If $\mathcal{E}$ is a locally free sheaf of rank n over S, then $f : \mathbb{V}(\mathcal{E}) \to S$ is a vector bundle of rank n over S. The sheaf of local sections over S (Example 4.13) is the locally free sheaf $\mathcal{E}^* = \underline{\mathrm{Hom}}_{\mathcal{O}_S}(\mathcal{E}, \mathcal{O}_S)$ of rank n. $\mathcal{E}^*$ is said to be the *dual sheaf* of $\mathcal{E}$.

CHAPTER 5

Proper and Projective Morphisms

We will provide a general theory of proper morphisms and projective morphisms, notions which are crucial in algebraic geometry.

In this chapter, the descriptions have a tendency to be longer than in the previous chapters. The reader may find it harder to follow the arguments, even though they consist of step by step fundamental facts. It is suggested that the reader take notes while checking the details at his/her own speed.

In this chapter, as we provide motivational background, we will give more general statements. The Noetherian assumption is not much loss of generality; however, we have decided to give a general result which might provide an insight on the assertions. When theorems and lemmas do not readily show their meaning, the reader is encouraged to examine the examples. If the examples give the reader sufficient insights into theorems and lemmas, the proofs can be skipped in the first reading.

The sheaf $\widetilde{M}$ associated with a graded S-module M over $\operatorname{Proj} S$ is sometimes denoted as $(M)^\sim$. When the expressions of M is long, $(M)^\sim$ will be used. In the localization M_f of the graded S-module M at an element f of a graded ring S, the part of degree 0 will be written as $M_{(f)}$.

5.1. Proper Morphisms

A continuous function f between topological spaces or manifolds is said to be a *proper map* if the inverse image of a compact set is always a compact set. The concept in algebraic geometry analogous to this proper map is a proper morphism. Since the image of a closed set under a proper map is closed, we will define a proper morphism using this property. We will begin with the notion of a closed morphism.

(a) Closed Morphisms

A scheme morphism $f : X \to Y$ is said to be a *closed morphism* if the image of a closed set of X under the map on underlying spaces is always a closed set. Moreover, for any morphism $g : Z \to Y$, if the base change $f_Z : X_Z = X \times_Y Z \to Z$ is always a closed morphism, then f is said to be a *universally closed morphism*, or just *universally closed*.

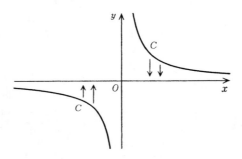

FIGURE 5.1. The origin is not contained in the image.

EXAMPLE 5.1. Consider an affine line $\mathbb{A}_k^1 = \operatorname{Spec} k[x]$ and an affine plane $\mathbb{A}_k^2 = \operatorname{Spec} k[x, y]$. The injective homomorphism $k[x] \to k[x, y]$ induces a scheme morphism $\varphi : \mathbb{A}_k^2 \to \mathbb{A}_k^1$. The image of the closed set $C = V(xy - 1)$ of $\mathbb{A}_k^2$ under φ is an open set $D(x) = \operatorname{Spec} k[x, \frac{1}{x}]$ of $\mathbb{A}_k^1$. Namely, φ is not a closed morphism. As one can observe from Figure 5.1, the point corresponding to the origin, i.e., a maximal ideal (x), of $\mathbb{A}_k^1$ is located at infinity. Since we can regard $\mathbb{A}_k^2 = \mathbb{A}_k^1 \times \mathbb{A}_k^1$, where

$$\mathbb{A}_k^1 = \operatorname{Spec} k[x] \quad \text{and} \quad \mathbb{P}_k^1 = \operatorname{Proj} k[y_0, y_1],$$

and $\overline{\varphi}$ is the projection onto the first component. We also can rewrite $\mathbb{P}_k^1 = \operatorname{Spec} k[\frac{y_1}{y_0}] \cup \operatorname{Spec} k[\frac{y_0}{y_1}]$. By defining $y = \frac{y_1}{y_0}$, we get

$$\mathbb{A}_k^2 = \operatorname{Spec} k[x, y] = \operatorname{Spec}\left(k[x] \otimes_k k\left[\frac{y_1}{y_0}\right]\right)$$

$$= \operatorname{Spec} k[x] \times_k \operatorname{Spec} k\left[\frac{y_1}{y_0}\right] \subset \mathbb{A}_k^1 \times_k \mathbb{P}_k^1.$$

Regard $k[x] \otimes_k k[y_0, y_1] = k[x, y_0, y_1]$ as a graded ring over $R = k[x]$. Then $\overline{\varphi}$ is considered as $\overline{\varphi} : \operatorname{Proj} R[y_0, y_1] \to \operatorname{Spec} R$, and the closed set $C = V(xy - 1)$ of $\mathbb{A}_k^2$ is extended to a closed set $\overline{C} = V(xy_1 - y_0)$ of X satisfying $\overline{\varphi}(\overline{C}) = \mathbb{A}_k^1$. The image of the point determined by the prime ideal (x, y_0) in X is the origin. Later we will show that $\mathbb{P}_k^1 \to \operatorname{Spec} k$ is a universally closed morphism.

LEMMA 5.2. (i) *A closed immersion is universally closed.*

(ii) *For a universally closed morphism $f : X \to Y$, the base change $f_Z : X_Z \to Z$ under any morphism $h : Z \to Y$ is also a universally closed morphism.*

(iii) *If morphisms of schemes $f : X \to Y$ and $g : Y \to Z$ are both universally closed, then the composition $g \circ f : X \to Z$ is a universally closed morphism.*

(iv) *If scheme morphisms $f : X \to Y$ over S and $f' : X' \to Y'$ over S are universally closed, then the fiber product over S*

$$f \times_S f' : X \times_S X' \to Y \times_S Y'$$

is a universally closed morphism.

(v) *For morphisms of schemes $f : X \to Y$ and $g : Y \to Z$, if $g \circ f$ is universally closed and g is separated, then f is universally closed.*

(vi) *For a universally closed $f : X \to Y$, $f_{\mathrm{red}} : X_{\mathrm{red}} \to Y_{\mathrm{red}}$ is also universally closed. (See Exercise 4.3.)*

PROOF. (i) A closed immersion $j : W \to X$ is a closed morphism. An arbitrary base change $j_Z : W \times_X Z \to Z$ is also a closed immersion. Hence j_Z is a closed morphism.

(ii) The base change of $f_Z : X_Z \to Z$ through $h' : Z' \to Z$ is precisely the base change of $f : X \to Z$ through $h \circ h' : Z' \to Y$. Hence $f_Z : X_Z \to Z$ is universally closed.

(iii) The composition of the base change $g_W : Y \times_Z W \to W$ of $g : Y \to Z$ via $h : W \to Z$ with the base change $f_{Y \times_Z W} : X \times_Y (Y \times_Z W) \to Y \times_Z W$ of f via $p : Y \times_Z W \to Y$ is a closed map. This is because this composition $g_W \circ f_{Y \times_Z W}$ consists of two closed morphisms. Then this composition is precisely the base change of $g \circ f : X \to Z$ via h.

(iv) For the fiber product $((Y \times_X Y'), (p_1, p_2))$ of Y and Y' over S and for a morphism $h : Z \to Y \times_S Y'$ over S, let $h_1 = p_1 \circ h : Z \to Y$ and $h_2 = p_2 \circ h : Z \to Y'$. Then h and h' are morphisms over S. Consider the base changes $f_Z : X \times_Y Z \to Z$ and $f'_Z : X' \times_{Y'} Z \to Z$. Denote the base change of f_Z by f'_Z as $\tilde{f} : (X \times_Y Z) \times_Z (X' \times_{Y'} Z) \to X' \times_{Y'} Z$. Since $(X \times_Y Z) \times_Z (X' \times_{Y'} Z) \tilde{\to} (X \times_S X') \times_{Y \times_S Y'} Z$, the base change $(f \times_S f')_Z$ of $f \times_S f'$ by h coincides with $f'_Z \circ \tilde{f}$.

As f'_Z and $\tilde{f}$ are closed morphisms, $f'_Z \circ \tilde{f}$ is also a closed morphism. Namely, $f \times_S f'$ is a universally closed morphism.

(v) From our assumption, $\Delta_{Y/Z} : Y \to Y \times_Z Y$ is a closed immersion. Hence, (i) implies universal closedness. By the natural projection on the first component $p_1 : Y \times_Z Y \to Y$, $\Delta_{Y/Z}$ can be considered as a morphism over Y. Also $\mathrm{id}_X : X \to X$ can be regarded as a morphism over Y via $f : X \to Y$, and id_X is a universally closed morphism. Then by (iv), $\mathrm{id}_X \times_Y \Delta_{Y/Z} : X \times_Y Y \to X \times_Y (Y \times_Z Y)$ is universally closed. This morphism is the morphism of the graph $\Gamma_f : X \to X \times_Z Y$ of f. (See the proof of Theorem 3.25(iv).) Notice also that the second projection $p_2 : X \times_Z Y \to Y$ coincides with the base change of $g \circ f : X \to Z$ by $g : Y \to Z$. Since $g \circ f$ is universally closed, p_2 is also universally closed. By (iii), $f = p_2 \circ \Gamma_f$ is also universally closed.

(vi) From Exercise 4.3, X_{red} is a closed subscheme of X whose underlying space coincides with that of X. Therefore, a closed subscheme F of $X_{\mathrm{red}} \times Z$ may be considered as a subscheme of $X \times Z$. For the base change $(f_{\mathrm{red}})_Z : X_{\mathrm{red}} \times Z \to Y_{\mathrm{red}} \times Z$, the image $(f_{\mathrm{red}})_Z(F)$ of F and $f_Z(f)$ are the same as sets. This is because the underlying spaces of $Y_{\mathrm{red}} \times Z$ and $Y \times Z$ are the same. Since $f_Z(F)$ is a closed set, $(f_{\mathrm{red}})_Z$ is a closed morphism. □

PROBLEM 1. Let $f : X \to Y$ and $g : Y \to Z$ be morphisms of schemes. If $g \circ f$ is universally closed and if the map induced by f on underlying spaces is surjective, then g is a universally closed morphism.

(b) **Proper Morphisms**

A morphism of schemes $f : X \to Y$ is said to be a *proper morphism* if f is a separated morphism of finite type and a universally closed morphism. We have the following proposition, corresponding to Lemma 5.2.

PROPOSITION 5.3. (i) *A closed immersion is a proper morphism.*

(ii) *The base change $f_W : X \times_Y W \to W$ of a proper morphism $f : X \to Y$ by a morphism $h : W \to Y$ is a proper morphism.*

(iii) *For proper morphisms $f : X \to Y$ and $g : Y \to Z$, the composition $g \circ f : X \to Z$ is also a proper morphism.*

(iv) *If morphisms of schemes $f : X \to Y$ and $f' : X' \to Y'$ over S are proper morphisms, then $f \times_S f' : X \times_S X' \to Y \times_S Y'$ is also a proper morphism.*

(v) *For a proper morphism $f : X \to Y$, $f_{\text{red}} : X_{\text{red}} \to Y_{\text{red}}$ is a proper morphism.*

PROOF. We proved the universal closedness in Lemma 5.2. We only need to show that these morphisms are separated morphisms of finite type. However, the separatedness is implied by Theorem 3.25. From the definition, one can show that those morphisms are of finite type. □

We will give some typical examples of a proper morphism.

THEOREM 5.4. *The morphism $\pi : \mathbb{P}^n_{\mathbb{Z}} = \operatorname{Proj} \mathbb{Z}[x_0, x_1, \ldots, x_n] \to \operatorname{Spec} \mathbb{Z}$ is a proper morphism.*

PROOF. For $0 \leq j \leq n$, let

$$R_j = \mathbb{Z}\left[\frac{x_0}{x_j}, \frac{x_1}{x_j}, \ldots, \frac{x_n}{x_j}\right], \qquad U_j = \operatorname{Spec} R_j.$$

Then $\{U_j\}_{0 \leq j \leq n}$ is an open covering of $\mathbb{P}^n_{\mathbb{Z}}$. The restriction of the diagonal morphism $\Delta : \mathbb{P}^n_{\mathbb{Z}} \to \mathbb{P}^n_{\mathbb{Z}} \times \mathbb{P}^n_{\mathbb{Z}}$ to $U_i \cap U_j$ is precisely the induced morphism $\Delta_{ij} : U_i \cap U_j \to U_i \times U_j$ by the natural immersions $U_i \cap U_j \to U_i$ and $U_i \cap U_j \to U_j$. Let $R = \mathbb{Z}[x_0, x_1, \ldots, x_n]$ and let $R_{(x_j)}$ be the subring of $R[\frac{1}{x_j}]$ consisting of elements of degree zero. Then $R_j = R_{(x_j)}$. For $i \neq j$, let $R_{ij} = R_{(x_i x_j)}$. We have $U_i \cap U_j = \operatorname{Spec} R_{ij}$. We can define a natural homomorphism

$$\sigma_{ij} : R_{(x_i)} \to R_{(x_i x_j)},$$
$$\frac{p(x_0, \ldots, x_n)}{x_i^m} \mapsto \frac{x_j^m p(x_0, \ldots, x_n)}{(x_i x_j)^m},$$

where $p(x_0, \ldots, x_n)$ is an integer coefficient homogeneous polynomial of degree m. The natural open immersion $U_i \cap U_j \to U_i$ is induced by $\sigma_{ij} : R_i \to R_{ij}$. Therefore, for $i \neq j$, a morphism $\Delta_{ij} : U_i \cap U_j \to U_i \times U_j$ is determined by the homomorphism

$$\psi_{ij} : R_i \otimes R_j \to R_{ij} = R_{(x_i x_j)},$$
$$f \otimes g \mapsto \sigma_{ij}(f) \cdot \sigma_{ij}(g).$$

Since R_{ij} is generated by $\frac{x_0}{x_i}, \ldots, \frac{x_{i-1}}{x_i}, \frac{x_{i+1}}{x_i}, \ldots, \frac{x_n}{x_i}, \frac{x_i}{x_j}$ over $\mathbb{Z}$, ψ_{ij} is a surjective homomorphism. Hence, Δ_{ij} is a closed immersion. Moreover, $\Delta_{ii} : U_i \to U_i \times U_i$ is a closed immersion. Consequently, π is a separated morphism since Δ is a closed immersion. Since R_j is finitely generated over $\mathbb{Z}$, π is a morphism of finite type.

We will show that π is a universally closed morphism. For an arbitrary scheme W, we need to show that $\pi_W : \mathbb{P}^n_{\mathbb{Z}} \times W \to W$ is a closed morphism. For an affine open covering $\{W_i\}_{i \in I}$ of W, it is enough to show that $\pi_{W_i} : \mathbb{P}^n_{\mathbb{Z}} \times W_i \to W_i$ is a closed morphism. Namely, we will prove that for an arbitrary commutative ring A, $\pi_A : \mathbb{P}^n_A = \operatorname{Proj} A[x_0, x_1, \ldots, x_n] \to \operatorname{Spec} A$ is a closed morphism. For simplicity, let

$$R = A[x_0, x_1, \ldots, x_n], \qquad R_j = R_{(x_j)} = A\left[\frac{x_0}{x_j}, \frac{x_1}{x_j}, \ldots, \frac{x_n}{x_j}\right],$$
$$U_i = \operatorname{Spec} R_j, \qquad j = 0, 1, 2, \ldots, n.$$

For a closed subscheme $Z \subset \mathbb{P}^n_A$, let $I(Z \cap U_i)$ be the ideal of R_i determined by $Z \cap U_i$. For $g \in I(Z \cap U_i)$, we will show that there is a homogeneous polynomial $G \in A[x_0, x_1, \ldots, x_n]$ of degree r satisfying $G/x_j^m \in I(Z \cap U_i)$, $0 \leq j \leq n$, and $G/x_i^m = g$ for a large m. Choose l so that $H = x_i^l g$ is a homogeneous polynomial. Then, since $H/x_j^l \in I(Z \cap U_j \cap U_i)$, one can find a positive integer n_{ij} to satisfy $(x_i^{n_{ij}} H)/x_j^{l+n_{ij}} \in I(Z \cap U_j)$. Let N be the largest such n_{ij}, and let $G = x_i^{l+N} g$. Then G is a homogeneous polynomial of degree $m = l + N$ so that $G/x_j^m \in I(Z \cap U_j)$ and $g = G/x_i^m$.

If $\pi(Z) = \operatorname{Spec} A$, then $\pi(Z)$ is a closed set. Suppose $\pi(Z) \neq \operatorname{Spec} A$. Let $\mathfrak{p}$ be a prime ideal of A satisfying $\mathfrak{p} \in \operatorname{Spec} A \setminus \pi(Z)$. Restricting to $\operatorname{Spec} A_{\mathfrak{p}} \subset \operatorname{Spec} A$, $\pi^{-1}(\mathfrak{p} A_{\mathfrak{p}})$ does not intersect with $Z \cap \pi^{-1}(\operatorname{Spec} A_{\mathfrak{p}})$. Then we have

$$(5.1) \quad I(Z \cap U_i) A_{\mathfrak{p}} + \mathfrak{p} A_{\mathfrak{p}} \left[\frac{x_0}{x_i}, \frac{x_1}{x_i}, \ldots, \frac{x_n}{x_i}\right] = A_{\mathfrak{p}} \left[\frac{x_0}{x_i}, \frac{x_1}{x_i}, \ldots, \frac{x_n}{x_i}\right].$$

Let R_m be the A-submodule of R consisting of homogeneous polynomials of degree m and let $I_m = I(Z) \cap R_m$. From (5.1), we obtain

$$(5.2) \qquad 1 = f_i + \sum_j p_{ij} f_{ij},$$

$$f_i \in I(Z \cap U_i) A_{\mathfrak{p}}, \qquad p_{ij} \in \mathfrak{p} A_{\mathfrak{p}}, \qquad f_{ij} \in A_{\mathfrak{p}} \left[\frac{x_0}{x_i}, \ldots, \frac{x_n}{x_i}\right].$$

Multiplying a sufficiently high power x_i^N of x_i with both sides of (5.2), we get

$$x_i^N = f_i' + \sum_j p_{ij} f_{ij}', \qquad f_i' \in I_N A_{\mathfrak{p}}, \qquad f_{ij}' \in A_{\mathfrak{p}}.$$

5.1. PROPER MORPHISMS

For an even larger N, we have
$$R_N A_\mathfrak{p} = I_N A_\mathfrak{p} + (\mathfrak{p} A_\mathfrak{p})(R_N A_\mathfrak{p}).$$
That is,
$$R_N/I_N \otimes_A A_\mathfrak{p}/\mathfrak{p} A_\mathfrak{p} = 0.$$
By Nakayama's lemma (see Problem 2 below), we have
$$(R_N/I_N) \otimes_A A_\mathfrak{p} = 0.$$
Namely, there exists $f \in A \setminus \mathfrak{p}$ satisfying $fR_N \subset I_N$. In particular, $fx_i^N \in I_N$. Consequently, we get $f \in I(Z \cap U_i)$. Hence,
$$Z \cap \mathbb{P}_\mathbb{Z}^n \times \operatorname{Spec} R_f = \varnothing.$$
We have proved that there is an open set $D(f)$ containing each point $\mathfrak{p}$ of $\operatorname{Spec} A \setminus \pi(Z)$ satisfying $D(f) \cap \pi(Z) = \varnothing$. □

PROBLEM 2. For a finite module M over a commutative ring R, if an ideal I of R satisfies $M = IM$, then there exists an element $f \in 1 + I$ to annihilate M, i.e., $fM = 0$. (This assertion is often called Nakayama's lemma; however, it was first proved by Krull and Azumaya.) In particular, when R is a local ring and I is the maximal ideal, then f is a unit of R. Consequently, we get $M = 0$.

COROLLARY 5.5. *For an arbitrary scheme Y, let $\pi_Y : \mathbb{P}_Y^n = \mathbb{P}_\mathbb{Z}^n \times Y \to Y$ be the base change of $\pi : \mathbb{P}_\mathbb{Z}^n \to \operatorname{Spec} \mathbb{Z}$. For a closed subscheme X of $\mathbb{P}_\mathbb{Z}^n \times Y$, let $\varphi : X \to Y$ be the restriction of π_Y to X. Then φ is a proper morphism.* □

The relevant statements to Corollary 5.5 about the properties of a projective scheme and a projective morphism will appear in the next section.

(c) **Valuative Criterion**

The notion of a proper morphism is extremely important for the finiteness of a coherent sheaf. However, it is not an easy task to tell whether a given scheme morphism is proper. We will describe a method using a valuation ring to determine whether a morphism is proper. This method is called the *valuative criterion*.

Let G be an additive group equipped with a total order $<$, i.e., for any elements x and y in G, we have $x = y$, $x < y$ or $y < x$. Furthermore, if we always have $x + x' \leq y + y'$ for $x \leq y$ and $x' \leq y'$, G is said to be a *totally ordered additive group*. Consider $\mathbb{Z}^n$ with the lexicographic order, namely, for a pair $(a_1, a_2, \ldots, a_n)$ and $(b_1, b_2, \ldots, b_n)$ in $\mathbb{Z}^n$, when $a_1 = b_1, a_2 = b_2, \ldots, a_l = b_l, a_{l+1} <$

b_{l+1}, then $(a_1, a_2, \ldots, a_n) < (b_1, b_2, \ldots, b_m)$, and when $a_1 = b_1, a_2 = b_2, \ldots, a_l = b_l, a_{l+1} > b_{l+1}$, then $(b_1, b_2, \ldots, b_n) < (a_1, a_2, \ldots, a_n)$. This example $\mathbb{Z}^n$ is a typical example of a totally ordered additive group.

Let K be a field and let $v : K \setminus \{0\} \to G$ be a map into a totally ordered additive group G. Then the map v is said to be a *valuation of K with values in G* if v satisfies the following two conditions.

(V1) For x and y in K, $x, y \neq 0$, we have $v(xy) = v(x) + v(y)$.

(V2) For x and y in K, $x, y \neq 0$ and $x + y \neq 0$, we have

$$v(x+y) \geq \min(v(x), v(y)).$$

If one assumes an element ∞ to be greater than all the elements in G, by defining $x + \infty = \infty$ and $v(0) = \infty$, then, without excluding 0, (V1) and (V2) hold for any x and y in K. In what follows, we will use ∞. For a valuation v of K with values in G, let

(5.3) $$R = \{x \in K | v(x) \geq 0\}.$$

From (V1) and (V2), R becomes a commutative ring. Also from (V1), we have $v(1) = v(1) + v(1)$, i.e., $v(1) = 0$. Namely, $1 \in R$. Then R is said to be the *valuation ring* of the valuation v. Furthermore,

(5.4) $$\mathfrak{m} = \{x \in K | v(x) > 0\}$$

is a maximal ideal of R.

PROBLEM 3. Prove that R as defined in (5.3) is a local ring with its maximal ideal $\mathfrak{m}$ as defined in (5.4).

An integral domain R is said to be a *valuation ring* if there exists a valuation of the quotient field K of R such that R is the valuation ring of this valuation. One often emphasizes the valuation ring more than the valuation itself.

PROBLEM 4. Let K be the quotient field of an integral domain R. Prove that the integral domain R is a valuation ring if and only if for an arbitrary element x in K, $x \neq 0$, either $x \in R$ or $x^{-1} \in R$.

Let R and S be local rings contained in a field K. For the maximal ideals $\mathfrak{m}_R$ and $\mathfrak{m}_S$ of R and S, respectively, if $R \subset S$ and $\mathfrak{m}_S \cap R = \mathfrak{m}_R$, then S is said to *dominate* R. Define an order: $S > R$ when S dominates R. One can characterize a valuation ring in terms of this order among local rings contained in K as follows.

5.1. PROPER MORPHISMS

PROPOSITION 5.6. *A local ring R contained in a field K and with the quotient field K is a valuation ring if and only if R is maximal with respect to the order defined above.*

PROOF. Let R be a valuation ring with the quotient field K. Then for an arbitrary element $x \neq 0$ in K, we have $x \in R$ or $x^{-1} \in R$. If a local ring $S \subset K$ dominates R, for $a \in S \setminus R$ we must have $a^{-1} \in R$. Hence $a^{-1} \in S$. If $a^{-1} \notin \mathfrak{m}_R$, from (5.3) and (5.4) we obtain $v(a^{-1}) = 0$. Then we would get $a = (a^{-1})^{-1} \in R$, contradicting our assumption. Therefore, $v(a^{-1}) > 0$, i.e., $a^{-1} \in \mathfrak{m}_R$. Since $\mathfrak{m}_S \cap R = \mathfrak{m}_R$, $a^{-1} \in \mathfrak{m}_S$. But since $a \in S$, we get $1 = a \cdot a^{-1} \in \mathfrak{m}_S$, a contradiction to our hypothesis. Consequently, we have $S = R$. We will not provide a proof for the converse. $\square$

PROBLEM 5. If $G = \mathbb{Z}$, a valuation v is said to be a *discrete valuation*. Then the valuation ring R of v is said to be the *discrete valuation ring*. Prove that the maximal ideal $\mathfrak{m}$ of the discrete valuation ring R is generated by one element π (π is said to be a *prime element* of R) so that an ideal of R has a form $(\pi^n) = \mathfrak{m}^n$.

EXERCISE 5.7. To define a morphism $q : \operatorname{Spec} R \to X$ from the affine scheme $\operatorname{Spec} R$ of a valuation ring R to a scheme X, it is necessary and sufficient to give a point x_1 of X and a point x_0 of the closure $Z = \overline{\{x_1\}}$ of x_1 in X satisfying the following conditions (1) and (2).

(1) An embedding $k(x_1) \subset K$ of the residue class field $k(x_1)$ at x_1 into K is given.

(2) With the reduced closed subscheme structure on Z, R dominates $\mathcal{O}_{Z,x_0}$.

PROOF. Let η_1 be the generic point of $\operatorname{Spec} R$ corresponding to the ideal (0) of R, and let η_0 be the point of $\operatorname{Spec} R$ corresponding to the maximal ideal $\mathfrak{m}_R$. Define $x_1 = q(\eta_1)$ and $x_0 = q(\eta_0)$. Then to give a local homomorphism $\mathcal{O}_{X,x_1} \to K = \mathcal{O}_{\operatorname{Spec} R, \eta_1}$ is to give an injective homomorphism of fields $k(x_1) = \mathcal{O}_{X,x_1}/\mathfrak{m}_{x_1} \subset K$. Notice also that the inverse image $q^{-1}(Z)$ of the closed set Z is a closed set. Then, since $q^{-1}(Z)$ contains the generic point of $\operatorname{Spec} R$, we get $q^{-1}(Z) = \operatorname{Spec} R$. Hence, in particular, $\eta_0 \in q^{-1}(Z)$. Namely, $x_0 = q(\eta_0) \in Z$. Furthermore, since $\operatorname{Spec} R$ is reduced, with the reduced closed subscheme structure on Z, one can obtain a scheme morphism $q : \operatorname{Spec} R \to Z$. A local homomorphism $\mathcal{O}_{Z,x_0} \to R = \mathcal{O}_{\operatorname{Spec} R, \eta_0}$ that is compatible with the embedding $k(x_1) \subset K$ satisfies $k(x_1) = \mathcal{O}_{Z,x_1}$ and $\mathcal{O}_{Z,x_0} \subset \mathcal{O}_{Z,x_1}$ (see Problem 6 following).

That is, R dominates $\mathcal{O}_{Z,x_0}$, in which case the scheme morphism $\operatorname{Spec} R \to \operatorname{Spec} \mathcal{O}_{Z,x_0}$, induced by $\mathcal{O}_{Z,x_0} \subset R$, composed with the natural morphism $\operatorname{Spec} \mathcal{O}_{Z,x_0} \to X$ gives $q : \operatorname{Spec} R \to X$. □

As we saw in the previous example, a point x_0 of a scheme X that is contained in the closure of $x_1 \in X$ is said to be a *specialization* of x_1.

PROBLEM 6. When for an arbitrary open set U of a scheme X, $\Gamma(U, \mathcal{O}_X)$ is an integral domain, X is said to be an *integral scheme*. Prove that X has a generic point x and that $\mathcal{O}_{X,x}$ is a field. Prove also that for any point y of X, $\mathcal{O}_{X,y}$ may be regarded as a subring of $\mathcal{O}_{X,x}$.

PROBLEM 7. Prove that the closure of x of the affine scheme $\operatorname{Spec} A$ determined by a prime ideal $\mathfrak{p}$ of a commutative ring A is given by

$$\overline{\{x\}} = \{\mathfrak{q} | \mathfrak{q} \text{ is a prime ideal of } A \text{ satisfying } \mathfrak{q} \supset \mathfrak{p}\},$$

and moreover, one can consider $\overline{\{x\}} = \operatorname{Spec} A/\mathfrak{p}$.

The following lemma characterizing a closed set in terms of a specialization will play a crucial role.

LEMMA 5.8. *Let* $f : X \to Y$ *be a quasicompact morphism of schemes. Then* $f(X)$ *is a closed set in* Y *if and only if* $f(X)$ *is closed under specialization, i.e.,* $x \in f(X)$ *and* $y \in \overline{\{x\}}$ *imply* $y \in f(X)$.

PROOF. It is clear from the definition that a point of a closed set is closed under specialization. Conversely, assume that $f(X)$ is closed under specialization. One may assume that X and Y are both reduced schemes, and if necessary, by introducing a reduced scheme structure on $f(X)$, one can assume that $Y = \overline{f(X)}$. Let y be an arbitrary point in Y. We need to show $y \in f(X)$. Let $\operatorname{Spec} A$ be any affine open set containing y. Instead of X and Y, consider $f^{-1}(\operatorname{Spec} A)$ and $\operatorname{Spec} A$ satisfying $Y = \operatorname{Spec} A$. Since f is a quasicompact morphism, $X = f^{-1}(\operatorname{Spec} A)$ is covered by finitely many affine open sets $\{X_j = \operatorname{Spec} B_j\}_{j \in J}$. Consider the restriction of f to X_j, i.e., $f|_{X_j} = f_j : X_j = \operatorname{Spec} B_j \to Y = \operatorname{Spec} A$. For simplicity, replace B_j with B. This morphism corresponds to a commutative ring homomorphism $g : A \to B$. Since X and Y are reduced, A and B have no nilpotent elements. If necessary, by replacing A by $\operatorname{Im} g$, one may assume g is injective. When Y is not irreducible, by taking

5.1. PROPER MORPHISMS

an irreducible component, one can assume Y is irreducible. That is, Y may be assumed to be irreducible and reduced. Namely, A is an integral domain. Let y' be the point corresponding to the prime ideal (0) of A. Then we have $Y \subset \overline{\{y'\}} = \operatorname{Spec} A$. Let K be the quotient field of A. Then the localization of A at (0) is K. Through g, regard B as an A-module. Then the localization at the prime ideal (0) of A is isomorphic to $B \otimes_A K$, which becomes a K-algebra. Since localization preserves exactness, $g_K : K \to B \otimes_A K$ is injective. Then for an arbitrary prime ideal $\mathfrak{q}_0$ of $B \otimes_A K$, we get $g_K^{-1}(\mathfrak{q}_0) = (0)$. Let $\mathfrak{q}$ be the inverse image of $\mathfrak{q}_0$ under the natural homomorphism $\lambda : B \to B \otimes_A K$. Then $\mathfrak{q}$ is a prime ideal and $g^{-1}(\mathfrak{q}) = \{0\}$. This is because if there is an element $a \in g^{-1}(\mathfrak{q})$, $a \neq 0$, then $1 = \frac{a}{a} \in K$ would imply $1 = g_K(\frac{a}{a}) \in \mathfrak{q}_0$. For x' in $X_j = \operatorname{Spec} B$ corresponding to the prime ideal $\mathfrak{q}$, we have $f_j(x') = y'$.

Consequently, we have obtained $y' \in f(X)$. Then y is a specialization of y', and $f(X)$ is closed under specialization. Hence, $y \in f(X)$. Therefore $\overline{f(X)} = f(X)$, i.e., $f(X)$ is a closed set. □

Based on the above argument, we will characterize a proper morphism in terms of a valuation ring.

THEOREM 5.9 (Valuative Criterion of Properness). *Let $f : X \to Y$ be a scheme morphism of finite type and let X be a Noetherian scheme. Then f is a proper morphism if and only if, for an arbitrary valuation ring R and a given morphism $s_1 : \operatorname{Spec} K \to X$ (K is the quotient field of R) and a given morphism $s : \operatorname{Spec} R \to Y$ making*

$$\begin{CD}
\operatorname{Spec} K @>{s_1}>> X \\
@V{\iota}VV @VV{f}V \\
\operatorname{Spec} R @>{s}>> Y
\end{CD}$$

commutative, there exists a unique morphism $g : \operatorname{Spec} R \to X$ making

(5.5)
$$\begin{CD}
\operatorname{Spec} K @>{s_1}>> X \\
@V{\iota}VV @VV{f}V \\
\operatorname{Spec} R @>>{s}> Y
\end{CD}$$

commute. Note that $\iota : \operatorname{Spec} K \to \operatorname{Spec} R$ is the natural open immersion induced by $R \subset K$.

PROOF. Let f be a proper morphism. First we will show that such a $g : \operatorname{Spec} R \to X$ exists. For the base change by $s : \operatorname{Spec} R \to Y$, i.e.,

$$\begin{array}{ccc} X_R = X \times_Y \operatorname{Spec} R & \longrightarrow & X \\ {\scriptstyle f_R} \downarrow & & \downarrow {\scriptstyle f} \\ \operatorname{Spec} R & \longrightarrow & Y \end{array}$$

f_R is a closed morphism. Then by $\iota : \operatorname{Spec} K \to \operatorname{Spec} R$ and $s_1 : \operatorname{Spec} K \to X$, we get a morphism $(s_1, \iota) : \operatorname{Spec} K \to X_R$. Let $\xi_1 \in X_R$ be the image of the unique point $\eta_1 \in \operatorname{Spec} K$ under the morphism (s_1, ι). Give the induced reduced closed subscheme structure on $Z = \overline{\{\xi_1\}}$ of X_R. Since f_R is closed, $Z' = f_R(Z)$ is a closed set of $\operatorname{Spec} R$. Moreover, since $f_R(\xi_1) = \iota(\eta_1)$ is the generic point of $\operatorname{Spec} R$, we have $Z' = \operatorname{Spec} R$. Let η_0 be the unique closed point of $\operatorname{Spec} R$. Then by Lemma 5.8, there exists ξ_0 in Z to satisfy $f_R(\xi_0) = \eta_0$. Then the restriction of f_R to Z determines a commutative ring homomorphism $\varphi : R \to \mathcal{O}_{Z,\xi_0}$. On the other hand, the residue class field $k(\xi_1)$ at the generic point ξ_1 of Z can be regarded as a subfield of K via (s_1, ι). As one can observe from Problem 6, $k(\xi_1)$ is the quotient field of $\mathcal{O}_{Z,\xi_0}$, i.e., the quotient field of $\mathcal{O}_{Z,\xi_0} \subset k(\xi_1) \subset K = R$. Thus, φ is an injective map, and furthermore, φ is a local homomorphism. Since, by Proposition 5.6, R is a maximal local ring contained in K, φ must be an isomorphism. In particular, R dominates $\mathcal{O}_{Z,\xi_0}$. We have shown that the condition in Example 5.7 is satisfied. Hence there exists a morphism $g_R : \operatorname{Spec} R \to X_R$. The composition of g_R with the natural morphism $X_R \to X$ is the morphism $g : \operatorname{Spec} R \to X$. It is clear from the definition of g that (5.5) is a commutative diagram.

Next we will prove that such a morphism g is unique. Let g' be another morphism from $\operatorname{Spec} R$ to X making (5.5) commutative. Let ξ_0 and ξ_0' be the images of the closed point η_0 of $\operatorname{Spec} R$ under g and g', respectively. Since $f : X \to Y$ is a separated morphism, the diagonal morphism $\Delta_{X/Y} : X \to X \times_Y X$ is a closed immersion. Then consider the morphism $(g, g') : \operatorname{Spec} R \to X \times_Y X$. Since g and g' coincide over $\operatorname{Spec} K$, the images (ξ_1, ξ_1') of the generic point of $\operatorname{Spec} R$ by (g, g') are on the diagonal set $\Delta_{X/Y}(X)$. Since $\Delta_{X/Y}(X)$ is a closed set, by Lemma 5.8 $\Delta_{X/Y}(X)$ is closed under specialization. On the other hand, (ξ_0, ξ_0') is the specialization of (ξ_1, ξ_1'). Hence, $(\xi_0, \xi_0') \in \Delta_{X/Y}(X)$. Namely, $\xi_0 = \xi_0'$. Example 5.7 implies $g = g'$.

5.1. PROPER MORPHISMS

Next we will prove the converse. By our assumption, f is a morphism of finite type. Then we will show that f is separated. That is, it is enough to show that the image $\Delta_{X/Y}(X)$ of the diagonal morphism $\Delta_{X/Y} : X \to X \times_Y X$ is closed under specialization. Let ζ_0 be a specialization of $\zeta_1 \in \Delta_{X/Y}(X)$. Induce the reduced closed subscheme structure on $W = \overline{\{\zeta_1\}}$. The residue class field K of ζ_1, i.e., $K = k(\zeta_1)$, may be regarded as the quotient field of $\mathcal{O}_{W,\zeta_0}$. Proposition 5.6 implies that there exists a valuation ring R for K dominating $\mathcal{O}_{W,\zeta_0}$. By Example 5.7, there is a morphism $h : \operatorname{Spec} R \to X \times_Y X$ that maps the generic point η_1 of $\operatorname{Spec} R$ to ζ_1 and the closed point η_0 to ζ_0. Let p_1 and p_2 be the projections from $X \times_Y X$ to the first and the second components, respectively. Then for $f : X \to Y$, we get $f \circ p_1 \circ h = f \circ p_2 \circ h$. For $\zeta_1 \in \Delta_{X/Y}(X)$, the restrictions of $p_1 \circ h$ and $p_2 \circ h$ to $\operatorname{Spec} R$ coincide. From the condition of this theorem, we must have $p_1 \circ h = p_2 \circ h$. By the definition of fiber product, the morphism h factors through the diagonal morphism $\Delta_{X/Y}$. Namely, there exists $\tilde{h} : \operatorname{Spec} R \to X$ so that $\operatorname{Spec} R \xrightarrow{\tilde{h}} X \xrightarrow{\Delta_{X/Y}} X \times_Y X$ may coincide with h. Therefore, $\zeta_0 \in \Delta_{X/Y}(X)$ and $\Delta_{X/Y}(X)$ is a closed set.

Finally we will prove that f is universally closed. For an arbitrary base change

$$\begin{array}{ccc} X' = Y' \times_Y X & \longrightarrow & X \\ {\scriptstyle f'}\downarrow & & \downarrow{\scriptstyle f} \\ Y' & \longrightarrow & Y \end{array}$$

and for a closed set Z of X', induce the reduced closed subscheme structure on Z. Since f is a morphism of finite type, f' is also of finite type. Hence the restriction f'_Z of f' to Z is a morphism of finite type and, in particular, is quasicompact. For simplicity, write f' for f'_Z. In order to show that $f'(Z)$ is a closed set, it is enough to show that $f'(Z)$ is closed under specialization. For $z_1 \in Z$, let y_0 be a specialization of $y_1 = f'(z_1) \in Y'$. Induce the reduced closed subscheme structure on $W = \overline{\{y_1\}}$. Then the quotient field of $\mathcal{O}_{W,y_0}$ is the residue class field $k(y_1)$ of y_1. Set $K = k(z_1)$. We have $k(y_1) \subset K$. Then by Proposition 5.6, there exists a valuation ring R of K dominating $\mathcal{O}_{W,y_0}$. Thus, we obtain the commutative diagram

$$\begin{array}{ccc} \operatorname{Spec} K & \xrightarrow{s_1'} & Z \\ {\scriptstyle \iota}\downarrow & & \downarrow{\scriptstyle f'} \\ \operatorname{Spec} R & \xrightarrow{s'} & Y' \end{array}$$

Compose s_1' and s' with $Z \to X' \to X$ and $Y' \to Y$, respectively, to obtain the morphisms $s_1 : \operatorname{Spec} K \to X$ and $s : \operatorname{Spec} R \to Y$. By the assumption of this theorem, there exists a morphism $g : \operatorname{Spec} R \to X$ making diagram (5.5) commutative. Taking the fiber product, g induces a morphism $g' : \operatorname{Spec} R \to X'$. By this construction of g' from g, the image $g'(\eta_1)$ of the generic point η_1 of $\operatorname{Spec} R$ is z_1 in Z. Since Z is a closed set, we get $g'(\operatorname{Spec} R) \subset Z$. For the closed point η_0 of $\operatorname{Spec} R$, let $z_0 = g'(\eta_0)$. Then $z_0 \in Z$ and $y_0 = f'(z_0)$. Therefore, $f'(Z)$ is closed under specialization, i.e., a closed set. We conclude that f is universally closed. $\square$

We can rephrase the existence of the unique morphism $g : \operatorname{Spec} R \to X$ to obtain the commutative diagram (5.5) as follows. As sets of morphisms over Y, there exists an isomorphism

(5.6)
$$\operatorname{Hom}_Y(\operatorname{Spec} R, X) \xrightarrow{\sim} \operatorname{Hom}_Y(\operatorname{Spec} K, X),$$
$$g \mapsto g \circ \iota.$$

The above proof also gives us the following proposition about a separated morphism.

PROPOSITION 5.10 (Valuative criterion of separatedness). *Let $f : X \to Y$ be a morphism of schemes. For a Noetherian scheme X, f is a separated morphism if and only if, for any morphisms s and s_1 making the diagram*

$$\begin{array}{ccc} \operatorname{Spec} K & \xrightarrow{s_1} & X \\ {\scriptstyle \iota}\downarrow & & \downarrow{\scriptstyle f} \\ \operatorname{Spec} R & \xrightarrow{s} & Y \end{array}$$

commutative, there exists at most one morphism $t : \operatorname{Spec} R \to X$ that can be added to the above diagram without losing commutativity.

EXAMPLE 5.11. Using the valuative criterion, we will show that the structure morphism of an n-dimensional projective space $\mathbb{P}^n_{\mathbb{Z}}$ over

$\mathbb{Z}$, i.e., $\pi : \mathbb{P}^n_{\mathbb{Z}} \to \mathbb{Z}$, is a proper morphism. For a given commutative diagram

$$\begin{array}{ccc} \operatorname{Spec} K & \xrightarrow{s_1} & X = \mathbb{P}^n_{\mathbb{Z}} \\ \iota \downarrow & & \downarrow f \\ \operatorname{Spec} R & \xrightarrow{s} & Y = \operatorname{Spec} \mathbb{Z} \end{array}$$

associated with a valuation ring R and its quotient field K, the morphism $s : \operatorname{Spec} R \to \operatorname{Spec} \mathbb{Z}$ is given by

$$\begin{cases} \mathbb{Z} \to R, \\ m \mapsto m \cdot 1, \\ \text{where 1 is a unit element of } R. \end{cases}$$

Since $U_j = D_+(x_j)$, $j = 0, 1, \ldots, n$, is an affine open covering of X, the image of s_1 belongs to one of these open sets. For the sake of simplicity, assume $s_1(\operatorname{Spec} K) \subset D_+(x_0)$. Then s_1 corresponds to the homomorphism

$$\varphi : \mathbb{Z}\left[\frac{x_1}{x_0}, \frac{x_2}{x_0}, \ldots, \frac{x_n}{x_0}\right] \to K.$$

Let $\varphi(\frac{x_j}{x_0}) = a_j$, $j = 1, 2, \ldots, n$, and let $a_j = \frac{b_j}{b_0}$, $b_0, b_j \in R$, $j = 1, 2, \ldots, n$. If $v(b_i)$ is the minimum among $v(b_0), v(b_1), \ldots, v(b_n)$, then we have $v(\frac{b_j}{b_i}) \geq 0$, i.e., $\frac{b_j}{b_i} \in R$. Then let

$$\varphi_i : \mathbb{Z}\left[\frac{x_0}{x_i}, \frac{x_1}{x_i}, \ldots, \frac{x_{i-1}}{x_i}, \frac{x_{i+1}}{x_i}, \ldots, \frac{x_n}{x_i}\right] \to R$$

be the homomorphism induced by $\varphi_i(\frac{x_j}{x_i}) = \frac{b_j}{b_i}$, $j \neq i$. Notice that φ_i determines $g : \operatorname{Spec} R \to D_+(x_i) \subset X$. The uniqueness of g follows from the uniqueness of the ratio $(b_0 : b_1 : \cdots : b_n)$. □

5.2. Quasicoherent Sheaves over a Projective Scheme

By applying the ideas in Chapter 4 to projective schemes, we will obtain some delicate results. A projective scheme is one of the most important objects to study in algebraic geometry.

(a) **Brief Review of Projective Schemes**

We touched on projective schemes in §2.3(c). We shall give a brief review.

For a graded ring $S = \bigoplus_{n=0}^{\infty} S_n$, S_0 is a commutative ring and S is regarded as an S_0-algebra. Then $S_+ = \bigoplus_{n \geq 1} S_n$ is a homogeneous ideal of S which is called an *irrelevant ideal*. Recall that the set of homogeneous ideals $\mathfrak{p} \not\supset S_+$ is $\operatorname{Proj} S$. We showed that $\operatorname{Proj} S$ is a scheme in §2.3(c). For a homogeneous element $f \in S_d$, the collection

$$D_+(f) = \{\mathfrak{p} \in \operatorname{Proj} S | f \notin \mathfrak{p}\}$$

forms a base for open sets. Then define the structure sheaf $\mathcal{O}_X$ of $X = \operatorname{Proj} S$ as

(5.7) $$\Delta(D_+(f), \mathcal{O}_X) = \left\{ \frac{g}{f^m} \,\bigg|\, g \in S_{md},\ m \geq 0 \right\}.$$

Note that in the right-hand side of (5.7) we have $\frac{g}{1} = \frac{fg}{f}$, where $g \in S_0$, $fg \in S_d$. Hence one may consider only the case when $m \geq 1$. The right-hand side of (5.7) is precisely the part $(S_f)_0$ of degree 0 of the localization S_f at $\{f^m\}_{m \geq 0}$ of S. Set $S_{(f)} = (S_f)_0$. That is, for a homogeneous element $f \in S_d$ of degree d, let

(5.8) $$S_{(f)} = \left\{ \frac{g}{f^m} \,\bigg|\, g \in S_{md},\ m \geq 0 \right\}.$$

We have $\Gamma(D_+(f), \mathcal{O}_X|D_+(f)) = \operatorname{Spec} S_{(f)}$. The natural homomorphism $S_0 \ni a \mapsto \frac{a}{1} \in S_{(f)}$ induces $\operatorname{Spec} S_{(f)} \to \operatorname{Spec} S_0$. Hence we obtain a scheme morphism $\pi : \operatorname{Proj} S \to \operatorname{Spec} S_0$, which is called the *structure morphism* of the scheme $\operatorname{Proj} S$.

PROBLEM 8. For a graded ring $S = \bigoplus_{m \geq 0} S_m$, $R = S_0$, prove that the structure morphism $\pi : \operatorname{Proj} S \to \operatorname{Spec} R$ is a separated morphism.

EXERCISE 5.12. Let S be a graded ring and let e be a positive integer. Define a subring of S as

$$S^{(e)} = \bigoplus_{m=0}^{\infty} S_{me}.$$

Then prove that there exists a natural isomorphism of schemes

$$\operatorname{Proj} S \overset{\sim}{\to} \operatorname{Proj} S^{(e)}.$$

In particular, let I be a homogeneous ideal of a polynomial ring $R[x_0, x_1, \ldots, x_n]$ over a commutative ring R. Consider the graded ring

$$S = R[x_0, x_1, \ldots, x_n]/I.$$

Let $I = \bigoplus_{n=1}^{\infty} I_n$, and for an arbitrary positive integer n_0, let $I' = \bigoplus_{n \geq n_0} I_n$ and $R' = R[x_0, x_1, \ldots, x_n]/I'$. Then for $d \geq n_0$ we have $S_d = S'_d$. Then show that $\operatorname{Proj} S \overset{\sim}{\to} \operatorname{Proj} S'$.

5.2. QUASICOHERENT SHEAVES OVER A PROJECTIVE SCHEME 69

PROOF. For a homogeneous prime ideal $\mathfrak{p}$ of S, it is clear that $\mathfrak{p}' = S^{(e)} \cap \mathfrak{p}$ is a homogeneous prime ideal of $S^{(e)}$. Unless $\mathfrak{p}$ is contained in the irrelevant ideal S_+ of S, there is a homogeneous element $a \notin \mathfrak{p}$ with $a \in S_d$. Then we have $a^e \in S_{de}$ and $a^e \notin \mathfrak{p}$. Therefore, $a^e \notin \mathfrak{p}'$, and $\mathfrak{p}'$ does not contain the irrelevant ideal $S_+^{(e)}$ of $S^{(e)}$. Namely, as a map of sets

$$\iota : \operatorname{Proj} S \to \operatorname{Proj} S^{(e)},$$

$$\mathfrak{p} \mapsto \mathfrak{p}' = S^{(e)} \cap \mathfrak{p},$$

is injective. Next consider a homogeneous ideal $\mathfrak{q}'$ of $S^{(e)}$. Then the ideal generated by $\mathfrak{q}'$ in S is homogeneous. Suppose that there are a and b satisfying $a \in S_{d_1}$ and $b \in S_{d_2}$ and $ab \in \mathfrak{q}$. Since $a^e b^e \in \mathfrak{q}'$, we have either $a^e \in \mathfrak{q}'$ or $b^e \in \mathfrak{q}'$. Let $\mathfrak{p} = \sqrt{\mathfrak{q}}$. Then $\mathfrak{p}$ is a prime ideal of S. It is also clear that $\mathfrak{p} \cap S^{(e)} = \mathfrak{q}'$. That is, the morphism ι is a set-theoretic bijection.

In $\operatorname{Proj} S$ we have $D_+(f) = D_+(f^e)$ and also $f^e \in S_{de}$. Hence, open sets of the type $D_+(f^e)$ constitute a base for open sets of $\operatorname{Proj} S^{(e)}$. Also in (5.8) we have $\frac{g}{f^m} = \frac{f^{m(e-1)}g}{f^{em}}$, and $f^{m(e-1)}g \in S_{mde}$. Namely, we get an isomorphism of commutative rings $S_{(f)} \xrightarrow{\sim} S_{(f^e)}^{(e)}$. Let $X = \operatorname{Proj} S$ and $X' = \operatorname{Proj} S^{(e)}$. We obtain a commutative ring isomorphism $\Gamma(D_+(f), \mathcal{O}_X) \xrightarrow{\sim} \Gamma(D_+(f^e), \mathcal{O}_{X'})$. Consequently, $(X, \mathcal{O}_X)$ and $(X', \mathcal{O}_{X'})$ are isomorphic as schemes.

The last part of the assertion is obtained from $S^{(e)} = S'^{(e)}$ for $e \geq n_0$. □

This example implies that one can replace a graded ring S with a subring $S^{(e)}$ to get a scheme $\operatorname{Proj} S$. In particular, if a graded ring S is finitely generated by $x_1, x_2, \ldots, x_n$ as an algebra over S_0, then a subring $S^{(d)} = \bigoplus_{m=0}^{\infty} S_{md}$ is generated by the elements in S_d as an S_0-algebra, where $d = d_1 \cdot d_2 \cdots d_n$ and $d_1 = \deg x_1, d_2 = \deg x_2, \ldots, d_n = \deg x_n$. Let $T_m = S_{md}$. Then $T = \bigoplus_{m=0}^{\infty} T_m$ is a graded ring, and T is generated by the elements in T_1 as a $T_0 = S_0$-algebra. From the above example, as schemes, $\operatorname{Proj} S$ and $\operatorname{Proj} T$ are isomorphic.

PROBLEM 9. Prove that a graded ring $S = \bigoplus_{m=0}^{\infty} S_m$ is a Noetherian ring if and only if S_0 is a Noetherian ring and S is finitely generated as an S_0-algebra.

In what will follow, we will mainly consider a graded ring S which is generated by S_1 as an S_0-algebra. In particular, we often consider

Proj S of a graded ring S that is Noetherian and generated by S_1 as an S_0-algebra. By Problem 9, namely, we consider the case where S is Noetherian. As was shown in the above example, even when $S \not\cong S'$, we can still have $\operatorname{Proj} S \xrightarrow{\sim} \operatorname{Proj} S'$.

From a homomorphism of commutative rings $\varphi : A \to B$, there is induced an affine scheme morphism $(\varphi^a, \varphi^\#) : \operatorname{Spec} B \to \operatorname{Spec} A$. (See Proposition 2.29.) Similarly, consider a homomorphism of graded rings $\varphi : S \to T$. Notice that we consider homogeneous ideals which do not contain the irrelevant ideal. This is a difference between $\operatorname{Proj} S$ and an affine scheme.

For graded rings S and T, if a homomorphism of commutative rings $\varphi : S = \bigoplus_{m=0}^\infty S_m \to T = \bigoplus_{m=0}^\infty T_m$ satisfies $\varphi(S_m) \subset T_m$, $m \geq 0$, then φ is said to be a *homomorphism of degree* 0. Then for a homogeneous ideal $\mathfrak{p}$ of T, $\varphi^{-1}(\mathfrak{p})$ is a homogeneous ideal of S. Since $\varphi^{-1}(\mathfrak{p}) \supset S_+$ is equivalent to $\mathfrak{p} \supset \varphi(S_+)$, we define

(5.9) $\qquad G(\varphi) = D_+(\varphi(S_+)) = \{\mathfrak{p} \in \operatorname{Proj} T | \mathfrak{p} \not\supset \varphi(S_+)\}.$

Then, for $\mathfrak{p} \in G(\varphi)$, we have $\varphi^{-1}(\mathfrak{p}) \in \operatorname{Proj} S$. Since $D_+(\varphi(S_+)) = \bigcup_{h \in \varphi(S_+)} D_+(h)$, $G(\varphi)$ is an open set of $\operatorname{Proj} T$. Induce an open subscheme structure on $G(\varphi)$ from $\operatorname{Proj} T$. We have the following.

PROPOSITION 5.13. *For a homomorphism φ of degree 0 of graded rings from S to T, there is induced a scheme morphism $\varphi^a : G(\varphi) \to \operatorname{Proj} S$. Furthermore, φ^a is an affine morphism.*

PROOF. For a homogeneous element $f \in S_d$, $d \geq 1$, let $g = \varphi(f)$. Then φ determines a commutative ring homomorphism $\varphi_f : S_{(f)} \to T_{(g)}$ inducing an affine scheme morphism $\varphi_f^a : D_+(g) = \operatorname{Spec} T_{(g)} \to D_+(f) = \operatorname{Spec} S_{(f)}$. Each point in $D_+(g)$ is a homogeneous ideal not containing $g = \varphi(f)$. Hence, $D_+(g) \subset G(\varphi)$. As f ranges over every homogeneous element of S_+, $\{D_+(g)\}$ is an affine open covering of $G(\varphi)$. Therefore, φ_f^a determines a scheme morphism $\varphi^a : G(\varphi) \to \operatorname{Spec} S$. Note also that $(\varphi^a)^{-1}(D_+(f)) = D_+(\varphi(f))$. That is, φ^a is an affine morphism. $\square$

EXAMPLE 5.14. Consider the graded ring $S = R[x_0, x_1, \ldots, x_n]$ generated by x_j of degree e_j, $j = 0, 1, \ldots, n$, over a commutative ring R. (Note that for $e_0 = e_1 = \cdots = e_n = 1$ we get the usual polynomial ring.) Then $\operatorname{Proj} S$ is said to be the *weighted projective space* over R of weights $e_0, e_1, \ldots, e_n$, denoted as $\mathbb{P}_R(e_0, e_1, \ldots, e_n)$. For the usual polynomial ring $R[y_0, y_1, \ldots, y_n]$, define a ring homomorphism

$$\varphi : S = R[x_0, x_1, \ldots, x_n] \to T = R[y_0, y_1, \ldots, y_n],$$
$$x_j \mapsto y_j^{e_j}.$$

Then we have $\varphi(S_m) \subset T_m$, i.e., a homomorphism of degree 0. We also get $\varphi(S_+) = (y_0^{e_0}, y_1^{e_1}, \ldots, y_n^{e_n})$ and $\sqrt{\varphi(S_+)} = T_+$. Therefore, $G(\varphi) = \operatorname{Proj} T = \mathbb{P}_R^n$. That is, we get a scheme morphism $\varphi^a : \mathbb{P}_R^n \to \mathbb{P}_R(e_0, \ldots, e_n)$. It is clear that φ^a is a finite morphism. □

EXAMPLE 5.15. Assume that a graded ring $S = \bigoplus_{m=0}^{\infty} S_m$ is finitely generated by S_1 over $R = S_0$ as an R-algebra. Choose a set of generators $z_0, z_1, \ldots, z_n \in S_1$ as an R-algebra. Then there exists a surjective homomorphism of degree 0 from the polynomial ring $R[y_0, y_1, \ldots, y_n]$ over R to S, i.e.,

$$\psi : R[y_0, y_1, \ldots, y_n] \to S,$$
$$y_j \mapsto z_j.$$

The kernel $I = \operatorname{Ker} \psi$ is a homogeneous ideal of $R[y_0, y_1, \ldots, y_n]$. Since the irrelevant ideal $(y_0, y_1, \ldots, y_n)$ of $R[y_0, y_1, \ldots, y_n]$ satisfies $\psi(y_0, y_1, \ldots, y_n) = S_+$, we have $G(\psi) = \operatorname{Proj} S$. Therefore, we obtain the scheme morphism $\psi^a : \operatorname{Proj} S \to \mathbb{P}_R^n$. Since ψ is an isomorphism from $R[y_0, y_1, \ldots, y_n]/I$ to S, ψ^a becomes an isomorphism from $\operatorname{Proj} S$ to the closed subscheme $V(I)$ determined by the homogeneous ideal I of $\mathbb{P}_R^n$. Namely, φ^a is an embedding. That is, one may consider $\operatorname{Proj} S$ as a closed subscheme of $\mathbb{P}_R^n$. Recall from Corollary 5.5 that for any commutative ring R the structure morphism $\mathbb{P}_R^n \to \operatorname{Spec} R$ is a proper morphism. Furthermore, the structure morphism $\pi : \operatorname{Proj} S \to \operatorname{Spec} R$ is the composition of the closed embedding $\operatorname{Proj} S \to \mathbb{P}_R^n$ and the structure morphism $\mathbb{P}_R^n \to \operatorname{Spec} R$. By Proposition 5.3(i), (iii), π is a proper morphism. □

In the above two examples, $G(\varphi) = \operatorname{Proj} T$ holds. There are many cases when $G(\varphi) \neq \operatorname{Proj} T$. We will give a simple example.

EXAMPLE 5.16. Consider polynomial rings $R[x_0, x_1, \ldots, x_n]$ and $R[x_0, x_1, \ldots, x_n, x_{n+1}, \ldots, x_{n+m}]$, $m \geq 1$. We have the natural injective homomorphism

$$\varphi : S = R[x_0, x_1, \ldots, x_n] \to T = R[x_0, x_1, \ldots, x_n, x_{n+1}, \ldots, x_{n+m}],$$
$$f(x_0, x_1, \ldots, x_n) \mapsto f(x_0, x_1, \ldots, x_n).$$

As graded rings, φ is a homomorphism of degree 0. We have $S_+ = (x_0, x_1, \ldots, x_n)$ and $\varphi(S_+) = (x_0, x_1, \ldots, x_n)$. Therefore,

$$G(\varphi) = D_+((x_0, x_1, \ldots, x_n)) \neq \mathbb{P}_R^{n+m}.$$

The scheme morphism $\varphi^a : G(\varphi) \to \mathbb{P}_R^n$ induced by φ is the morphism from the open subscheme to $\mathbb{P}_R^n$.

(b) **Quasicoherent Sheaves**

For a graded ring $S = \bigoplus_{m=0}^\infty S_m$, an S-module $M = \bigoplus_{n \in \mathbb{Z}} M_n$ is said to be a graded S-module if, for arbitrary m and n, $S_m \cdot M_n \subset M_{m+n}$. For a homogeneous element $f \in S_d$, define

(5.10) $$M_{(f)} = \left\{ \frac{\alpha}{f^l} \; \middle| \; l \geq 0, \; \alpha \in M_{ld} \right\}.$$

Then $M_{(f)}$ is an $S_{(f)}$-module, and $\widetilde{M_{(f)}}$ is a quasicoherent sheaf over $\operatorname{Spec} S_{(f)} = D_+(f)$. For $g \in S_e$, one can also define the quasicoherent sheaf $\widetilde{M_{(g)}}$ over $D_+(g) = \operatorname{Spec} S_{(g)}$. One can identify $D_+(fg) = D_+(f) \cap D_+(g)$ with the open set $D(\frac{g^d}{f^e})$ of $\operatorname{Spec} S_{(f)}$. Notice also that $M_{(fg)}$ can be identified with the localization $(M_{(f)})_h$ of $M_{(f)}$ at $h = \frac{g^d}{f^e}$. Namely, since for $\frac{\alpha}{(fg)^l}$, $\alpha \in M_{l(d+e)}$, we have $\frac{\alpha}{(fg)^l} = \frac{\alpha(fg)^{l(d-1)}}{(fg)^{ld}}$, we get an isomorphism from $M_{(fg)}$ to $(M_{(f)})_h$ by assigning $\frac{\alpha}{(fg)^l} \in M_{(fg)}$ to $\frac{\alpha(fg)^{l(d-1)}}{f^{l(d+e)}}/h^l \in (M_{(f)})_h$. Hence, we obtain $\widetilde{M_{(f)}}|D_+(fg) = \widetilde{M_{(fg)}}$. That is, one has the quasicoherent sheaf $\widetilde{M}$ over $X = \operatorname{Proj} S$. Note that $\widetilde{S} = \mathcal{O}_X$.

For a graded S-module $M = \bigoplus_{m \in \mathbb{Z}} M_m$ and an integer l, one can define another graded S-module $M(l)$, where

(5.11) $$M(l) = \bigoplus_{m \in \mathbb{Z}} M(l)_m, \qquad M(l)_m = M_{m+l}.$$

Notice that $M(0) = M$. Furthermore, for integers l_1 and l_2, we have $M(l_1 + l_2) = (M(l_1))(l_2)$. The quasicoherent sheaf $\widetilde{M(l)}$ can be constructed from $M(l)$, which we write as $\widetilde{M}(l)$. We also write $\mathcal{O}_X(l)$ for $\widetilde{S}(l)$. Then the S-module $S(l)$ is a free S-module generated by $1 \in S(l)_{-l}$ as an S-module.

For the study of projective schemes, we often assume that a graded ring $S = \bigoplus_{m=0}^\infty S_m$ is generated by S_1 as an S_0-algebra. The reasoning for this assumption is the following.

5.2. QUASICOHERENT SHEAVES OVER A PROJECTIVE SCHEME 73

LEMMA 5.17. *If a graded ring* $S = \bigoplus_{m=0}^{\infty} S_m$ *is generated by* S_1 *as an* S_0-*algebra, then* $\{D_+(f)\}_{f \in S_1}$ *is an affine open covering of* $\operatorname{Proj} S$.

PROOF. Consider a homogeneous prime ideal $\mathfrak{p}$ not containing the irrelevant ideal S_+ of S. If $\mathfrak{p} \supset S_1$, then, since S_1 generates S_+, we get $\mathfrak{p} \supset S_+$. This contradicts our assumption. Hence there exists $f \in S_1$ satisfying $f \notin \mathfrak{p}$. Then $\mathfrak{p} \in D_+(f)$. Consequently, we get $\operatorname{Proj} S \subset \bigcup_{f \in S_1} D_+(f)$. □

PROBLEM 10. Prove that for a set $\{f_\lambda\}_{\lambda \in \Lambda}$ of homogeneous elements of a graded ring S, $\{D_+(f_\lambda)\}_{\lambda \in \Lambda}$ is an open affine covering of $\operatorname{Proj} S$ if and only if the homogeneous ideal J of S generated by $\{f_\lambda\}_{\lambda \in \Lambda}$ satisfies $\sqrt{J} \supset S_+$.

EXERCISE 5.18. When a graded ring S is generated by S_1 as an S_0-algebra, the quasicoherent sheaf $\mathcal{O}_X(l)$, $l \in \mathbb{Z}$, over $X = \operatorname{Proj} S$ is an invertible sheaf.

PROOF. From Lemma 5.17, $\{D_+(f)\}_{f \in S_1}$ is an open covering of $X = \operatorname{Proj} S$. Since $S(l)_m = S_{m+l}$, for $f \in S_1$ we have

$$S(l)_{(f)} = \left\{ \frac{\beta}{f^n} \,\middle|\, n \geq 0,\ \beta \in S(l)_n = S_{n+l} \right\}.$$

Then the $S_{(f)}$-module homomorphism

$$\varphi_{(f)} : S_{(f)} \to S(l)_{(f)},$$
$$\frac{\alpha}{f^n} \mapsto \frac{\alpha f^l}{f^n},$$

is an isomorphism. It is easy to see that the inverse of $\varphi_{(f)}$ is given by the $S_{(f)}$-homomorphism

$$\psi_{(f)} : S(l)_{(f)} \to S_{(f)},$$
$$\frac{\beta}{f^n} \mapsto \frac{\beta}{f^{n+l}}.$$

This isomorphism $\varphi_{(f)}$ induces an $\mathcal{O}_{D_+(f)}$-module isomorphism

$$\mathcal{O}_{D_+(f)} \xrightarrow{\sim} \mathcal{O}_X(l)|D_+(f)$$

on $D_+(f)$. Therefore $\mathcal{O}_X(l)$ is an invertible sheaf. □

Next, for graded S-modules M and N, we will examine the relationship among the tensor product $M \otimes_S N$, $\operatorname{Hom}_S(M, N)$, and the

quasicoherency. For $a \in M_{d_1}$ and $b \in N_{d_2}$, define
$$\deg(a \otimes b) = d_1 + d_2 = \deg a + \deg b.$$
Then $M \otimes_S N$ becomes a graded module over S. This is because, for $f \in S_d, a \in M_{d_1}$ and $b \in N_{d_2}$, we have $f(a \otimes b) = (fa) \otimes b + a \otimes (fb)$, and consequently $S_d \cdot (M \otimes N)_n \subset (M \otimes N)_{n+d}$. For $f \in S_d$, one can define an $S_{(f)}$-module homomorphism

(5.12)
$$\lambda_{(f)} : M_{(f)} \otimes_{S_{(f)}} N_{(f)} \to (M \otimes_S N)_{(f)},$$
$$\frac{a}{f^m} \otimes \frac{b}{f^n} \mapsto \frac{a \otimes b}{f^{m+n}}.$$

This homomorphism induces an $\mathcal{O}_X$-module homomorphism over $X = \operatorname{Proj} S$,

(5.13)
$$\tilde{\lambda} : \widetilde{M} \otimes_{\mathcal{O}_X} \widetilde{N} \to (M \otimes_S N)^\sim.$$

Later we will show that $\tilde{\lambda}$ is an isomorphism when S is generated by S_1 as an S_0-algebra.

Next we will define the graded S-module $\operatorname{Hom}_S(M, N)$. An S-homomorphism $\varphi : M \to N$ is said to be an *S-homomorphism of degree n* if for an arbitrary integer m, we always have $\varphi(M_m) \subset N_{m+n}$. We denote the totality of S-homomorphisms of degree n by $\operatorname{Hom}_S(M, N)_n$. Then define $\operatorname{Hom}_S(M, N)$ as

(5.14)
$$\operatorname{Hom}_S(M, N) = \bigoplus_{n \in \mathbb{Z}} \operatorname{Hom}_S(M, N)_n.$$

We obtain a graded S-module having the elements in $\operatorname{Hom}_S(M, N)_n$ as the homogeneous elements of degree n. For $f \in S_d$, define the $S_{(f)}$-module homomorphism

(5.15)
$$\mu_{(f)} : \operatorname{Hom}_S(M, N)_{(f)} \to \operatorname{Hom}_{S_{(f)}}(M_{(f)}, N_{(f)})$$

by

(5.16)
$$\mu_{(f)}\left(\frac{\varphi}{f^m}\right)\left(\frac{a}{f^l}\right) = \frac{\varphi(a)}{f^{m+l}}, \qquad \varphi \in \operatorname{Hom}_S(M, N)_{md}, \qquad a \in M_{ld}.$$

For $g \in S_e$ we have $D_+(f) \cap D_+(g) = D_+(fg)$. The natural homomorphism
$$\operatorname{Hom}_S(M, N)_{(f)} \to \operatorname{Hom}_S(M, N)_{(fg)},$$
$$\frac{\varphi}{f^m} \mapsto \frac{g^m \cdot \varphi}{(fg)^m},$$

induces the commutative diagram

$$\begin{array}{ccc} \mathrm{Hom}_S(M,N)_{(f)} & \xrightarrow{\mu_{(f)}} & \mathrm{Hom}_{S_{(f)}}(M_{(f)}, N_{(f)}) \\ \downarrow & & \downarrow \\ \mathrm{Hom}_S(M,N)_{(fg)} & \xrightarrow{\mu_{(fg)}} & \mathrm{Hom}_{S_{(fg)}}(M_{(fg)}, N_{(fg)}) \end{array}$$

Therefore, we obtain an $\mathcal{O}_X$-module homomorphism

(5.17) $$\tilde{\eta} : \mathrm{Hom}_S(M,N)^\sim \to \underline{\mathrm{Hom}}_{\mathcal{O}_X}(\widetilde{M}, \widetilde{N}).$$

One can wonder when $\tilde{\eta}$ may be an isomorphism. We need to define the notion of finite presentation. A graded S-module M is said to be *finitely presented* if there exist integers $l_1, \ldots, l_s, m_1, \ldots, m_t$ to make the sequence of degree 0 homomorphisms

(5.18) $$\bigoplus_{i=1}^s S(l_i) \to \bigoplus_{j=1}^t S(m_j) \to M \to 0$$

exact. This definition is a generalization of a finitely presented S-module M to the case of a graded S-module.

PROPOSITION 5.19. *Let S be a graded ring generated by S_1 as an S_0-algebra. Then, for graded S-modules M and N, the $\mathcal{O}_X$-homomorphism $\tilde{\lambda}$ in (5.13) is an isomorphism. Furthermore, if M is finitely presented, the $\mathcal{O}_X$-homomorphism $\tilde{\mu}$ in (5.17) is an isomorphism.*

PROOF. From our assumption on S, $\{D_+(f)\}_{f \in S_1}$ is an open covering of $X = \mathrm{Proj}\, S$. Therefore, it is enough to prove our assertions over $D_+(f)$. First we will show that for $f \in S_1$, the homomorphism $\lambda_{(f)}$ in (5.12) is an isomorphism. Suppose

$$\lambda_{(f)}\left(\sum_i \frac{a_i}{f^{m_i}} \otimes \frac{b_i}{f^{n_i}}\right) = 0, \qquad a_i \in M_{m_i}, \qquad b_i \in N_{n_i}.$$

Then, in $(M \otimes_S N)_{(f)}$ we get $\sum_i \frac{a_i \otimes b_i}{f^{m_i+n_i}} = 0$. Namely, there exists a positive integer l such that $f^l(\sum_i a_i \otimes b_i) = 0$ in $M \otimes_S N$. Hence we have $\sum_i (f^l a_i) \otimes b_i = 0$, and in $M_{(f)} \otimes_{S_{(f)}} N_{(f)}$ we obtain

$$\sum_i \frac{f^l a_i}{f^{m_i+l}} \otimes \frac{b_i}{f^{n_i}} = 0.$$

Namely, $\lambda_{(f)}$ is injective. Since for $\frac{a \otimes b}{f^n} \in (M \otimes_S N)_{(f)}$, where $a \in S_d, b \in S_{n-d}$, we have $\frac{a}{f^d} \in M_{(f)}$ and $\frac{b}{f^{n-d}} \in N_{(f)}$, we obtain

$$\lambda_{(f)}\left(\frac{a}{f^d} \otimes \frac{b}{f^{n-d}}\right) = \frac{a \otimes b}{f^n}.$$

Namely, $\lambda_{(f)}$ is surjective. Therefore, over $D_+(f)$, $\tilde{\lambda}$ is an isomorphism. Since $\{D_+(f)\}_{f \in S_1}$ is an open covering of $X = \operatorname{Proj} S$, $\tilde{\lambda}$ is an isomorphism.

Next we assume that M is finitely presented. That is, we have an exact sequence as in (5.18). Then we get an exact sequence of S-modules

$$0 \to \operatorname{Hom}_S(M, N) \to \bigoplus_{j=1}^{t} \operatorname{Hom}_S(S(m_j), N) \to \bigoplus_{i=1}^{s} \operatorname{Hom}_S(S(l_i), N).$$

Therefore, for $f \in S$ we obtain the following commutative diagram:

(5.19)
$$\begin{array}{c}
0 \longrightarrow \operatorname{Hom}_S(M, N)_{(f)} \longrightarrow \bigoplus_{j=1}^{t} \operatorname{Hom}_S(S(m_j), N)_{(f)} \\
\downarrow \mu_{(f)} \qquad\qquad\qquad \downarrow \\
0 \longrightarrow \operatorname{Hom}_{S_{(f)}}(M_{(f)}, N_{(f)}) \longrightarrow \bigoplus_{j=1}^{t} \operatorname{Hom}_{S_{(f)}}(S(m_j)_{(f)}, N_{(f)})
\end{array}$$

$$\begin{array}{c}
\longrightarrow \bigoplus_{i=1}^{s} \operatorname{Hom}_S(S(l_i), N)_{(f)} \\
\downarrow \\
\longrightarrow \bigoplus_{i=1}^{s} \operatorname{Hom}_{S_{(f)}}(S(l_i)_{(f)}, N_{(f)})
\end{array}$$

The two horizontal sequences are exact. In order to show $\mu_{(f)}$ to be an isomorphism, we need to prove the vertical homomorphisms are isomorphic. That is, for an arbitrary integer l, it is enough to prove that the $S_{(f)}$-homomorphism

(5.20) $\qquad \nu_{(f)} : \operatorname{Hom}_S(S(l), N)_{(f)} \to \operatorname{Hom}_{S_{(f)}}(S(l)_{(f)}, N_{(f)})$

is an isomorphism. Note that each element φ of $\operatorname{Hom}_S(S(l), N)_n$ is determined by $\varphi(1) \in N_{n-l}$, $1 \in S(l)_{-l}$. Hence, $\operatorname{Hom}_S(S(l), N)_n \xrightarrow{\sim}$

5.2. QUASICOHERENT SHEAVES OVER A PROJECTIVE SCHEME

N_{n-l}. Therefore, as S-modules, $\mathrm{Hom}_S(S(l), N) \stackrel{\sim}{\to} N(-l)$. Now we can rewrite (5.20) as

$$\nu_{(f)} : N(-l)_{(f)} \to \mathrm{Hom}_{S_{(f)}}(S(l)_{(f)}, N_{(f)}).$$

For $\frac{a}{f^n}$, $a \in N(-l)_n = N_{n-l}$, and for $\alpha \in S(l)_m = S_{m+l}$, the $S_{(f)}$-homomorphism $\varphi = \nu_{(f)}(\frac{a}{f^n})$ is defined by $\varphi(\frac{\alpha}{f^m}) = \frac{\alpha a}{f^{n+m}}$. Note that $\alpha a \in N_{m+n}$. Suppose $\varphi = 0$. For $1 \in S(l)_{-l}$, we have $f^l \cdot 1 \in S(l)_{(f)}$ and $\varphi(f^l \cdot 1) = \frac{f^l a}{f^n} = 0$. Therefore, for a certain positive integer k we get $f^k a = 0$. Hence $\frac{a}{f^n} = \frac{f^k a}{f^{n+k}} = 0$. Namely, $\nu_{(f)}$ is an injection. Conversely, for a given $\psi \in \mathrm{Hom}_{S_{(f)}}(S(l)_{(f)}, N_{(f)})$, we can write $\psi(f^l \cdot 1) = \frac{b}{f^m}$, $b \in N_m$. Then $b \in N(-l)_{m+l}$ and $\frac{b}{f^{m+l}} \in N(-l)_{(f)}$. For $\beta \in S(l)_k$,

$$\nu_{(f)}\left(\frac{b}{f^{m+l}}\right)\left(\frac{\beta}{f^k}\right) = \frac{\beta b}{f^{m+k+l}}.$$

On the other hand, we have

$$\psi\left(\frac{\beta}{f^k}\right) = \psi\left(\frac{\beta}{f^{k+l}} \cdot (f^l \cdot 1)\right) = \frac{\beta}{f^{k+l}} \cdot \frac{b}{f^m} = \frac{\beta b}{f^{m+k+l}}.$$

Namely, $\psi = \nu_{(f)}(\frac{b}{f^{m+l}})$. Hence $\nu_{(f)}$ is surjective. Consequently, $\nu_{(f)}$ is an isomorphism. From ((5.19)), $\mu_{(f)}$ is an isomorphism. That is, the homomorphism $\tilde{\mu}$ in (5.17) becomes an isomorphism when restricted to $D_+(f)$, $f \in S_1$. Since $\{D_+(f)\}_{f \in S_1}$ is an open covering, we conclude that $\tilde{\mu}$ is an isomorphism. □

COROLLARY 5.20. *Assume a graded ring S is generated by S_1 as an S_0-algebra. For arbitrary integers l and m, we have the isomorphisms over $X = \mathrm{Proj}\, S$ as*

$$\mathcal{O}_X(l) \otimes_{\mathcal{O}_X} \mathcal{O}_X(m) \stackrel{\sim}{\to} \mathcal{O}_X(l+m),$$
$$\mathcal{O}_X(-l) \stackrel{\sim}{\to} \underline{\mathrm{Hom}}_{\mathcal{O}_X}(\mathcal{O}_X(l), \mathcal{O}_X).$$

For a graded S-module M, we have

$$\widetilde{M(l)} \stackrel{\sim}{\to} \widetilde{M} \otimes_{\mathcal{O}_X} \mathcal{O}_X(l). \quad \square$$

We have been considering the quasicoherent sheaf $\widetilde{M}$ over $\mathrm{Proj}\, S$ induced by a graded S-module M. We will find a sufficient condition for a given quasicoherent sheaf to be the associated quasicoherent sheaf with a graded S-module. In what will follow, we will assume that a graded ring S is generated by S_1 as an S_0-algebra. For a quasicoherent sheaf $\mathcal{F}$ over $X = \operatorname{Proj} S$, define $\mathcal{F}(n) = \mathcal{F} \otimes_{\mathcal{O}_X} \mathcal{O}_X(n)$ and

(5.21) $$\Gamma_*(\mathcal{F}) = \bigoplus_{n \in \mathbb{Z}} \Gamma(X, \mathcal{F}(n)).$$

Then $\Gamma_*(\mathcal{F})$ is a graded $\Gamma_*(\mathcal{O}_X)$-module, where the degree of an element in $\Gamma(X, \mathcal{F}(n))$ is n. For a homogeneous element $f \in S_d$, one can define a homomorphism

$$\alpha_n(f) : S_n \to S(n)_{(f)} = \Gamma(D_+(f), \mathcal{O}_X(n)),$$
$$a \mapsto \frac{a}{1}.$$

Furthermore, for a homogeneous element $g \in S_e$, we have the following commutative diagram:

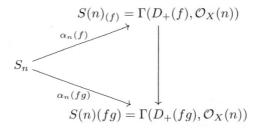

Then, by glueing $\{\alpha_n(f)(a) | f$ is a homogeneous element of $S_+\}$, we get an element of $\Gamma(X, \mathcal{O}_X(n))$. Thus, we can define the homomorphism of additive groups

$$\alpha_n : S_n \to \Gamma(X, \mathcal{O}_X(n)).$$

Also by Corollary 5.20, one gets the graded ring homomorphism

(5.22) $$\alpha = \bigoplus_{n=0}^{\infty} \alpha_n : S = \bigoplus_{n=0}^{\infty} S_n \to \Gamma_*(\mathcal{O}_X).$$

For the quasicoherent sheaf $\mathcal{F}$ over X, the homomorphism α induces a graded S-module structure on $\Gamma_*(\mathcal{F})$. For $f \in S_d$ and $x \in \Gamma_*(\mathcal{F})_{nd}$, we have $\frac{x}{f^n} \in \Gamma_*(\mathcal{F})_{(f)}$. Denote the restriction of x to $D_+(f)$ by $x|D_+(f)$. We have

$$\frac{x|D_+(f)}{(\alpha_d(f)|D_+(f))^n} \in \Gamma(D_+(f), \mathcal{F}).$$

5.2. QUASICOHERENT SHEAVES OVER A PROJECTIVE SCHEME

By assigning $\frac{x}{f^n} \in \Gamma_*(\mathcal{F})_{(f)}$ to

$$\frac{x|D_+(f)}{(\alpha_d(f)|D_+(f))^n} \in \Gamma(D_+(f), \mathcal{F}),$$

an $S_{(f)}$-module homomorphism

$$\beta_{(f)} : \Gamma_*(\mathcal{F})_{(f)} \to \Gamma(D_+(f), \mathcal{F})$$

can be defined. Furthermore, for $g \in S_e$, we have a commutative diagram as follows:

$$\begin{array}{ccc} \Gamma_*(\mathcal{F})_{(f)} & \xrightarrow{\beta_{(f)}} & \Gamma(D_+(f), \mathcal{F}) \\ \downarrow & & \downarrow \\ \Gamma_*(\mathcal{F})_{(fg)} & \xrightarrow{\beta_{(fg)}} & \Gamma(D_+(fg), \mathcal{F}) \end{array}$$

Thus, we can define an $\mathcal{O}_X$-module homomorphism

(5.23) $$\beta_{\mathcal{F}} : \widetilde{\Gamma_*(\mathcal{F})} \to \mathcal{F}.$$

The necessary and sufficient conditions for α and β to be isomorphisms will be given in the following theorem.

THEOREM 5.21. *Suppose a graded ring S is generated by finitely many elements $f_1, f_2, \ldots, f_n$ of S_1 as an S_0-algebra.*

(i) *If S is an integral domain, then the homomorphism α in (5.22) is injective. Furthermore, if for all i, f_i is a prime element (i.e., the ideal (f_i) is a prime ideal), then α is an isomorphism. When S is the polynomial ring over a commutative ring R, i.e., $S = R[x_0, x_1, \ldots, x_m]$, α is also an isomorphism.*

(ii) *If $\mathcal{F}$ is a quasicoherent $\mathcal{O}_X$-module, then $\beta_{\mathcal{F}}$ in (5.23) is an isomorphism.*

PROOF. If $\alpha_m(a) = 0$ for $a \in S_m$, we get

$$\alpha(a)|D_+(f_i) = \frac{a}{f_i^m} \cdot \frac{f_i^m}{1} = 0.$$

Namely, for each i, we have $\frac{a}{f_i^m} = 0$. Hence for a large integer N, $f_i^N a = 0$ must hold, $i = 1, 2, \ldots, n$. Since S is generated by $f_1, f_2, \ldots, f_n$ as an S_0-algebra, and S is an integral domain, we obtain $a = 0$. That is, α is an injection. Assume also that f_i is a prime element of S. For $h \in \Gamma(X, \mathcal{O}_X(m))$, we can write

$$h|D_+(f_i) = \frac{b_i}{f_i^{m_i}} \cdot \frac{f_i^m}{1} \in \Gamma(D_+(f_i), \mathcal{O}_X(m)) = S(m)_{(f_i)}, \qquad b_i \in S_m.$$

We may assume $m_i \geq m$. Then $h|D_+(f_i)$ and $h|D_+(f_j)$ must coincide when restricted to $D_+(f_i) \cap D_+(f_j) = D_+(f_i f_j)$. That is, in $S(m)_{(f_i f_j)}$ we have

$$\frac{b_i}{f_i^{m_i}} \cdot \frac{f_i^m}{1} = \frac{b_j}{f_j^{m_j}} \cdot \frac{f_j^m}{1}.$$

Since S is an integral domain, this condition is equivalent to $b_i f_j^{m_j - m} = b_j f_i^{m_i - m}$. For $m_i > m$, since f_i is a prime element, either $b_i \in (f_i)$ or $f_j \in (f_i)$. If $f_j \in (f_i)$, we get $f_j = g f_i$, $g \in S_0$. Since f_j is a prime element, g is a unit, i.e., $D_+(f_i) = D_+(f_j)$. Let $b_i \in (f_i)$. There is an element $c_i \in S_{m_i - 1}$ to satisfy $b_i = c_i f_i$. Since

$$\frac{c_i}{f_i^{m_i - 1}} = \frac{b_i}{f_i^{m_i}},$$

by applying the above argument to c_i, we can assume $m_i = m$. That means we may assume $b_i f_j^{m_j - m} = b_j$. If $m_j > m$, the above argument implies $m_j = m$. Therefore, for arbitrary $1 \leq i, j \leq n$, we can assume $b_i = b_j$. This implies that there exists $b \in S_m$ satisfying $\alpha_m(b) = h$. Namely, α is surjective. Consequently, α is an isomorphism. When S is the polynomial ring over R, $D_+(x_j)$, $j = 0, 1, \ldots, n$, form an affine open covering of $X = \mathbb{P}_R^n$. For this covering we can apply the above method. Note that x_j is not a zero divisor in this case.

(ii) In order to show that $\beta_{\mathcal{F}}$ is an isomorphism, it is enough to prove that $\beta_{\mathcal{F}_{(f)}}$ is an isomorphism. Let us write $\beta_{(f)}$ for $\beta_{\mathcal{F}_{(f)}}$. For $s \in \Gamma(X, \mathcal{F}(md))$, we have

$$\beta_{(f)}(s|D_+(f)) = \frac{s|D_+(f)}{(\alpha_d(f)|D_+(f))^m}.$$

If $\beta_{(f)}(s|D_+(f)) = 0$, then for each i, we must have $\beta_{(ff_i)}(s|D_+(ff_i)) = 0$. Namely, we have

$$\frac{s|D_+(ff_i)}{(\alpha_d(f)|D_+(ff_i))^m} = 0.$$

Hence, for a sufficiently large integer N,

$$(\alpha_d(f)|D_+(ff_i))^N \cdot s|D_+(ff_i) = (\alpha_d(f)^N s)|D_+(ff_i) = 0$$

for all i. We can consider $\alpha_d(f)^N s \in \Gamma(X, \mathcal{F}((N+m)d))$, and $\{D_+(ff_i)\}_{1 \leq i \leq n}$ is an open covering of the affine scheme $D_+(f)$. Hence, $\alpha_d(f)^N s|D_+(f) = 0$ must hold. Then we get

$$s|D_+(f) = \frac{s}{f^m} = \frac{\alpha_d(f)^N s}{f^{m+N}} = 0.$$

That is, $\beta_{(f)}$ is injective. Conversely, let $g \in \Gamma(D_+(f), \mathcal{F})$. Then we can consider

$$g|D_+(ff_i) \in \Gamma(D_+(ff_i), \mathcal{F}) = \Gamma(D_+(f_i), \mathcal{F})_{h_i}, \qquad h_i = \frac{f}{f_i^d} \in S_{(f_i)}.$$

By choosing a large enough integer N, we can extend $h_i^N(g|D_+(ff_i))$ to an element in $\Gamma(D_+(f_i), \mathcal{F})$, and hence extend $f^N(g|D_+(ff_i))$ to an element s_i in $\Gamma(D_+(f_i), \mathcal{F}(Nd))$. Namely, there is a sufficiently large N so that for all i, $f^N(g|D_+(ff_i))$ can be extended to $s_i \in \Gamma(D_+(f_i), \mathcal{F}(Nd))$. Note that the restrictions of $s_i|D_+(ff_i)$ and $s_j|D_+(ff_j)$ to $D_+(ff_if_j)$ are the same. Therefore, there exists a positive integer M so that, as an element of $\Gamma(D_+(f_if_j), \mathcal{F}((N+M)d))$, $(f^M s_i - f^M s_j)|D_+(f_if_j)$ becomes 0. That is, for a sufficiently large M, we have $(f^M s_i - f^M s_j)|D_+(f_if_j) = 0$. Therefore, $\{f^M s_i\}_{1 \le i \le n}$ determines an element $s \in \Gamma(X, \mathcal{F}((N+M)d))$. From the construction of s, we have $\beta_{(f)}(\frac{s}{f^{N+M}}) = g$. Namely, $\beta_{(f)}$ is surjective, and so $\beta_{(f)}$ is an isomorphism. □

The above proof required a lengthy general theory. The following example will play a significant role in what will follow.

EXAMPLE 5.22. Let us consider the polynomial ring $S = R[x_0, x_1, \ldots, x_n]$ over a commutative ring R. From the above theorem, for the invertible sheaf $\mathcal{O}(m)$ over the n-dimensional projective space $\mathbb{P}_R^n = \operatorname{Proj} R[x_0, x_1, \ldots, x_n]$, we have the following fact:

$$(5.24) \qquad \Gamma(\mathbb{P}_R^n, \mathcal{O}_{\mathbb{P}_R^n}(m)) = \begin{cases} 0, & m < 0, \\ R[x_0, x_1, \ldots, x_n]_m, & m \geq 0. \end{cases}$$

Notice that by α_m, we are identifying the right-hand side with the left-hand side of (5.24). In particular, $\Gamma(\mathbb{P}_R^n, \mathcal{O}_{\mathbb{P}_R^n}) = R$. For the n-dimensional projective space $\mathbb{P}_k^n$ over a field k, we have

$$(5.25) \qquad \dim_k \Gamma(\mathbb{P}_k^n, \mathcal{O}_{\mathbb{P}_k^n}(m)) = \begin{cases} 0, & m < 0, \\ \binom{m+n}{m}, & m \geq 0, \end{cases}$$

where $\binom{n}{0} \stackrel{\text{def}}{=} 1$. Namely, $\Gamma(\mathbb{P}_k^n, \mathcal{O}_{\mathbb{P}_k^n}(m))$ is a finite-dimensional vector space. This fact differs from the affine case. For example, if $Y = \operatorname{Spec} k[x]$, then $\Gamma(Y, \mathcal{O}_Y) = k[x]$, which is an infinite-dimensional vector space over the field k. □

EXERCISE 5.23. Let $M = \bigoplus_{m \in \mathbb{Z}} M_m$ be a graded S-module. For an arbitrary integer n_0, let $N = \bigoplus_{m \geq m_0} M_m$. Then N is a graded

S-module as well. Then the quasicoherent sheaves $\widetilde{M}$ and $\widetilde{N}$ over $X = \operatorname{Proj} S$ are isomorphic as $\mathcal{O}_X$-modules.

PROOF. Let
$$0 \to A \to B \to C \to 0$$
be an exact sequence of S-homomorphisms among the graded S-modules of degree 0. Then, for a homogeneous element $f \in S_d$,
$$0 \to A_{(f)} \to B_{(f)} \to C_{(f)} \to 0$$
is exact. From this, we obtain the exact sequence
$$0 \to \widetilde{A_{(f)}} \to \widetilde{B_{(f)}} \to \widetilde{C_{(f)}} \to 0$$
of $\mathcal{O}_{D_+(f)}$-modules over $D_+(f)$. For another homogeneous element $g \in S_e$, we get the corresponding exact sequence of $\mathcal{O}_{D_+(g)}$-modules. Those two exact sequences over $D_+(f)$ and $D_+(g)$ coincide over $D_+(f \cdot g)$. Consequently, we obtain an exact sequence
$$0 \to \widetilde{A} \to \widetilde{B} \to \widetilde{C} \to 0$$
of $\mathcal{O}_X$-modules. Since N is an S-submodule of M, there is an exact sequence of homomorphisms of degree 0
$$0 \to N \to M \to L \to 0.$$
Note that $L_m = M_m/N_m$. By the assumption, for $m > n_0$ we have $L_m = 0$. Hence, we will show that $\widetilde{L} = 0$. For a homogeneous element $f \in S_d$, consider $L_{(f)}$. For $a \in L_{ld}$, $\frac{a}{f^l} \in L_{(f)}$ can be written as $\frac{a}{f^l} = \frac{f^n a}{f^{l+n}}$. For a large n, $(n+l)d > n_0$ can hold, i.e., $f^n a \in L_{(n+l)d}$. Namely, $f^n a = 0$. Then $\frac{a}{f^l} = 0$, i.e., $L_{(f)} = 0$. We obtain $\widetilde{L} = 0$. □

In Theorem 5.21 we showed that for a finitely generated graded ring S by S_1 as an S_0-algebra, a quasicoherent sheaf $\mathcal{F}$ over $X = \operatorname{Proj} S$ can be written as $\mathcal{F} = \widetilde{M}$, where $M = \Gamma_*(\mathcal{F})$, the graded S-module. As the above example shows, unlike the affine case, M is not uniquely determined by $\mathcal{F}$. We will discuss the case where S is a Noetherian ring, and we will discuss coherent sheaves over $X = \operatorname{Proj} S$. Note that $N = \bigoplus_{m \geq n_0} M_m$ may be a finite S-module even when the graded S-module $M = \bigoplus_{m \in \mathbb{Z}} M_m$ is not a finite S-module.

5.2. QUASICOHERENT SHEAVES OVER A PROJECTIVE SCHEME

PROBLEM 11. Let L, M, and N be graded S-modules, and let $\varphi : L \to M$ and $\psi : M \to N$ be S-homomorphisms of degree 0. Prove that if the sequence induced by φ and ψ,

$$0 \to \bigoplus_{m \geq n_0} L_m \to \bigoplus_{m \geq n_0} M_m \to \bigoplus_{m \geq n_0} N_m \to 0,$$

is exact, then the sequence of $\mathcal{O}_X$-module homomorphisms over $X = \operatorname{Proj} S$,

$$0 \to \widetilde{L} \to \widetilde{M} \to \widetilde{N} \to 0,$$

is exact.

THEOREM 5.24. (i) *If a graded ring S is Noetherian, then $\operatorname{Proj} S$ is a Noetherian scheme.*

(ii) *If a graded ring S is finitely generated over $S_0 = R$, then the structure morphism $\pi : \operatorname{Proj} S \to \operatorname{Spec} R$ is a morphism of finite type.*

(iii) *Suppose that a graded ring S is a Noetherian ring, and is generated by S_1 as an S_0-algebra. If M is a graded S-module and a finite S-module (such an M is said to be a finite graded S-module), then $\widetilde{M}$ is a coherent sheaf over $X = \operatorname{Proj} S$. Conversely, a coherent sheaf $\mathcal{F}$ over X can be written as $\mathcal{F} = \widetilde{M}$, where M is a finite graded S-module.*

PROOF. For a positive integer d, let $S^{(d)} = \bigoplus_{m=0}^{\infty} S_{md}$. Since S is Noetherian, S_0 is a Noetherian ring. Hence S is generated by finitely many homogeneous elements $f_1, f_2, \ldots, f_l$ as an S_0-algebra (Problem 9). Let d_j be the degree of f_j. Define $G = \{f_1^{n_1} f_2^{n_2} \cdots f_l^{n_l} \mid 0 \leq n_j \leq d, j = 1, \ldots, n, \sum_{j=1}^{l} n_j d_j \equiv 0 \pmod{d}\}$. Then G is a finite set consisting of homogeneous elements of $S^{(d)}$. We will prove that, as an S_0-algebra, $S^{(d)}$ is generated by G. An element of S_{md} has a form $\sum \alpha_{m_1, \ldots, m_l} f_1^{m_1} \cdots f_l^{m_l}$ satisfying $\sum m_j d_j = md$. Therefore, $f_1^{m_1} \cdots f_l^{m_l}$ can be written as a product of elements in G. Hence, as an S_0-algebra, $S^{(d)}$ is generated by G. Since S_0 is a Noetherian ring, $S^{(d)}$ is also a Noetherian ring (Problem 9). For $f \in S_d$, one can define the homomorphism

(5.26)
$$\psi : S^{(d)}/(f-1) \to S_{(f)},$$
$$[x] \mapsto \frac{x}{f^m}, \quad \text{where } [x] \text{ is the image of } x \in S_{md}.$$

Obviously, ψ is surjective. Suppose $\psi(\sum_{i=1}^{k}[x_i]) = 0$, $x_i \in S_{m_i d}$. Choosing $m \geq m_i$, $i = 1, \ldots, k$, we can write

$$0 = \psi\left(\sum_{i=1}^{k}[x_i]\right) = \sum_{i=1}^{k} \frac{x_i}{f^{m_i}} = \sum_{i=1}^{k} \frac{f^{m-m_i} x_i}{f^m}.$$

Therefore, one can choose a positive integer N so that

$$f^N \left(\sum_{i=1}^{k} f^{m-m_i} x_i\right) = 0.$$

That is, $\sum_{i=1}^{k}[x_i] = 0$ in $S^{(d)}/(f-1)$. Hence, ψ is injective. Consequently, ψ is an isomorphism. Since $S^{(d)}/(f-1)$ is Noetherian, $S_{(f)}$ is also a Noetherian ring. On the other hand, S is generated by $f_1, \ldots, f_l$ as an S_0-algebra. Namely, $\operatorname{Proj} S = \bigcup_{j=1}^{l} D_+(f_j)$. Since $S_{(f_j)}$ is Noetherian, $D_+(f_j) = \operatorname{Spec} S_{(f_j)}$ is a Noetherian scheme. Therefore, $\operatorname{Proj} S$ is a Noetherian scheme as well.

(ii) In (i) we showed that, if S is finitely generated over S_0, then $S^{(d)}$ is also finitely generated over S_0. (Note that the Noetherianness of S was not used to prove this assertion.) We also proved the isomorphism of (5.26) without the Noetherianness assumption. Since $S^{(d)}$ is finitely generated as an S_0-algebra, $S^{(d)}/(f-1)$ is also finitely generated as an S_0-algebra. Therefore $S_{(f)}$ too is finitely generated as an S_0-algebra. Let $f_1, f_2, \ldots, f_l$ be generators for S as an S_0-algebra. Then $\operatorname{Proj} S = \bigcup_{j=1}^{l} D_+(f_j)$ is of finite type over $\operatorname{Spec} S_0$. This is because $S_{(f_j)}$ is a finitely generated S_0-algebra.

(iii) Choose a set of generators $f_1, \ldots, f_l \in S_1$ of the S_0-algebra S. Then $\operatorname{Proj} S = \bigcup_{j=1}^{l} D_+(f_j)$. (i) implies that $S_{(f_j)}$ is a Noetherian ring, and $\operatorname{Proj} S$ is a Noetherian scheme. In order to show the coherency of $\widetilde{M}$, it is enough to prove that $M_{(f_j)}$ is a finite $S_{(f_j)}$-module. For a homogeneous element $f \in S$, as in (i) let $M^{(d)} = \bigoplus_{m \in \mathbb{Z}} M_{md}$. Then the homomorphism

$$\psi_M : M^{(d)}/(f-1)M^{(d)} \to M_{(f)}$$

is an isomorphism. By regarding $M^{(d)}/(f-1)M^{(d)}$ as an $S^{(d)}/(f-1)$-module, the isomorphism ψ in (5.26) gives that ψ_M is an isomorphism between the $S^{(d)}/(f-1)$-module and the $S_{(f)}$-module. In particular, for $f \in S_1$, $M/(f-1)M$ is an $S/(f-1)$-module. Since M is finitely generated as an S-module, $M/(f-1)M$ is finitely generated as an $S/(f-1)$-module. Therefore, as an $S_{(f)}$-module, $M_{(f)}$ is finitely

5.2. QUASICOHERENT SHEAVES OVER A PROJECTIVE SCHEME 85

generated. Namely, $\widetilde{M_{(f)}}$ is a coherent sheaf over $\operatorname{Spec} S_{(f)} = D_+(f)$. Since $\widetilde{M}|D_+(f) = \widetilde{M_{(f)}}$, $\widetilde{M}$ is a coherent sheaf.

Conversely, for a coherent sheaf $\mathcal{F}$ over $X = \operatorname{Proj} S$, let $N = \Gamma_*(\mathcal{F})$. Then Theorem 5.21 implies $\widetilde{N} = \mathcal{F}$. Since

$$\operatorname{Proj} S = \bigcup_{j=1}^{l} D_+(f_j)$$

and $\mathcal{F}|D_+(f_j)$ is coherent over $D_+(f_j) = \operatorname{Spec} S_{(f_j)}$, it follows that $\Gamma(D_+(f_j), \mathcal{F}) = N_{(f_j)}$ is a finitely generated $S_{(f)}$-module. Let $\frac{\alpha_{ji}}{f_j^{d_{ij}}}$, $i = 1, 2, \ldots, s_j$, $\alpha_{ji} \in N_{d_{ij}}$, be generators for $N_{(f_j)}$. Let d be the maximum of $\{d_{ij}, 1 \leq j \leq l, 1 \leq i \leq s_j\}$. Rewrite

$$\frac{\alpha_{ji}}{f_j^{d_{ij}}} = \frac{f_i^{d - d_{ji}} \alpha_{ji}}{f_j^d} = \frac{\beta_{ji}}{f_j^d}.$$

We have $\beta_{ji} = N_d$. Define an S-submodule $M = \sum_{j,i} S\beta_{ji}$. Then M is a finitely generated S-module. Since $M_{(f_j)} = N_{(f_j)}$, we get $\widetilde{M} = \widetilde{N}$. That is, $\mathcal{F} = \widetilde{M}$. □

EXAMPLE 5.25. Let X be the closed subscheme determined by a quasicoherent sheaf of ideals over an n-dimensional projective space $P = \mathbb{P}_R^n$ over a commutative ring R. Then $\Gamma_*(\mathcal{J})$ is an ideal of $\Gamma_*(\mathcal{O}_P)$. By Example 5.22, we can regard $\Gamma_*(\mathcal{O}_P) = R[x_0, \ldots, x_n]$. Hence $J = \Gamma_*(\mathcal{J})$ is a homogeneous ideal of $R[x_0, \ldots, x_n]$. Let $S = R[x_0, \ldots, x_m]/J$. From Theorem 5.21(ii), we get $\widetilde{J} = \mathcal{J}$ and $X = \operatorname{Proj} S$.

Conversely, if R is a field k, then let J be the homogeneous ideal generated by a homogeneous polynomial $F \in k[x_0, x_1, \ldots, x_m]$ of degree m. Then $V(F) = \operatorname{Proj} k[x_0, x_1, \ldots, x_n]/J$ is a closed subscheme of $\mathbb{P}_k^n$. When $V(F)$ is reduced and irreducible, $V(F)$ is said to be a *hypersurface of degree m*. For $n = 2$, $V(F)$ is called a plane curve of degree m, and for $n = 3$, $V(F)$ is called a surface of degree m. The defining ideal $\mathcal{J}_X$ of $X = V(F)$ is precisely $\widetilde{J}$. Over each open set $D_+(x_j)$, $\mathcal{J}_X$ is the sheaf of ideals generated by $f_j = F(x_0, x_1, \ldots, x_n)/x_j^m$. Therefore, there exists an isomorphism

$$\mathcal{J}_X|D_+(x_j) \xrightarrow{\sim} \mathcal{O}_X|D_+(x_j),$$
$$f_j h \mapsto h,$$

such that $\mathcal{J}_X$ is an invertible sheaf. From this isomorphism, we get $\mathcal{J}_X \xrightarrow{\sim} \mathcal{O}_X(m)$. We often write $\mathcal{O}_F(-X)$ for $\mathcal{J}_X$.

PROBLEM 12. Let a graded ring S be the graded ring over a commutative ring R, i.e., $S_0 = R$. For a commutative ring homomorphism $\varphi : R \to A$, let $T = S \otimes_R A$. Then T is a graded ring over A. Prove that $\operatorname{Proj} T = \operatorname{Proj} S \times_{\operatorname{Spec} R} \operatorname{Spec} A$.

(c) $\operatorname{Proj} \mathcal{S}$

Consider a sheaf $\mathcal{S} = \bigoplus_{n=0}^{\infty} \mathcal{S}_n$ of quasicoherent graded $\mathcal{O}_X$-algebras over a scheme X. A product in the quasicoherent $\mathcal{O}_X$-module $\mathcal{S}$ is defined as

$$\mathcal{S}_m \cdot \mathcal{S}_n \subset \mathcal{S}_{m+n},$$

and this product is $\mathcal{O}_X$-bilinear. Then we have the projection $p_n : \mathcal{S} \to \mathcal{S}_n$ and the natural injective homomorphism $\iota_n : \mathcal{S}_n \to \mathcal{S}$. Those two maps are $\mathcal{O}_X$-module homomorphisms. Then $\mathcal{S}_n$ is isomorphic to the image of $\iota_n \circ p_n$. By Corollary 4.24, $\mathcal{S}_n$ is a quasicoherent sheaf. In particular, since $\mathcal{S}_0 \cdot \mathcal{S}_0 \subset \mathcal{S}_0$, $\mathcal{S}_0$ is a sheaf of $\mathcal{O}_X$-algebras. We will study the case where $\mathcal{S}_0 = \mathcal{O}_X$.

Let U be an affine open set of X. Then we have $\mathcal{S}_n|U = \widetilde{\Gamma(U, \mathcal{S}_n)}$ and also

$$\left(\bigoplus_{n=0}^{\infty} \Gamma(U, \mathcal{S}_n) \right)^{\sim} = \bigoplus_{n=0}^{\infty} (\Gamma(U, \mathcal{S}_n))^{\sim}.$$

Hence, we get

$$\mathcal{S}|U = \bigoplus_{n=0}^{\infty} \widetilde{\Gamma(U, \mathcal{S}_n)}.$$

Therefore,

$$\Gamma(U, \mathcal{S}) = \bigoplus_{n=0}^{\infty} \Gamma(U, \mathcal{S}_n).$$

For $U = \operatorname{Spec} R$, $\bigoplus_{n=0}^{\infty} \Gamma(U, \mathcal{S}_n)$ is a graded R-algebra. That is, we can define a projective scheme and the structure morphism

(5.27) $\qquad \pi_U : \operatorname{Proj} \left(\bigoplus_{n=0}^{\infty} \Gamma(U, \mathcal{S}_n) \right) \to \operatorname{Spec} R = U.$

Let $U' = \operatorname{Spec} R' \subset U = \operatorname{Spec} R$ be an affine open set contained in U. We have $\Gamma(U', \mathcal{S}) = \Gamma(U, \mathcal{S}) \otimes_R R'$. From Problem 12, we get

$\operatorname{Proj}\Gamma(U',\mathcal{S}) = \operatorname{Proj}\Gamma(U,\mathcal{S}) \times_{\operatorname{Spec} R} \operatorname{Spec} R'$. Then

$$\pi_{U'} : \operatorname{Proj}\Gamma(U',\mathcal{S}) \to \operatorname{Spec} R'$$

is obtained by the base change $\operatorname{Spec} R' \to \operatorname{Spec} R$ of π_U. Hence, for an affine covering $\{U_\lambda\}_{\lambda \in \Lambda}$ of X, construct $\pi_{U_\lambda} : \operatorname{Proj}\Gamma(U_\lambda, \mathcal{S}) \to U_\lambda$ as in (5.27). Then, over $U_\lambda \cap U_\mu \neq \varnothing$, π_{U_λ} and π_{U_μ} coincide, so we obtain the scheme $\operatorname{Proj}\mathcal{S}$ by glueing $\bigcup_{\lambda \in \Lambda} \operatorname{Proj}\Gamma(U_\lambda, \mathcal{S})$. Furthermore, by glueing π_{U_λ}, one obtains $\pi : \operatorname{Proj}\mathcal{S} \to X$. Then, $\operatorname{Proj}\mathcal{S}$ (to be precise, with the structure morphism $\pi : \operatorname{Proj}\mathcal{S} \to X$) is said to be the *projective scheme* associated with the sheaf $\mathcal{S}$ of quasicoherent graded $\mathcal{O}_X$-algebras.

EXAMPLE 5.26. Let $\mathcal{S} = \mathcal{O}_X[x_0, x_1, \ldots, x_n]$, where $\mathcal{S}_m$ is the totality of homogeneous polynomials of degree m with coefficients in $\mathcal{O}_X$, i.e.,

$$\mathcal{S}_m = \bigoplus_{a_0 + a_1 + \cdots + a_n = m} \mathcal{O}_X x_0^{a_0} x_1^{a_1} \cdots x_n^{a_n}.$$

Then $\mathcal{S}$ is a sheaf of quasicoherent graded $\mathcal{O}_X$-algebras. It is clear that for an affine open set $U = \operatorname{Spec} R$, we have

$$\Gamma(U, \mathcal{S}) = R[x_0, x_1, \ldots, x_n].$$

Therefore, $\operatorname{Proj}\Gamma(U, \mathcal{S}) = \mathbb{P}^n_R$. Write $\mathbb{P}^n_X$ for $\operatorname{Proj}\mathcal{S}$, which is called the projective space of dimension n over X, i.e., precisely speaking, $\mathbb{P}^n_X \to X$. Since $R[x_0, x_1, \ldots, x_n] = \mathbb{Z}[x_0, x_1, \ldots, x_n] \otimes_\mathbb{Z} R$, it follows that $\mathbb{P}^n_X = \mathbb{P}^n_\mathbb{Z} \times_{\operatorname{Spec}\mathbb{Z}} X$. □

EXAMPLE 5.27. For a quasicoherent $\mathcal{O}_X$-module $\mathcal{E}$ over a scheme X, the symmetric algebra $\mathbb{S}(\mathcal{E})$ (see Exercise **4.5**) is a sheaf of quasicoherent graded algebras. One can define $\mathbb{S}(\mathcal{E})_n$ as the image of $\underbrace{\mathcal{E} \otimes \cdots \otimes \mathcal{E}}_{n}$. Write $\mathbb{P}(\mathcal{E})$ for $\operatorname{Proj}\mathbb{S}(\mathcal{E})$. The existence of the structure morphism $\pi : \mathbb{P}(\mathcal{E}) \to X$ is clear. In particular, for $\mathcal{E} = \mathcal{O}_X^{\oplus(n+1)}$, we get $\mathbb{S}(\mathcal{E}) \xrightarrow{\sim} \mathcal{O}_X[x_0, x_1, \ldots, x_n]$ and, therefore, $\mathbb{P}(\mathcal{O}_X^{\oplus(n+1)}) = \mathbb{P}^n_X$. □

We have defined $\operatorname{Proj}\mathcal{S}$ by glueing the projective schemes obtained from a graded ring over a commutative ring via (5.27). Hence, almost all the arguments in (a) and (b) can be applied to this case. We will not describe all the corresponding results, but the reader is encouraged to modify the results in (b) to the $\operatorname{Proj}\mathcal{S}$ case.

PROBLEM 13. Let $\mathcal{S} = \bigoplus_{n=0}^{\infty} \mathcal{S}_n$ be a sheaf of quasicoherent graded algebras over a scheme X. Assume that there exists a non-zero section $f \in \Gamma(X, \mathcal{S}_d)$, $d \geq 1$. Prove that there exists an open set Z_f of $Z = \operatorname{Proj} \mathcal{S}$ satisfying the following conditions (1) and (2).

(1) For an affine open set U of X, we have $Z_f \cap \pi^{-1}(U) = D_+(f|U)$, where $\pi : Z \to X$ is the structure morphism.

(2) For a non-zero $g \in \Gamma(X, \mathcal{S}_e)$, $e \geq 1$, we have $Z_f \cap Z_g = Z_{fg}$.

PROBLEM 14. Let $\mathcal{S} = \bigoplus_{n=0}^{\infty} \mathcal{S}_n$ be a sheaf of quasicoherent graded algebras over a scheme X. Let $g^*\mathcal{S} = \bigoplus_{n=0}^{\infty} g^*\mathcal{S}_n$ be the pull-back of a scheme morphism $g : X' \to X$. Prove that

$$\operatorname{Proj} g^*\mathcal{S} \overset{\sim}{\to} \operatorname{Proj} \mathcal{S} \times_X X'.$$

EXAMPLE 5.28. Let $X = \mathbb{A}_k^n = \operatorname{Spec} k[x_1, x_2, \ldots, x_n]$ be an n-dimensional affine spaces over a field k and let $\mathcal{J}$ be the sheaf of ideals of $\mathcal{O}_X$ determined by the defining ideal $J = (x_1, x_2, \ldots, x_n)$ of the origin.

Then consider the sheaf $\mathcal{S} = \bigoplus_{d=0}^{\infty} \mathcal{J}^d$ of graded algebras over $\mathcal{O}_X$, where $\mathcal{J}^0 = \mathcal{O}_X$. The sheaf $\mathcal{S}$ is the quasicoherent sheaf over X determined by the graded algebra $S = \bigoplus_{d=0}^{\infty} J^d$ over $R = k[x_1, x_2, \ldots, x_n]$. If you let $\widetilde{X} = \operatorname{Proj} \mathcal{S}$, then you have $\widetilde{X} = \operatorname{Proj} S$. The R-homomorphism of degree 0

$$R[y_0, y_1, \ldots, y_{n-1}] \to S,$$
$$f(y_0, y_1, \ldots, y_{n-1}) \mapsto f(x_1, x_2, \ldots, x_n),$$

is an onto map. Hence, $\widetilde{X}$ can be regarded as a closed subscheme of $\mathbb{P}_R^{n-1} = \operatorname{Proj} R[y_0, y_1, \ldots, y_{n-1}]$. The restriction of the structure morphism $\mathbb{P}_R^{n-1} \to \operatorname{Spec} R = X$ to $\widetilde{X}$ is the natural morphism $\pi : \widetilde{X} \to X$. Then $\pi : \widetilde{X} \to X$, or simply $\widetilde{X}$, is said to be a blowing up of X at the origin. Let $U = X \setminus \{(0, \ldots, 0)\}$. We have $\mathcal{S}|U \overset{\sim}{\to} \bigoplus_{d=0}^{\infty} \mathcal{O}_U^{\otimes d} \overset{\sim}{\to} \mathcal{O}_U[T]$. Therefore, $\pi^{-1} \overset{\sim}{\to} \operatorname{Proj} \mathcal{O}_U[T] = U$, i.e., namely, π is an isomorphism from $\pi^{-1}(U)$ to U. On the other hand, we have $S \otimes_R (R/J) \overset{\sim}{\to} k[y_0, y_1, \ldots, y_{n-1}]$ over the origin, and we get $\operatorname{Proj} k[y_0, y_1, \ldots, y_{n-1}] = \mathbb{P}_k^{n-1}$. □

Next we will consider quasicoherent sheaves over $\operatorname{Proj} \mathcal{S}$. For $\mathcal{S} = \bigoplus_{n=0}^{\infty} \mathcal{S}_n$, let $\mathcal{M} = \bigoplus_{n \in \mathbb{Z}} \mathcal{M}_n$ be a quasicoherent graded $\mathcal{S}$-module. That is, $\mathcal{M}$ is an $\mathcal{S}$-module satisfying $\mathcal{S}_m \cdot \mathcal{M}_n \subset \mathcal{M}_{n+m}$. Then $\mathcal{M}_n$ is a quasicoherent $\mathcal{O}_X$-module and, for an affine open U of X, we have $\Gamma(U, \mathcal{M}) = \bigoplus_{n \in \mathbb{Z}} \Gamma(U, \mathcal{M}_n)$. Therefore, $\Gamma(U, \mathcal{M})$ is a graded $\Gamma(U, \mathcal{S})$-module, and hence, we can define the quasicoherent sheaf

5.2. QUASICOHERENT SHEAVES OVER A PROJECTIVE SCHEME 89

$\widetilde{\Gamma(U,\mathcal{M})}$ over $\operatorname{Proj}\Gamma(U,\mathcal{S})$. By glueing these sheaves, one obtains a quasicoherent sheaf $\widetilde{\mathcal{M}}$ over $Z = \operatorname{Proj}\mathcal{S}$. For an integer n, define a graded $\mathcal{S}$-module $\mathcal{M}(n) = \bigoplus_{m\in\mathbb{Z}}\mathcal{M}(n)_m$ by

$$\mathcal{M}(n)_m = \mathcal{M}_{n+m}, \quad m \in \mathbb{Z}.$$

In particular, define $\mathcal{O}_Z(n) = \widetilde{\mathcal{S}(m)}$. A graded $\mathcal{S}$-module $\mathcal{M}$ over a scheme X is said to be finitely presented if for an arbitrary point x in X, there exists an open neighborhood U of x to satisfy an exact sequence

$$\bigoplus_{i=1}^{k}\mathcal{S}(m_i)|U \to \bigoplus_{j=1}^{l}\mathcal{S}(n_j)|U \to \mathcal{M}|U \to 0,$$

where m_i and n_j are integers. Then Proposition 5.19 and Corollary 5.20 imply the following.

PROPOSITION 5.29. *Let $\mathcal{S} = \bigoplus_{n=0}^{\infty}\mathcal{S}_n$ be a quasicoherent sheaf of graded algebra generated by $\mathcal{S}_1$ as an $\mathcal{S}_0 = \mathcal{O}_X$-algebra over a scheme X. Let $\mathcal{M}$ and $\mathcal{N}$ be quasicoherent graded $\mathcal{S}$-modules and let $Z = \operatorname{Proj}\mathcal{S}$. Then:*

(i) There exists an $\mathcal{O}_Z$-homomorphism $\tilde{\lambda}: \widetilde{\mathcal{M}} \otimes_{\mathcal{O}_Z} \tilde{h} \to \widetilde{\mathcal{M} \otimes_{\mathcal{S}} \mathcal{N}}$, and this homomorphism is an isomorphism.

(ii) There exists an $\mathcal{O}_Z$-homomorphism $\tilde{\mu}: (\underline{\operatorname{Hom}}_{\mathcal{S}}(\mathcal{M},\mathcal{N}))^{\sim} \to \operatorname{Hom}_{\mathcal{O}_Z}(\widetilde{\mathcal{M}}, \widetilde{\mathcal{N}})$. When $\mathcal{M}$ is finitely presented, $\tilde{\mu}$ is an isomorphism.

(iii) For any integer m, $\mathcal{O}_Z(m) = \widetilde{\mathcal{S}(m)}$ is an invertible sheaf. We have isomorphisms

$$\mathcal{O}_Z(l) \otimes \mathcal{O}_Z(m) \xrightarrow{\sim} \mathcal{O}_Z(l+m), \qquad \mathcal{O}_Z(-l) \xrightarrow{\sim} \underline{\operatorname{Hom}}_{\mathcal{O}_Z}(\mathcal{O}_Z(l), \mathcal{O}_Z).$$

Furthermore, for a quasicoherent graded $\mathcal{S}$-module $\mathcal{M}$ and for any natural number l, we have an isomorphism

$$\widetilde{\mathcal{M}(l)} \xrightarrow{\sim} \widetilde{\mathcal{M}} \otimes_{\mathcal{O}_Z} \mathcal{O}_Z(l). \qquad \square$$

In what follows, we assume that a quasicoherent sheaf $\mathcal{S} = \bigoplus_{n=0}^{\infty}\mathcal{S}_n$ is generated by $\mathcal{S}_1$ as an $\mathcal{S}_0$-algebra. Let $\pi: Z = \operatorname{Proj}\mathcal{S} \to X$ be the structure morphism. For an $\mathcal{O}_Z$-module $\mathcal{F}$, let $\mathcal{F}(n) = \mathcal{F} \otimes_{\mathcal{O}_Z} \mathcal{O}_Z(n)$, and let

(5.28) $$\Gamma_*(\mathcal{F}) = \bigoplus_{n\in\mathbb{Z}} \pi_*(\mathcal{F}(n)).$$

For arbitrary $\mathcal{O}_Z$-modules $\mathcal{A}$ and $\mathcal{B}$, there exists the natural $\mathcal{O}_X$-homomorphism $\pi_*(\mathcal{A}) \otimes \pi_*(\mathcal{B}) \to \pi_*(\mathcal{A} \otimes \mathcal{B})$. Hence, $\Gamma_*(\mathcal{O}_Z)$ becomes a graded ring over X, and $\Gamma_*(\mathcal{F})$ has the structure of a graded $\Gamma_*(\mathcal{O}_Z)$-module. In particular, if U is an affine open set of X, from (5.22) we can define a homomorphism

$$\alpha(U) : \Gamma(U, \mathcal{S}) = \bigoplus_{n=0}^{\infty} \Gamma(U, \mathcal{S}_n) \to \bigoplus_{n \in \mathbb{Z}} \Gamma(U, \mathcal{O}_Z(n)).$$

Using this map, one can easily define a homomorphism of sheaves of graded $\mathcal{O}_Z$-modules

(5.29) $$\alpha : \mathcal{S} = \bigoplus_{n=0}^{\infty} \mathcal{S}_n \to \Gamma_*(\mathcal{O}_Z).$$

Through α, we can regard $\Gamma_*(\mathcal{O}_Z)$ as a graded $\mathcal{S}$-module sheaf. Therefore, the sheaf $\widetilde{\Gamma_*(\mathcal{F})}$ can be defined over Z. For an affine open set U, (5.25) implies

$$\beta(U) : \bigoplus_{n \in \mathbb{Z}} (\Gamma(\pi^{-1}(U), \mathcal{F}(n)))^{\sim} \to \mathcal{F}|\pi^{-1}(U).$$

By glueing these, one constructs an $\mathcal{O}_Z$-module homomorphism

(5.30) $$\beta : \widetilde{\Gamma_*(\mathcal{F})} \to \mathcal{F}.$$

The following theorem rephrases Theorem 5.21.

THEOREM 5.30. *Assume a quasicoherent sheaf $\mathcal{S} = \bigoplus_{n=0}^{\infty} \mathcal{S}_n$ of graded algebras is generated by $\mathcal{S}_1$ as an $\mathcal{S}_0 = \mathcal{O}_X$-algebra, and also assume $\mathcal{S}_1$ is finitely generated as an $\mathcal{O}_X$-module. Then, for $\pi : Z = \operatorname{Proj} \mathcal{S} \to X$ and for an arbitrary quasicoherent sheaf $\mathcal{F}$ over Z, we have an isomorphism $\beta : \widetilde{\Gamma_*(\mathcal{F})} \to \mathcal{F}$.* □

For a given homomorphism $f : \mathcal{E} \to \mathcal{F}$ of quasicoherent sheaves over X, let us study the relation between the projective schemes $\mathbb{P}(\mathcal{E})$ and $\mathbb{P}(\mathcal{F})$ over X. The above f induces a homomorphism $\mathbb{S}(f) : \mathbb{S}(\mathcal{E}) \to \mathbb{S}(\mathcal{F})$ of symmetric algebras which is a graded algebra homomorphism of degree 0. Over an affine open set of X, one can define the open subscheme $G(\mathbb{S}(f))$ of $\mathbb{P}(\mathcal{F})$ as in (5.9). Hence, we have the scheme morphism $\mathbb{S}(f)^a : G(\mathbb{S}(f)) \to \mathbb{P}(\mathcal{E})$. Then we have the following fact.

PROPOSITION 5.31. *If a homomorphism of quasicoherent sheaves $f : \mathcal{E} \to \mathcal{F}$ over X is surjective, then $G(\mathbb{S}(f)) = \mathbb{P}(\mathcal{F})$, and $\mathbb{S}(f)^a : \mathbb{P}(\mathcal{F}) \to \mathbb{P}(\mathcal{E})$ is a closed immersion.*

PROOF. It is enough to consider the case when X is an affine scheme $\operatorname{Spec} R$, $\mathcal{E} = \widetilde{M}, \mathcal{F} = \widetilde{N}$, where M and N are R-modules. Then f is determined by an R-module homomorphism $\varphi : M \to N$. The surjectivity of f is equivalent to that of φ. Then $\mathbb{S}(\varphi) : \mathbb{S}(M) \to \mathbb{S}(N)$ is a surjective homomorphism satisfying $\mathbb{S}(\varphi)(\mathbb{S}(M)_1) = \mathbb{S}(N)_1$. Hence, we can define $\mathbb{S}(\varphi)^a : \operatorname{Proj} \mathbb{S}(N) \to \operatorname{Proj} \mathbb{S}(M)$. By Proposition 5.13 and Example 4.39(1), $\mathbb{S}(\varphi)^a$ is a closed immersion. □

PROBLEM 15. For a quasicoherent sheaf $\mathcal{E}$ and an invertible sheaf $\mathcal{L}$ over a scheme X, prove that $\mathbb{P}(\mathcal{E})$ and $\mathbb{P}(\mathcal{E} \otimes \mathcal{L})$ are isomorphic as schemes over X.

5.3. Projective Morphisms

(a) **Categorical Characterization of $\mathbb{P}(\mathcal{E})$**

Let $\mathbb{P}(\mathcal{E}) = \operatorname{Proj} \mathbb{S}(\mathcal{E})$ be the projective scheme determined by a quasicoherent finitely generated $\mathcal{O}_X$-module $\mathcal{E}$ over a scheme X. We will characterize $\mathbb{P}(\mathcal{E})$ in terms of a representable functor. That is, for an arbitrary scheme (Y, f) over X, i.e., $Y \xrightarrow{f} X$, we will capture $\operatorname{Hom}_X(Y, \mathbb{P}(\mathcal{E}))$ as the value of a functor at Y. For the sake of simplicity, put $P = \mathbb{P}(\mathcal{E})$ and let $\pi : P \to X$ be the structure morphism. If we put $\mathcal{S} = \mathbb{S}(\mathcal{E})$, then $\mathcal{S}_1 = \mathcal{E}$, and by (5.29) we get an $\mathcal{O}_X$-module homomorphism $\alpha_1 : \mathcal{S}_1 = \mathcal{E} \to \pi_*\mathcal{O}_P(1)$. By Lemma 4.37, α_1 induces the $\mathcal{O}_P$-module homomorphism $\alpha_1^{\#} : \pi^*\mathcal{E} \to \mathcal{O}_P(1)$. This homomorphism has the following interpretation. By the property of a symmetric algebra, we have the surjection of $\mathcal{O}_X$-modules $\mathcal{E} \otimes_{\mathcal{O}_X} \mathbb{S}(\mathcal{E}) \to \mathbb{S}(\mathcal{E})(1)$. By regarding $\mathcal{E} \otimes \mathbb{S}(\mathcal{E})$ and $\mathbb{S}(\mathcal{E})(1)$ as graded $\mathbb{S}(\mathcal{E})$-modules, we obtain the $\mathcal{O}_P$-module homomorphism $(\mathcal{E} \otimes_{\mathcal{O}_X} \mathbb{S}(\mathcal{E}))^{\sim} \to (\mathbb{S}(\mathcal{E})(1))^{\sim}$. Since $(\mathcal{E} \otimes_{\mathcal{O}_X} \mathbb{S}(\mathcal{E}))^{\sim} = \pi^*\mathcal{E}$ and $(\mathbb{S}(\mathcal{E})(1))^{\sim} = \mathcal{O}_P(1)$, the above $\mathcal{O}_P$-module homomorphism is exactly α_1. Therefore, $\alpha_1 : \pi^*\mathcal{E} \to \mathcal{O}_P(1)$ is surjective.

Let $\varphi : Y \to P$ be a morphism over X. Namely, we have the commutative diagram

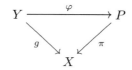

where $g = \pi \circ \varphi$. Pulling back the $\mathcal{O}_P$-module surjective homomorphism $\alpha_1 : \pi^*\mathcal{E} \to \mathcal{O}_P(1)$ by φ over Y, we get an $\mathcal{O}_Y$-homomorphism
$$\gamma_\varphi : g^*\mathcal{E} \to \varphi^*\mathcal{O}_P(1).$$
Let $\mathcal{L}_\varphi = \varphi^*\mathcal{O}_P(1)$. Then $\mathcal{L}_\varphi$ is an invertible sheaf, and
$$\gamma_\varphi : g^*\mathcal{E} \to \mathcal{L}_\varphi$$
is surjective.

Conversely, assume that an invertible sheaf $\mathcal{L}$ over Y and a surjective $\mathcal{O}_Y$-homomorphism $\gamma : g^*\mathcal{E} \to \mathcal{L}$ are given. Then the surjective homomorphism of graded algebras
$$\mathbb{S}(\gamma) : \mathbb{S}(g^*\mathcal{E}) = g^*(\mathbb{S}(\mathcal{E})) \to \mathbb{S}(\mathcal{L}) = \bigoplus_{n=0}^{\infty} \mathcal{L}^{\otimes n}$$
is determined. Hence we obtain the scheme morphism
$$\mathbb{S}(\gamma)^a : \operatorname{Proj} \mathbb{S}(\mathcal{L}) \to \operatorname{Proj} g^*(\mathbb{S}(\mathcal{E})).$$
Problem 15 implies $\operatorname{Proj} \mathbb{S}(\mathcal{L}) = \mathbb{P}(\mathcal{L}) = \mathbb{P}(\mathcal{O}_Y \otimes \mathcal{L}) = \mathbb{P}(\mathcal{O}_Y) = Y$, and Problem 14 implies $\operatorname{Proj} g^*(\mathbb{S}(\mathcal{E})) = \mathbb{P}(\mathcal{E}) \times_X Y$. Therefore we get $\mathbb{S}(\gamma)^a : Y \to \mathbb{P}(\mathcal{E}) \times_X Y$. That is, we have the scheme morphism over X
$$\varphi_{(\mathcal{L}, \gamma)} : Y \to \mathbb{P}(\mathcal{E}).$$

Let $\mathcal{L}$ and $\mathcal{L}'$ be invertible sheaves over Y and let $\gamma : g^*\mathcal{E} \to \mathcal{L}$ and $\gamma' : g^*\mathcal{E} \to \mathcal{L}'$ be surjective $\mathcal{O}_Y$-homomorphisms. Furthermore, let $h : \mathcal{L} \xrightarrow{\sim} \mathcal{L}'$ be an $\mathcal{O}_Y$-isomorphism satisfying $\gamma' = h \circ \gamma$. Namely,

(5.31)
$$\begin{array}{ccc} & & \mathcal{L} \\ & \nearrow^{\gamma} & \\ g^*\mathcal{E} & & \downarrow h \\ & \searrow_{\gamma'} & \\ & & \mathcal{L}' \end{array}$$

is commutative. Since $\mathbb{S}(h)^a : \mathbb{P}(\mathcal{L}) = Y \to \mathbb{P}(\mathcal{L}') = Y$ is an identity morphism, we get $\varphi_{(\mathcal{L}, \gamma)} = \varphi_{(\mathcal{L}', \gamma')}$. We will define an equivalence relation among the pairs $(\mathcal{L}, \gamma)$, where $\mathcal{L}$ is an invertible sheaf over Y, and $\gamma : g^*\mathcal{E} \to \mathcal{L}$ is a surjective $\mathcal{O}_Y$-homomorphism:

$(\mathcal{L}, \gamma) \sim (\mathcal{L}', \gamma') \Leftrightarrow$ there exists an $\mathcal{O}_Y$-isomorphism $h : \mathcal{L} \xrightarrow{\sim} \mathcal{L}'$

so that (5.31) is commutative.

5.3. PROJECTIVE MORPHISMS

Also define a contravariant functor $P_{\mathcal{E}}$ from the category $(\mathrm{Sch})/X$ of schemes over X to the category (Set) of sets by

(5.32)
$$P_{\mathcal{E}}((Y,g)) = \left\{ (\mathcal{L},\gamma) \;\middle|\; \begin{array}{l} \mathcal{L} \text{ is an invertible sheaf over } Y, \text{ and} \\ \gamma : g^*\mathcal{E} \to \mathcal{L} \text{ is a surjective} \\ \mathcal{O}_Y\text{-homomorphism} \end{array} \right\} \Big/ \sim .$$

Note that the right-hand side of (5.32) is the totality of equivalence classes. Let $f : Z \to Y$ be a morphism over X, i.e., the diagram

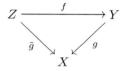

is commutative. Then for $(\mathcal{L},\gamma)$ over Y, $f^*\mathcal{L}$ is invertible over Z, and $f^*\gamma : f^*g^*(\mathcal{E}) = \tilde{g}^*(\mathcal{E}) \to f^*\mathcal{L}$ is a surjective $\mathcal{O}_Z$-homomorphism. Hence $(f^*\mathcal{L}, f^*\gamma)$ determines an element of $P_{\mathcal{E}}((Z,\tilde{g}))$. Therefore, we have $P_{\mathcal{E}}(f) : P_{\mathcal{E}}((Y,g)) \to P_{\mathcal{E}}((Z,\tilde{g}))$. That is, $P_{\mathcal{E}}$ is a contravariant functor. The next theorem characterizes $P_{\mathcal{E}}$.

THEOREM 5.32. *For a quasicoherent sheaf $\mathcal{E}$ over a scheme X, let $P_{\mathcal{E}}$ be the contravariant functor as defined in (5.32). Then $P_{\mathcal{E}}$ is represented by a scheme over X, i.e., $\pi : \mathbb{P}(\mathcal{E}) \to X$. That is, for any $g : Y \to X$ we have an isomorphism of sets*

(5.33)
$$\mathrm{Hom}_X(Y, \mathbb{P}(\mathcal{E})) \xrightarrow{\sim} P_{\mathcal{E}}((Y,g)),$$
$$\varphi \mapsto [(\mathcal{L}_\varphi, \gamma_\varphi)], \quad \mathcal{L}_\varphi = \varphi^*\mathcal{O}_{\mathbb{P}(\mathcal{E})}(1),$$

where $[(\mathcal{L},\gamma)]$ is the equivalence class of $(\mathcal{L},\gamma)$.

PROOF. We have already shown the correspondence from $\varphi \in \mathrm{Hom}_X(Y,\mathbb{P}(\mathcal{E}))$ to the equivalence class of $(\mathcal{L}_\varphi, \gamma_\varphi)$ and from the equivalence class of $(\mathcal{L},\gamma)$ to $\varphi_{(\mathcal{L},\gamma)} \in \mathrm{Hom}_X(Y,\mathbb{P}(\mathcal{E}))$. We will show that those correspondences are set-theoretic isomorphisms. For $(\mathcal{L}_\varphi, \gamma_\varphi)$ given by $\varphi \in \mathrm{Hom}_X(Y,\mathbb{P}(\mathcal{E}))$, we will show that $\varphi_{(\mathcal{L}_\varphi,\gamma_\varphi)} = \varphi$. By considering affine coverings of X and Y, it is sufficient to consider the case when $X = \mathrm{Spec}\,R, Y = \mathrm{Spec}\,A, \mathcal{E} = \widetilde{E}$ for an R-module E satisfying $\varphi(Y) \subset D_+(a) \subset \mathbb{P}(\mathcal{E})$, $a \in E = \mathbb{S}(E)_1$. Let $B = \mathbb{S}(E)$ and $P = \mathbb{P}(\mathcal{E})$. The morphism $\varphi : Y \to D_+(a) \subset P$ is determined by a commutative ring homomorphism $\psi : B_{(a)} \to A$. We have $g^*\mathcal{E} = (E \otimes_R A)^\sim$ and $\mathcal{L}_\varphi = \varphi^*\mathcal{O}_P(1) = (B(1)_{(a)} \otimes_{G_{(a)}} A)^\sim$. Let

$L_\varphi = B(1)_{(a)} \otimes_{B_{(a)}} A$. The surjective homomorphism of A-modules
$$\eta : E \otimes_R A \to L_\varphi = B(1)_{(a)} \otimes_{B_{(a)}} A,$$
$$e \otimes \gamma \mapsto \frac{a}{1} \otimes \psi\left(\frac{e}{a}\right) \gamma,$$
induces $\gamma_\varphi : g^*\mathcal{E} \to \mathcal{L}_\varphi$, where we note that
$$\frac{e}{1} \otimes \gamma = \left(\frac{e}{a}\right) \cdot \frac{a}{1} \otimes \gamma = \frac{a}{1} \otimes \psi\left(\frac{e}{a}\right) \gamma.$$
If necessary, choose a finer affine covering for Y so that we may have $\mathcal{L}_\varphi = \mathcal{O}_Y$. That is, we may assume that L_φ is a free A-module of rank 1. Then we get $L_\varphi = A \cdot \left(\frac{a}{1} \otimes 1\right)$. Let $b = \frac{a}{1} \otimes 1$. The homomorphism η induces
$$\mathbb{S}^n(\eta) : \mathbb{S}^n(E \otimes_R A) = \mathbb{S}^n(E) \otimes_R A \to L_\varphi^{\otimes n} = A \cdot b^n,$$
$$(e_1 \otimes \cdots \otimes e_n) \otimes \gamma \mapsto \psi\left(\frac{e_1}{a}\right) \cdots \psi\left(\frac{e_n}{a}\right) \gamma \cdot b^n.$$
Then an R-homomorphism $\hat{\psi} : \mathbb{S}(E) \to A$ can be defined: $\hat{\psi}(v) = \psi(\frac{v}{a^n})$, $v \in \mathbb{S}^n(E)$. For the natural homomorphism $\hat{\iota} : B = \mathbb{S}(E) \to B_{(a)}$, we have $\hat{\psi} = \psi \circ \hat{\iota}$. Therefore, the scheme morphism $\varphi_{(\mathcal{L}_\varphi, \gamma_\varphi)} : Y \to \mathbb{P}_\mathcal{E}$ induced by $\mathbb{S}(\eta) : \mathbb{S}(E \otimes_R A) \to \mathbb{S}(L_\varphi)$ coincides with the scheme morphism determined by $\hat{\psi} : \mathbb{S}(E) \to A$. Namely, $\varphi_{(\mathcal{L}_\varphi, \gamma_\varphi)} = \varphi : Y \to \mathbb{P}(\mathcal{E})$.

Next, for $(\mathcal{L}, \gamma)$ we will construct $\varphi = \varphi_{(\mathcal{L},\gamma)} \in \mathrm{Hom}_X(Y, \mathbb{P}(\mathcal{E}))$ and $(\mathcal{L}_\varphi, \gamma_\varphi)$. Then we will show that in the sense of (5.31), $(\mathcal{L}, \gamma)$ and $(\mathcal{L}_\varphi, \gamma_\varphi)$ are equivalent. Assume that X is an affine scheme $\mathrm{Spec}\, R$. Let $\mathcal{E} = \widetilde{E}$ for an R-module E. Let $V = \mathrm{Spec}\, A$ be an affine open set in Y. By taking a small enough V, we may assume $\mathcal{L}|V \stackrel{\sim}{\to} \mathcal{O}_V$. Let $\mathcal{L}|V = \widetilde{L}$ for an A-module L. Then L may be assumed to be a free A-module of rank one. Write $L = A \cdot b$. The restriction of $\gamma : g^*\mathcal{E} \to \mathcal{L}$ to V corresponds to the surjective A-module homomorphism $\tilde{\eta} : E \otimes_R A \to L$. For $e \in E$, define $\tilde{\psi}_1(e) \in A$ by $\tilde{\eta}(e \otimes 1) = \tilde{\psi}_1(e)b$. Then we get an R-homomorphism $\tilde{\psi}_1 : E \to A$. Since $\tilde{\eta}$ is surjective, we can find $e_1, e_2, \ldots, e_n \in E$ so that we may have $\tilde{\psi}_1(e_1)A + \tilde{\psi}_1(e_2)A + \cdots + \tilde{\psi}_1(e_n)A = A$. By replacing A by the localized A, one can choose $a \in E$ to satisfy $\tilde{\psi}_1(a)A = A$. (This means that by choosing an affine covering of $V = \mathrm{Spec}\, A$, replace V by one of those.) Then $\varphi|V : V \to \mathbb{P}(\mathcal{E})$ is determined by the R-algebra homomorphism $\tilde{\psi} : B = \mathbb{S}(E) \to A$ induced by $\tilde{\psi}_1(a)$. Since $\tilde{\psi}_1(a)$ is invertible in A, we have $\varphi(V) \subset D_+(a)$. Therefore, $\tilde{\psi}$ is a composition of $\psi : B_{(a)} \to A$ and the natural homomorphism $B \to$

$B_{(a)}$. Furthermore, $\mathcal{L}_\varphi|V = \varphi^*\mathcal{O}_{\mathbb{P}(\mathcal{E})}(1)|V = (B(1)_{(a)} \otimes_{B_{(a)}} A)^\sim$ and $\gamma_\varphi|V : g^*\mathcal{E}|V \to \mathcal{L}_\varphi|V$ is determined by the A-module homomorphism

$$\eta : E \otimes_R A \to B(1)_{(a)} \otimes_{B_{(a)}} A,$$
$$e \otimes \gamma \mapsto \frac{a}{1} \otimes \psi\left(\frac{e}{a}\right)\gamma.$$

Hence the $\mathcal{O}_V$-isomorphism $h_V : \mathcal{L}_\varphi|V \xrightarrow{\sim} \mathcal{L}|V$ is determined by the A-module homomorphism

(5.34)
$$\theta_V : B(1)_{(a)} \otimes_{B_{(a)}} A \to L = A \cdot b$$
$$\frac{a}{1} \otimes \gamma \mapsto \tilde{\psi}_1(a)\gamma \cdot b.$$

If $\varphi(V) \subset D_+(a')$, $a' \in E$, then $\mathcal{L}_\varphi|V = (B(1)_{(a')} \otimes_{b_{(a')}} A)^\sim$. As in (5.34), θ'_V can be defined as $\theta'_V(\frac{a'}{1} \otimes \gamma) = \tilde{\psi}_1(a')\gamma \cdot b$. Regarding $\frac{a'}{1} \otimes \gamma \in B(1)_{(a)} \otimes_{B_{(a)}} A$, we have

$$\frac{a'}{1} \otimes \gamma = \frac{a}{1} \otimes \psi\left(\frac{a'}{a}\right)\gamma.$$

Consequently,

$$\theta'_V\left(\frac{a}{1} \otimes \psi\left(\frac{a'}{a}\right)\gamma\right) = \tilde{\psi}_1(a)\psi\left(\frac{a'}{a}\right)\gamma \cdot b = \tilde{\psi}_1(a')\gamma \cdot b.$$

That is, h_V is uniquely determined for any choice of $a \in E$. Hence, by varying an affine open set V in Y, one can define an $\mathcal{O}_X$-isomorphism $h : \mathcal{L}_\varphi \to \mathcal{L}$. By the construction of h_V, we obtain $\gamma = h \circ \gamma_\varphi$. Namely, $(\mathcal{L}, \gamma)$ and $(\mathcal{L}_\varphi, \gamma_\varphi)$ are equivalent. □

The above proof is long but not difficult. The above theorem describes many properties of a projective scheme. We will provide a few examples.

EXERCISE 5.33. Let X be a scheme over a field k and let $\mathcal{L}$ be an invertible sheaf over X. Assume that $\Gamma(X, \mathcal{L})$ is a finite-dimensional vector space over k and that $\mathcal{L}$ is generated by $\Gamma(X, \mathcal{L})$ as an $\mathcal{O}_X$-module. Namely, for basis elements $f_0, f_1, \ldots, f_n$ of $\Gamma(X, \mathcal{L})$ over k, we are assuming the $\mathcal{O}_X$-homomorphism

(5.35)
$$\gamma : \mathcal{O}_X^{\oplus(n+1)} \to \mathcal{L},$$
$$(a_0, a_1, \ldots, a_n) \mapsto \sum_{j=0}^n a_j f_j,$$

to be surjective (i.e., for each point $x \in X$, there is f_j satisfying $f_j(x) \neq 0$). Then one can find a scheme morphism $\varphi : X \to \mathbb{P}^n$ over k so that $\varphi^* \mathcal{O}_{\mathbb{P}^n_k}(1) \xrightarrow{\sim} \mathcal{L}$.

PROOF. Since $\gamma : \mathcal{O}_X^{\oplus(n+1)} \to \mathcal{L}$ is a surjective $\mathcal{O}_X$-homomorphism, by Theorem 5.32, there exists a scheme morphism $\varphi : X \to \mathbb{P}(\mathcal{O}_X^{\oplus(n+1)}) = \mathbb{P}^n_k$ so that $(\mathcal{L}, \gamma)$ and $(\mathcal{L}_\varphi = \varphi^* \mathcal{O}_{\mathbb{P}^n_k}(1), \gamma_\varphi)$ are equivalent. □

EXERCISE 5.34. For an $(n+1)$-dimensional vector space E over a field k, the totality of k-valued points of $\mathbb{P}(E)$ over k, which is denoted as $\mathbb{P}(E)(k)$, is isomorphic to $(k^{n+1} \setminus \{(0,\ldots,0)\})/\sim$ as sets, where the equivalence relation $\sim$ is defined as follows:

$$(a_0, a_1, \ldots, a_n) \sim (b_0, b_1, \ldots, b_n) \Leftrightarrow \text{ for some } \alpha \in k^\times = k \setminus \{0\},$$
$$(a_0, a_1, \ldots, a_n) = (\alpha b_0, \alpha b_1, \ldots, \alpha b_n).$$

PROOF. Since an invertible sheaf over $\operatorname{Spec} k$ is k, Theorem 5.32 implies

$\operatorname{Hom}_{\operatorname{Spec} k}(\operatorname{Spec} k, \mathbb{P}(\mathcal{E}))$
$= \{\psi : E \to k | \varphi \text{ is a non-zero } k\text{-linear map}\}/\sim .$

Note that k-linear maps $\psi : E \to k$ and $\psi' : E \to k$ are equivalent when $\psi = \alpha \psi'$, $\alpha \in k^\times = k \setminus \{0\}$. Choose a base $\{e_0, e_1, \ldots, e_n\}$ of E over k. For an arbitrary element $v = x_0 e_0 + x_1 e_1 + \cdots + x_n e_n$, a k-linear map ψ can be written as $\psi(v) = a_0 x_0 + a_1 x_1 + \cdots + a_n x_n$. Hence, there is a one-to-one correspondence between the non-zero k-linear map ψ and $(a_0, a_1, \ldots, a_n) \in k^{n+1} \setminus \{(0,\ldots,0)\}$, and the equivalence relation of ψ is exactly the one given in the assertion. □

In §1.5(a), we defined the n-dimensional projective space $\mathbb{P}^n_k$ over an algebraically closed field k as the quotient space $W/\sim$ of $W = k^{n+1} \setminus \{(0,\ldots,0)\}$. The equivalence relation $\sim$ is defined above. We notice that $\mathbb{P}^n_k$ is precisely the totality $\mathbb{P}^n_k(k)$ of k-valued points of $\mathbb{P}^n_k$ over k.

The quotient space $W/\sim$ can be regarded as the totality of one-dimensional vector subspaces of k^{n+1} over the field k. This pre-Grothendieck viewpoint can be replaced by the more categorical viewpoint. That is, as shown in Exercise 5.34, we can regard each element of W as a linear form (i.e., a k-linear map $E^* \to k$). Then $W/\sim$ is the totality of equivalence classes of non-zero linear forms. For a quasicoherent sheaf $\mathcal{E}$, there are many surjective $\mathcal{O}_Y$-homomorphisms from

$g^*\mathcal{E}$ to an invertible sheaf $\mathcal{L}$. But, conversely, there are not many injective homomorphisms from $\mathcal{L}$ to $g^*\mathcal{E}$. Namely, Grothendieck's definition is more natural. One needs to be aware that the definitions of the projective space $\mathbb{P}^n_k$ and $\mathbb{P}(\mathcal{E})$ for a local free sheaf $\mathcal{E}$ over a scheme X are dual to the classical counter-definitions.

Let k be an algebraically closed field. For an invertible sheaf $\mathcal{L}$ over a scheme over k, assume $H^0(X,\mathcal{L})$ to be a finite-dimensional k-vector space. Let $\{f_0, f_1, \ldots, f_n\}$ be a base for $H^0(X,k)$. If there are no common zeros on X for $f_0, f_1, \ldots, f_n$, from Exercise 5.33 and Exercise 5.34, one can define a map from the totality of k-valued points $X(k)$ to $\mathbb{P}^n_k(k)$:

$$\Phi_{|\mathcal{L}|} : X(k) \to \mathbb{P}^n_k(k),$$
$$x \mapsto (f_0(x) : f_1(x) : f_2(x) : \cdots : f_n(x)).$$

We write $\mathbb{P}^n(k)$ for $\mathbb{P}^n_k(k)$. The above map is affected by the choice of basis elements for $H^0(X,\mathcal{L})$, but differs only by a projective transformation.

(b) **The Segre Morphism**

Let $\mathbb{P}^n(k)$ be the totality of k-valued points of the projective space $\mathbb{P}^n_k$ over an algebraically closed field k. By Exercise 1.6, the image of

(5.36)
$$\mathbb{P}^1(k) \times \mathbb{P}^1(k) \to \mathbb{P}^3(k)$$
$$((a_0 : a_1), (b_0 : b_1)) \mapsto ((a_0b_0 : a_0b_1 : a_1b_0 : a_1b_1))$$

is a quadratic surface in $\mathbb{P}^3(k)$. Let us study this surface in terms of schemes.

In order to distinguish the two $\mathbb{P}^1(k)$'s, let $X_1 = \operatorname{Proj} k[x_0, x_1] = \mathbb{P}^1_k$ and $X_2 = \operatorname{Proj} k[y_0, y_1] = \mathbb{P}^1_k$, and let $Z = X_1 \times_{\operatorname{Spec} k} X_2$. Let $f_1 : X_1 \to \operatorname{Spec} k$, $f_2 : X_2 \to \operatorname{Spec} k$ and $g : Z \to \operatorname{Spec} k$ be the structure morphisms. Also let $p_j : Z = X_1 \times_{\operatorname{Spec} k} X_2 \to X_j$ be the projection for $j = 1, 2$. From Example 5.22, we have

$$f_*\mathcal{O}_{X_1}(1) = \Gamma(X_1, \mathcal{O}_{X_1}(1)) = kx_0 \oplus kx_1 = V_1,$$
$$f_*\mathcal{O}_{X_2}(1) = \Gamma(X_2, \mathcal{O}_{X_2}(1)) = ky_0 \oplus ky_1 = V_2.$$

Then the natural $\mathcal{O}_X$-homomorphism $\gamma_1 : f_1^*V_1 = \mathcal{O}_{X_1,x_0} \oplus \mathcal{O}_{X_1,x_1} \to \mathcal{O}_{X_1}(1)$ is surjective, and $\varphi_{(\mathcal{O}_{X_1}(1),\gamma_1)} : X_1 \to \mathbb{P}(V_1)$ is an isomorphism over k. Let $\mathcal{L} = p_1^*\mathcal{O}_{X_1}(1) \otimes p_2^*\mathcal{O}_{X_2}(1)$. Then $\mathcal{L}$ is an invertible sheaf over Z. Notice that $\gamma : g^*(V_1 \otimes V_2) = p_1^*(f_1^*(V_1)) \otimes p_2^*(f_2^*(V_2)) \to \mathcal{L}$ is a surjective homomorphism. Therefore, we obtain a scheme morphism $\varphi_{(\mathcal{L},\gamma)} : Z \to \mathbb{P}(V_1 \otimes V_2)$. Since $V_1 \otimes V_2$ is isomorphic to the 4-dimensional vector space k^4 over k, we get $\mathbb{P}(V_1 \otimes V_2) \xrightarrow{\sim} \mathbb{P}^3_k$. Then we can show that the map $Z(k) = \mathbb{P}(V_1 \otimes V_2)(k) = \mathbb{P}^3(k)$ between k-valued points coincides with the map in (5.36).

The k-valued points on X_1 are in one-to-one correspondence with the equivalence classes of non-zero k-linear maps $\psi_1 : V_1 \to k$. Then ψ_1 is uniquely determined by $\psi_1(x_0) = a_0$ and $\psi_1(x_1) = a_1$, and the equivalence class corresponds to $(a_0 : a_1)$ in one-to-one fashion. Similarly, the k-valued points on X_2 are in one-to-one correspondence with the equivalence classes of non-zero k-linear maps $\psi_2 : V_2 \to k$ which are uniquely determined by $\psi_2(y_0) = b_0$ and $\psi_2(y_1) = b_1$. The equivalence class corresponds to $(b_0 : b_1)$. Therefore, $((a_0 : a_1), (b_0 : b_1))$ corresponds to an element of $X_1(k) \times X_2(k)$ in one-to-one fashion. Namely, the image of $((a_0 : a_1), (b_0 : b_1))$ under the morphism $\varphi_{(\mathcal{L},\gamma)} : Z \to \mathbb{P}(V_1 \otimes V_2)$ corresponds to the equivalence class of $\psi = \psi_1 \otimes \psi_2 : V_1 \otimes V_2 \to k$. Then the k-linear map ψ is uniquely determined by $\psi(x_0 \otimes y_0), \psi(x_0 \otimes y_1), \psi(x_1 \otimes y_0)$ and $\psi(x_1 \otimes y_1)$. In this case, we get $\psi(x_0 \otimes y_0) = a_0 b_0, \psi(x_0 \otimes y_1) = a_0 b_1, \psi(x_1 \otimes y_0) = a_1 b_0, \psi(x_1 \otimes y_1) = a_1 b_1$. That is, the equivalence class of ψ corresponds to $(a_0 b_0 : a_0 b_1 : a_1 b_0 : a_1 b_1)$. Namely, the map of k-valued points $Z(k) \to \mathbb{P}(V_1, V_2)(k)$ determined by $\varphi_{(\mathcal{L},\gamma)}$ is exactly the map in (5.36).

The above argument can be generalized. Let $\mathbb{P}(\mathcal{E}^1)$ and $\mathbb{P}(\mathcal{E}_2)$ be the projective schemes obtained from quasicoherent sheaves $\mathcal{E}_1$ and $\mathcal{E}_2$ over a scheme X. Let $\pi_j : \mathbb{P}(\mathcal{E}_j) \to X$, $j = 1, 2$, be the structure morphisms. Consider $P = \mathbb{P}(\mathcal{E}_1) \times_X \mathbb{P}(\mathcal{E}_2)$ and their projections $p_j : P \to \mathbb{P}(\mathcal{E}_j)$, $j = 1, 2$. We have $\pi = \pi_j \circ p_j$. Let $\mathcal{L} = p_1^* \mathcal{O}_{\mathbb{P}(\mathcal{E}_1)}(1) \otimes_{\mathcal{O}_P} p_2^* \mathcal{O}_{\mathbb{P}(\mathcal{E}_2)}(1)$. Since $\pi_j^* \mathcal{E}_j \to \mathcal{O}_{\mathbb{P}(\mathcal{E}_j)}(1)$ is a surjective homomorphism, $\gamma : \pi^*(\mathcal{E}_1 \otimes_{\mathcal{O}_X} \mathcal{E}_2) = p_1^*(\pi_1^* \mathcal{E}_1) \otimes p_2^*(\pi_2^* \mathcal{E}_2) \to \mathcal{L}$ is also a surjective homomorphism. Therefore, one can define a morphism $\varphi_{(\mathcal{L},\gamma)} : P = \mathbb{P}(\mathcal{E}_1) \times_X \mathbb{P}(\mathcal{E}_2) \to \mathbb{P}(\mathcal{E}_1 \otimes_{\mathcal{O}_X} \mathcal{E}_2)$. This morphism $\varphi_{(\mathcal{L},\gamma)}$ is said to be a *Segre morphism*. A proof of the following proposition is notationally complicated but easy. The proof is left as an exercise.

PROPOSITION 5.35. *The Segre morphism*

$$\mathbb{P}(\mathcal{E}_1) \times_X \mathbb{P}(\mathcal{E}_2) \to \mathbb{P}(\mathcal{E}_1 \otimes_{\mathcal{O}_X} \mathcal{E}_2)$$

is a closed immersion.

Similarly, for m quasicoherent sheaves $\mathcal{E}_1, \mathcal{E}_2, \ldots, \mathcal{E}_m$ over X, a closed immersion

$$\mathbb{P}(\mathcal{E}_1) \times_X \mathbb{P}(\mathcal{E}_2) \times_X \cdots \times_X \mathbb{P}(\mathcal{E}_m) \to \mathbb{P}(\mathcal{E}_1 \otimes_{\mathcal{O}_X} \mathcal{E}_2 \otimes_{\mathcal{O}_X} \cdots \otimes_{\mathcal{O}_X} \mathcal{E}_m)$$

can be defined. This closed immersion is also called a Segre morphism.

5.3. PROJECTIVE MORPHISMS

(c) Ample Invertible Sheaves

A scheme morphism $\mu : Z \to W$ is said to be an *immersion* if $\mu(Z)$ is contained in an open set W_0 of W and $\mu : Z \to W_0$ is a closed immersion. For a given scheme morphism $f : Y \to X$, an invertible sheaf $\mathcal{L}$ over Y is said to be an f-*very ample invertible sheaf* if the following condition is satisfied.

(VA) There exists a quasicoherent sheaf $\mathcal{E}$ over X so that $\varphi_{(\mathcal{L},\gamma)} : Y \to \mathbb{P}(\mathcal{E})$ is an immersion and the $\mathcal{O}_Y$-homomorphism $\gamma : f^*\mathcal{E} \to \mathcal{L}$ is surjective.

LEMMA 5.36. *Let $f : Y \to X$ be a scheme morphism. For an f-very ample invertible sheaf $\mathcal{L}$ over Y, we have:*

(i) *f is a separated morphism.*
(ii) *If f is a quasicompact morphism, then $f_*\mathcal{L}$ is a quasicoherent $\mathcal{O}_X$-module, the natural $\mathcal{O}_Y$-homomorphism $\gamma : f^*f_*\mathcal{L} \to \mathcal{L}$ is surjective, and $\varphi_{(\mathcal{L},\gamma)} : Y \to \mathbb{P}(f_*\mathcal{L})$ is an immersion.*

PROOF. (i) Note that for a quasicoherent $\mathcal{O}_X$-module $\mathcal{E}$ over X, $\pi : \mathbb{P}(\mathcal{E}) \to X$ is a separated morphism. This can be shown by proving that $\pi^{-1}(U) \to U$ is separated for an affine open set U. (But this is the very statement of Problem 8.) Since for an open set V of $\mathbb{P}(\mathcal{E})$, the natural morphism $\iota_V : V \to \mathbb{P}(\mathcal{E})$ is an open immersion, ι_V is a separated morphism (i.e., Theorem 3.25). Hence their composition $\pi \circ \iota_V : V \to X$ is a separated morphism (see Theorem 3.25). Furthermore, one can choose V so that $j : Y \to V$ is a closed immersion. Since a closed immersion is separated (see Theorem 3.25), $f \circ \iota_V \circ j : Y \to X$ is a separated morphism.

(ii) By Proposition 4.36, $f_*\mathcal{L}$ is a quasicoherent $\mathcal{O}_X$-module. Since $\mathcal{L}$ is f-very ample, there exist a quasicoherent $\mathcal{O}_X$-module $\mathcal{E}$ and a surjective $\mathcal{O}_Y$-homomorphism $\gamma' : f^*\mathcal{E} \to \mathcal{L}$. Moreover, $\varphi_{(\mathcal{L},\gamma')} : Y \to \mathbb{P}(\mathcal{E})$ is a closed immersion. Then γ' is the composition of $f^*\mathcal{E} \xrightarrow{f^*(\gamma')} f^*f_*\mathcal{L} \xrightarrow{\gamma} \mathcal{L}$, and γ' is surjective. Hence γ is also surjective. Furthermore, there are induced $\mathcal{O}_X$-algebra homomorphisms

$$\mathbb{S}(f^*(\mathcal{E})) \xrightarrow{\mathbb{S}(f^*\gamma')} \mathbb{S}(f^*f_*\mathcal{L}) \xrightarrow{\mathbb{S}(\gamma)} \bigoplus_{n=0}^{\infty} \mathcal{L}^{\otimes n}.$$

Let $\psi = \mathbb{S}(f^*(\gamma'))$. We get a commutative diagram:

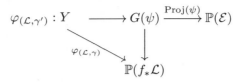

From Proposition 5.13, $\mathrm{Proj}(\psi)$ is an affine morphism. By the assumption, $\varphi_{(\mathcal{L},\gamma')}$ is a closed immersion into an open set U of $\mathbb{P}(\mathcal{E})$. Let $V = \mathrm{Proj}(\psi)^{-1}(U)$. Then $Y \to V$ is also a closed immersion. Since $G(\psi)$ is an open set of $\mathbb{P}(f_*\mathcal{L})$, we conclude that $\varphi_{(\mathcal{L},\gamma)} : Y \to V$ is a closed immersion. □

COROLLARY 5.37. *Let $f : Y \to X$ be a quasicompact morphism. An invertible sheaf $\mathcal{L}$ over Y is an f-very ample sheaf if and only if the natural $\mathcal{O}_X$-homomorphism $\gamma : f^*f_*\mathcal{L} \to \mathcal{L}$ is surjective and $\varphi_{(\mathcal{L},\gamma)}$ is an immersion.* □

For a morphism $f : Y \to X$ and for f-very ample invertible sheaves $\mathcal{L}_1$ and $\mathcal{L}_2$ over Y, we will ask whether $\mathcal{L}_1 \otimes \mathcal{L}_2$ is f-very ample or not. We will find that even a weaker condition is enough for f-very ampleness.

EXERCISE 5.38. For a morphism $f : Y \to X$ and for an f-very ample invertible sheaf $\mathcal{L}$ over Y, assume that for an invertible sheaf $\mathcal{L}_2$ over Y there exists a quasicoherent sheaf $\mathcal{E}_2$ over X to obtain a surjective homomorphism $\gamma_2 : f^*\mathcal{E}_2 \to \mathcal{L}_2$.

PROOF. The hypothesis implies that for a quasicoherent sheaf $\mathcal{E}_1$ and a surjective homomorphism $\gamma_1 : f^*\mathcal{E}_1 \to \mathcal{L}_1$, $\varphi_1 = \varphi_{(\mathcal{L}_1,\gamma_1)} : Y \to \mathbb{P}(\mathcal{E}_1)$ is an immersion. Then $\gamma = \gamma_1 \otimes \gamma_2 : f^*(\mathcal{E}_1 \otimes_{\mathcal{O}_X} \mathcal{E}_2) = f^*\mathcal{E}_1 \otimes_{\mathcal{O}_Y} f^*\mathcal{E}_2 \to \mathcal{L}_1 \otimes_{\mathcal{O}_Y} \mathcal{L}_2$ is also surjective. Let $\mathcal{L} = \mathcal{L}_1 \otimes_{\mathcal{O}_Y} \mathcal{L}_2$. We can define $\varphi = \varphi_{(\mathcal{L},\gamma)} : Y \to \mathbb{P}(\mathcal{E}_1 \otimes_{\mathcal{O}_X} \mathcal{E}_2)$. On the other hand, the morphism $\varphi_2 = \varphi_{(\mathcal{L}_2,\gamma_2)} : Y \to \mathbb{P}(\mathcal{E}_2)$ is defined from $\gamma_2 : f^*\mathcal{E}_2 \to \mathcal{L}_2$. Hence we get $\varphi_1 \times_X \varphi_2 : Y \to \mathbb{P}(\mathcal{E}_1) \times_X \mathbb{P}(\mathcal{E}_2)$. Since φ_1 is an immersion, the base change $\varphi_1 \times_X \varphi_2$ of φ_1 through $\mathbb{P}(\mathcal{E}_2) \to X$ is also an immersion (by applying Theorem 3.25(iii) to $\varphi_1 : Y \to U$ for an open set U of $\mathbb{P}(\mathcal{E}_1)$). Furthermore, φ is the composition of $\varphi_1 \times_X \varphi_2$ and the Segre morphism $\psi : \mathbb{P}(\mathcal{E}_1) \times \mathbb{P}(\mathcal{E}_2) \to \mathbb{P}(\mathcal{E}_1 \otimes_{\mathcal{O}_X} \mathcal{E}_2)$, and ψ is an immersion. We conclude that φ is an immersion. □

PROBLEM 16. For $f : Y \to X$, let $\mathcal{L}$ be f-very ample. Then prove that $\mathcal{L}^{\otimes m}$ is also f-very ample, where m is any positive integer.

5.3. PROJECTIVE MORPHISMS

We have discussed the existence of a quasicoherent $\mathcal{O}_X$-module $\mathcal{E}$ so that for an f-very ample invertible sheaf $\mathcal{L}$, $Y \to \mathbb{P}(\mathcal{E})$ is an immersion. In algebraic geometry, we often consider the case where X is a Noetherian scheme and f is a morphism of finite type. Then we would like $\mathcal{E}$ to be a coherent $\mathcal{O}_X$-module. We will prove a more general statement than the above.

THEOREM 5.39. *Let X be a quasicompact and separated scheme (i.e., the structure morphism $X \to \operatorname{Spec} \mathbb{Z}$ is quasicompact and separated), let $f : Y \to X$ be a morphism of finite type, and let $\mathcal{L}$ be an f-very ample invertible sheaf over Y. Then one can choose a quasicoherent finitely generated $\mathcal{O}_X$-module $\mathcal{F}$ and a surjective homomorphism $\delta : f^*\mathcal{F} \to \mathcal{L}$ so that $\varphi_{(\mathcal{L},\delta)} : Y \to \mathbb{P}(\mathcal{F})$ is an immersion.*

PROOF. Since $\mathcal{L}$ is f-very ample, we have a quasicoherent $\mathcal{O}_X$-module $\mathcal{E}$ over X and a surjective homomorphism $\gamma : \mathcal{E} \to \mathcal{L}$ to obtain an immersion $\varphi_{(\mathcal{L},\gamma)} : X \to \mathbb{P}(\mathcal{E})$. We will show that there exists a quasicoherent finitely generated $\mathcal{O}_X$-submodule $\mathcal{F}$ to satisfy the conclusion of this theorem.

Since X is quasicompact, there is a finite affine open covering $\{U_i\}_{i=1}^m$. Let $\mathbb{P}(\mathcal{E})|U_i$ be the inverse image $\pi^{-1}(U_i)$ of U_i under $\pi : \mathbb{P}(\mathcal{E}) \to X$. We have $\mathbb{P}(\mathcal{E})|U_i = \mathbb{P}(\mathcal{E}) \times_X U_i$. Since $\varphi = \varphi_{(\mathcal{L},\gamma)}$ is an immersion, one can choose an open set V of $\mathbb{P}(\mathcal{E})$ so that $\varphi = \varphi_{(\mathcal{L},\gamma)} : X \to V$ is a closed immersion. Set $V_i = \pi^{-1}(U_i) \cap V$. Then $\varphi_i = \varphi|f^{-1}(U_i) : f^{-1}(U_i) \to V_i$ is a closed immersion. For $U_i = \operatorname{Spec} R_i$, there exists an R_i-module E_i to satisfy $\mathcal{E}|U_i = \widetilde{E}_j$. For an affine open covering $\{D_+(x_{ij})\}_{j=1}^{n_i}$, $x_{ij} \in \mathbb{S}(E_i)$, of V_i, we get an affine open covering $V_{ij} = \varphi^{-1}(D_+(x_{ij}))$ of $f^{-1}(U_i)$. Since f is a morphism of finite type, $f^{-1}(U_i)$ is quasicompact. Hence one may choose a finite affine covering $\{V_{ij}\}_{j=1}^{n_i}$. Let $V_{ij} = \operatorname{Spec} A_{ij}$. The R_i-algebra A_{ij} is finitely generated, and then $\varphi_{ij} = \varphi|V_{ij} : V_{ij} \to D_+(x_{ij})$ is a closed immersion. One may assume $\mathcal{L}|V_{ij} \xrightarrow{\sim} \mathcal{O}_X|V_{ij}$ by taking a finer affine covering if necessary. Since $\varphi_{ij} : V_{ij} \to D_+(x_{ij})$ is a closed immersion, we get the corresponding R_i-algebra surjective homomorphism $\psi_{ij} : \mathbb{S}(E_i)_{(x_{ij})} \to A_{ij}$. Since A_{ij} is a finitely generated R_i-algebra, one can choose a finite R_i-submodule E'_{ij} of E_i so that $x_{ij} \in \mathbb{S}(E'_{ij})$ and $\psi_{ij}|\mathbb{S}(E'_{ij}) : \mathbb{S}(E'_{ij}) \to A_{ij}$ is surjective. For a finite R_i-module $E'_i = \sum_{j=1}^{n_i} E'_{ij}$, we will show that there exists a finitely generated $\mathcal{O}_X$-submodule $\mathcal{F}_i$ of $\mathcal{E}$ satisfying $\mathcal{F}_i|U_i = \widetilde{E'_i}$.

First, for a quasicompact open set W of X and an affine open set U, we will show that a quasicoherent finitely generated $\mathcal{O}_{W\cap U}$-submodule $\mathcal{G}'$ of $\mathcal{E}|W \cap U$ may be extended to a quasicoherent finitely generated $\mathcal{O}_U$-submodule $\mathcal{G}''$ of $\mathcal{E}|U$. After proving the above statement, for a given quasicoherent finitely generated $\mathcal{O}_W$-submodule $\mathcal{G}_1$ of $\mathcal{E}|W$ over W, $\mathcal{G}' = \mathcal{G}_1|W \cap U$ can be extended to a quasicoherent finitely generated $\mathcal{O}_U$-submodule $\mathcal{G}''$ of $\mathcal{E}|U$. Since $\mathcal{G}|W \cap U = \mathcal{G}''|W \cap U$, by gluing $\mathcal{G}_1$ and $\mathcal{G}''$ over $W \cap U$, we get a quasicoherent finitely generated $\mathcal{O}_{W\cup U}$-submodule $\mathcal{G}_2$ of $\mathcal{E}|W \cup U$. Then $\mathcal{G}_2|W = \mathcal{G}_1$. Namely, $\mathcal{G}_1$ can be extended to a quasicoherent finitely generated $\mathcal{O}_{W\cup U}$-submodule over $W \cup U$. By this construction, beginning from $\widetilde{E'_i}$, extend the sheaf from U_{j_1}, $U_i \cup U_{j_1} \cup U_{j_2}, \ldots$ to obtain a finitely generated $\mathcal{O}_X$-submodule $\mathcal{F}_i$ of $\mathcal{E}$.

Note that by Theorem 3.25, for a quasicompact open set W, the natural open immersion $\lambda : W \to X$ is a quasicompact separated morphism. Hence $\mu = \lambda|U : W \cap U \to U$ is also a quasicompact separated morphism. By Proposition 4.36, $\mu_*\mathcal{G}'$ is a quasicoherent $\mathcal{O}_U$-submodule of $\mu_I(\mathcal{E}|W \cap U)$. Lemma 4.37 implies

$$\mathrm{Hom}_{\mathcal{O}_{W\cap U}}(\mu^*(\mathcal{E}|U), \mathcal{E}|W \cap U) \xrightarrow{\sim} \mathrm{Hom}_{\mathcal{O}_U}(\mathcal{E}|U, \mu_*\mu^*(\mathcal{E}|U)).$$

Then consider the corresponding homomorphism

$$\nu : \mathcal{E}|U \to \mu_*\mu^*(E|U) = \mu_*(\mathcal{E}|W \cap U)$$

to the identity map on the left-hand side. Let $\mathcal{H}$ be the inverse image $\nu^{-1}(\mu_*\mathcal{G}')$ of $\mu_*\mathcal{G}' \subset \mu_*(\mathcal{E}|W \cap U)$. Since U is an affine scheme, $\mathcal{H} = \widetilde{H}$, where $H = \Gamma(U, \mathcal{H})$. Let $U = \mathrm{Spec}\, R$. Then H is an R-module, and H can be expressed as the direct limit $\varinjlim_{j\in J} H_j = H$ of finite R-submodules $\{H_j\}_{j\in J}$. Then the corresponding sheaf $\mathcal{H}$ can be expressed as the direct limit $\varinjlim_{j\in J} \widetilde{H_j}$ of finitely generated $\mathcal{O}_U$-submodules $\{\widetilde{H_j}\}_{j\in J}$. On the other hand, $W \cap U$ is quasicompact, and $\mathcal{G}' = \mathcal{H}|W \cap U$ is a finitely generated $\mathcal{O}_{W\cap U}$-submodule. One finds a finitely generated $\mathcal{O}_U$-submodule $\mathcal{H}_j$ satisfying $\mathcal{H}_j|W \cap U = \mathcal{G}'$. Hence, we take $\mathcal{F}_i = \mathcal{H}_j$.

For the quasicoherent finitely generated $\mathcal{O}_X$-submodules $\mathcal{F}_i$ of $\mathcal{E}$ constructed from $\widetilde{E'_i}$, let $\mathcal{F} = \sum_{i=1}^m \mathcal{F}_i$. Then $\mathcal{F}$ is a quasicoherent finitely generated $\mathcal{O}_X$-submodule of $\mathcal{E}$. For a finite R_i-module F_i, let $\mathcal{F}|U_i = \widetilde{F_i}$. Then, by the definition of $\mathcal{F}$, F_i is an R_i-submodule of E_i and $\psi_{ij}(\mathbb{S}(F_i)_{(x_{ij})}) = A_{ij}$. Let δ be the natural homomorphism

5.3. PROJECTIVE MORPHISMS

$f^*\mathcal{F} \to f^*\mathcal{E} \to \mathcal{L}$. Then, since $\psi_{ij}(\mathbb{S}(F_i)_{(x_{ij})}) = A_{ij}$, the morphism

$$\delta|V_{ij} : f^*\mathcal{F}|V_{ij} = (F_i \otimes_{R_i} A_{ij})^\sim \to (E_i \otimes_{R_i} A_{ij})^\sim \to \mathcal{L}|V_{ij} = \widetilde{A_{ij}}$$

is surjective. Hence δ is surjective. That is, we can define $\varphi_{(\mathcal{L},\varphi)} : Y \to \mathbb{P}(\mathcal{F})$. Since $\varphi_{(\mathcal{L},\varphi)}|V_{ij} : V_{ij} \to D_+(x_{ij}) \subset \mathbb{P}(\mathcal{F}|U_i)$ is a closed immersion, $\varphi_{(\mathcal{L},\varphi)}$ is an immersion. □

Applying this theorem to a Noetherian scheme, we get the following.

COROLLARY 5.40. *For a separated Noetherian scheme X and for a morphism $f : Y \to X$ of finite type, if $\mathcal{L}$ is an f-very ample invertible sheaf, then one can find a coherent sheaf $\mathcal{F}$ over X and a surjective homomorphism $\delta : \mathcal{F} \to \mathcal{L}$ so that $\varphi_{(\mathcal{L},\delta)} : Y \to \mathbb{P}(\mathcal{F})$ is an immersion.* □

A Noetherian scheme X is quasicompact, and $\mathcal{O}_X$ is coherent. Since a finitely generated $\mathcal{O}_X$-module is also coherent (see Proposition 4.27), this corollary holds.

There is a notion of an *ample invertible sheaf*. Rather than describing a general theory, we will give a more restricted definition.

For a quasicompact separated scheme X and a quasicompact separated morphism $f : X \to W$, an invertible sheaf $\mathcal{L}$ over X is said to be f-*ample* if for an arbitrary quasicoherent finitely generated $\mathcal{O}_X$-module $\mathcal{F}$, there is an n_0 to get a surjective $\mathcal{O}_X$-homomorphism $f^*f_*(\mathcal{F} \otimes \mathcal{L}^{\otimes n}) \to \mathcal{F} \otimes \mathcal{L}^{\otimes n}$ for $n \geq n_0$.

The following theorem explains the relationship between f-ampleness and f-very ampleness.

THEOREM 5.41. *For a given quasicompact scheme and a given separated morphism of finite type $f : X \to W$, an invertible sheaf $\mathcal{L}$ over X is f-ample if and only if for a positive integer n, $\mathcal{L}^{\otimes n}$ is f-very ample. Moreover, there is a positive integer n_0 such that for $n \geq n_0$, $\mathcal{L}^{\otimes n}$ is f-very ample.* □

We omit the proof of this theorem.

DEFINITION 5.42. (i) For a quasicompact scheme W, a morphism $f : X \to W$ is said to be a *quasiprojective morphism* if f is a morphism of finite type and if there exists an f-ample invertible sheaf over X. Then X over W is said to be a *quasiprojective scheme*.

(ii) A morphism $f : X \to W$ is said to be a *projective morphism* if there exists a quasicoherent finitely generated $\mathcal{O}_W$-module $\mathcal{E}$ that has a closed immersion $\varphi : X \to \mathbb{P}(\mathcal{E})$ with the commutative diagram

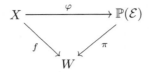

Then X is said to be *projective over* W.

We will give several remarks on the above definitions. The condition that $f : X \to W$ is quasiprojective is not local with respect to W. Namely, even if for an affine open covering $\{V_i\}_{i \in I}$ of W, $f^{-1}(V_i) \to V_i$ is quasiprojective, f need not be quasiprojective, for the following reason. When $q_i = f|f^{-1}(V_i) : f^{-1}(V_i) \to V_i$ is quasiprojective, there exists a q_i-ample invertible sheaf $\mathcal{L}_i$ over $f^{-1}(V_i)$. But there is no guarantee that there exists an invertible sheaf over X to satisfy $\mathcal{L}|f^{-1}(V_i) = \mathcal{L}_i$. We will give an example in Chapter 7 where a quasiprojective morphism $f^{-1}(V_i) \to V_i$ is not quasiprojective. Note also that by the definition, an ample invertible sheaf demands the separatedness of the scheme. Namely, a quasiprojective morphism $f : X \to W$ is separated.

In particular, when W is quasicompact, by Theorem 5.41, there is an f-ample sheaf $\mathcal{L}$ for a quasiprojective morphism $f : X \to W$ so that $\mathcal{L}^{\otimes n}$ is f-very ample. Hence we may assume that $\mathcal{L}$ is f-very ample.

EXERCISE 5.43. A scheme X over a field k is quasiprojective (i.e., the structure morphism $f : X \to \operatorname{Spec} k$ is quasiprojective) if and only if there exists an immersion morphism λ from X to a projective space $\mathbb{P}^n_k$ over k that completes the commutative diagram

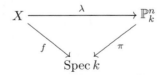

Note that $\pi : \mathbb{P}^n_k \to \operatorname{Spec} k$ is the structure morphism. That is, X can be considered as a closed subscheme of an open subscheme of $\mathbb{P}^n_k$.

PROOF. Let $f : X \to \operatorname{Spec} k$ be quasiprojective. Then there exists an f-very ample invertible sheaf $\mathcal{L}$ over X. (See the remark in the paragraph before Exercise 5.43.) Since f is a morphism of finite type over k, Theorem 5.39 implies that there exists a finitely generated

$\mathcal{O}_{\mathrm{Spec}\,k}$-module over k, i.e., a finite-dimensional k-vector space V and an immersion $\lambda : X \to \mathbb{P}(V) \cong \mathbb{P}^n_k$. Since λ is a morphism over $\mathrm{Spec}\,k$, the commutativity of the diagram follows.

Conversely, for an immersion $\lambda : X \to \mathbb{P}^n_k$, define $f = \pi \circ \lambda : X \to \mathrm{Spec}\,k$. Then f is a morphism of finite type, and $\mathcal{L} = \lambda^* \mathcal{O}_{\mathbb{P}^n_k}(1)$ is f-very ample. □

From Definition 5.42, a projective morphism $f : X \to W$ is quasiprojective. Under what condition can the converse be true? This naive question leads to the connection with a proper morphism.

THEOREM 5.44. (i) *A projective morphism is a quasiprojective and proper morphism.*

(ii) *Conversely, if W is a quasicompact separated scheme, then a proper and quasiprojective morphism is a projective morphism.*

PROOF. (i) Since a projective morphism $f : X \to W$ is a morphism of finite type, it is enough to show the universal closedness. First, we will prove that for a quasicoherent finitely generated $\mathcal{O}_W$-module $\mathcal{F}$, $\pi : \mathbb{P}(\mathcal{F}) \to W$ is universally closed. For a scheme morphism $g : Z \to W$, $g^*\mathcal{F}$ is a quasicoherent $\mathcal{O}_Z$-module and $\mathbb{P}(g^*\mathcal{F}) = \mathbb{P}(\mathcal{F}) \times_W Z$ holds. Then, since the structure morphism is the base change of π through g, it is sufficient to prove the closedness of $\pi : \mathbb{P}(\mathcal{F}) \to W$. Furthermore, for an affine covering $\{U_i\}_{i \in I}$ of W, if $\mathbb{P}(\mathcal{F}|U_i) = \mathbb{P}(\mathcal{F}) \times_W U_i \to U_i$ is a closed morphism, then π is a closed morphism. Hence we may assume that W is an affine scheme $\mathrm{Spec}\,R$. Then $\mathcal{F} = \widetilde{M}$, where M is a finite R-module. When M is generated by finitely many elements $a_0, a_1, \ldots, a_n$ as an R-module, we can define a surjective homomorphism of graded algebras of degree 0,

$$\psi : R[x_0, x_1, \ldots, x_n] \to \mathbb{S}(M) = \bigoplus_{n=0}^{\infty} \mathbb{S}^n(M),$$

$$x_j \mapsto a_j.$$

From this π, a closed immersion

$$\mu : \mathbb{P}(\mathcal{F}) = \mathrm{Proj}\,\mathbb{S}(M) \to \mathbb{P}^n_R$$

is induced. By Theorem 5.4, $\tilde{\pi} : \mathbb{P}^n_R \to \mathrm{Spec}\,R$ is a proper morphism, and by Proposition 5.3, a closed immersion is a proper morphism. Hence $\pi = \tilde{\pi} \circ \mu$ is a proper morphism, i.e., a closed morphism.

By the definition of a projective morphism $f : X \to W$, we can find a quasicoherent finitely generated $\mathcal{O}_W$-module $\mathcal{F}$ and a closed immersion $\lambda : X \to \mathbb{P}(\mathcal{F})$ over W. For the structure morphism $\pi : \mathbb{P}(\mathcal{F}) \to W$, we have $f = \pi \circ \lambda$. Since λ and π are universally closed, by Lemma 5.2, f is also a universally closed morphism.

(ii) Since W is a quasicompact separated morphism, Theorem 5.41 implies that for $f : X \to W$, there exists an f-very ample invertible sheaf $\mathcal{L}$ over X. Then by Theorem 5.39, there exists a quasicoherent finitely generated $\mathcal{O}_W$-module $\mathcal{F}$ so that an immersion $\lambda : X \to \mathbb{P}(\mathcal{F})$ over W may exist. For the structure morphism $\pi : \mathbb{P}(\mathcal{F}) \to W$, we have $f = \pi \circ \lambda$. From (i), π is a proper morphism, and in particular, a separated morphism. Since f is proper, Proposition 5.3 implies that λ is a proper morphism. Hence, $\lambda(X)$ is a closed subscheme of $\mathbb{P}(\mathcal{F})$. That is, λ is a closed immersion, and f is a projective morphism. □

A reduced projective scheme over a field is said to be a *projective variety*, and a reduced proper scheme over k (i.e., the structure morphism $f : X \to \operatorname{Spec} k$ is a proper morphism, and X is a reduced scheme) is said to be a *complete variety*. The notion of a proper morphism is more general than the notion of a projective morphism; an example will be given in Chapter 7. By Exercise 5.43, a projective variety is precisely a reduced closed scheme of $\mathbb{P}_k^n$.

Summary

5.1. Definitions of a closed morphism and a universally closed morphism, and their fundamental properties.

5.2. When a scheme morphism $f : X \to Y$ is a separated morphism of finite type and universally closed, f is said to be a proper morphism. Then X is said to be a proper scheme over Y. A proper morphism can be characterized in terms of valuation rings (valuative criterion).

5.3. A graded S-module M corresponds to a quasicoherent sheaf over the projective scheme $X = \operatorname{Proj} S$ of a graded ring S. Then for an arbitrary integer m_0, $M = \bigoplus_{n \in \mathbb{Z}} M_n$ and $N = \bigoplus_{n \geq m_0} M_n$ define the same quasicoherent $\mathcal{O}_X$-module. This correspondence is not uniquely determined, contrary to the case of the correspondence between quasicoherent sheaves and R-modules for an affine scheme $\operatorname{Spec} R$.

5.4. If a graded ring S is a Noetherian ring, then a coherent sheaf over $X = \operatorname{Proj} S$ corresponds to a finite graded S-module M.

5.5. For a sheaf $\mathcal{S}$ of quasicoherent graded $\mathcal{O}_X$-algebras, one can define a scheme $\operatorname{Proj}\mathcal{S}$. For a quasicoherent $\mathcal{O}_X$-module $\mathcal{E}$, the scheme $\mathbb{P}(\mathcal{E})$ and the structure morphism $\pi : \mathbb{P}(\mathcal{E}) \to X$ can be defined.

5.6. For a quasicoherent finitely generated $\mathcal{O}_X$-module $\mathcal{E}$ over X, $\pi : \mathbb{P}(\mathcal{E}) \to X$ can be characterized in terms of representable functors. That is, there exists a one-to-one correspondence between morphisms from schemes $g : W \to X$ over X to $\mathbb{P}(\mathcal{E})$ and equivalence classes of pairs $(\mathcal{L}, \gamma)$ of invertible sheaves $\mathcal{L}$ over W and $\mathcal{O}_W$-module surjective homomorphisms $\gamma : g^*\mathcal{E} \to \mathcal{L}$.

5.7. For a scheme morphism $f : W \to X$ and an invertible sheaf $\mathcal{L}$, we can find a quasicoherent sheaf $\mathcal{E}$ over X and a surjective $\mathcal{O}_W$-homomorphism $\gamma : f^*\mathcal{E} \to \mathcal{L}$ so that the corresponding morphism $W \to \mathbb{P}(\mathcal{E})$ over X is an immersion. Then $\mathcal{L}$ is said to be f-very ample.

5.8. Let X be a quasicompact separated scheme, let $f : X \to W$ be a quasicompact separated morphism, and let $\mathcal{L}$ be an invertible sheaf over X. For an arbitrary quasicoherent $\mathcal{O}_X$-module $\mathcal{F}$, if there exists m_0 yielding a surjective $\mathcal{O}_X$-homomorphism $f^*f_*(\mathcal{F} \otimes \mathcal{L}^{\otimes m}) \to \mathcal{F} \otimes \mathcal{L}^{\otimes m}$, $m \geq m_0$, then the invertible sheaf $\mathcal{L}$ is said to be f-ample. For a quasicompact separated morphism of finite type, $\mathcal{L}$ is f-ample if and only if for some positive integer n, $\mathcal{L}^{\otimes n}$ is f-very ample.

5.9. A scheme morphism $f : X \to W$ is said to be a projective morphism if f can be written as the composition of the closed immersion $\varphi : X \to \mathbb{P}(\mathcal{E})$, for a quasicoherent finitely generated $\mathcal{O}_W$-module $\mathcal{E}$, and the structure morphism $\pi : \mathbb{P}(\mathcal{E}) \to W$. A projective morphism is a proper morphism. When the morphism $f : X \to W$ is of finite type and an f-ample invertible sheaf $\mathcal{L}$ exists over X, the morphism f is said to be quasiprojective.

Exercises

5.1. Let $\mathcal{S} = \bigoplus_{n \in \mathbb{Z}}^{\infty} \mathcal{S}_n$ be a sheaf of quasicoherent graded algebras satisfying $\mathcal{S}_0 = \mathcal{O}_X$ and such that $\mathcal{S}$ is generated by $\mathcal{S}_1$ as an $\mathcal{O}_X$-algebra and $\mathcal{S}_1$ is a finitely generated $\mathcal{O}_X$-module. For $\pi : Z = \operatorname{Proj}\mathcal{S} \to X$, let Y be the closed subscheme of Z defined by a quasicoherent sheaf $\mathcal{J}$ of ideals. For $\alpha : \mathcal{S} \to \Gamma_*(\mathcal{O}_Z)$, let $\mathcal{I} = \alpha^{-1}(\Gamma_*(\mathcal{J}))$ and $\mathcal{S}' = \mathcal{S}/\mathcal{J}'$. Then show that Y is isomorphic to $\operatorname{Proj}\mathcal{S}'$ over X.

5.2. Let $\mathbb{P}^n(k)$ be the totality of k-valued points of the n-dimensional projective space $\mathbb{P}^n_k$ over a field k. For $A = (a_{ij}) \in GL(n+1, k)$, define an automorphism φ_A of the polynomial ring $R = k[x_0, x_1, \ldots, x_n]$ over k by $\varphi_A(x_i) = \sum_{j=0}^n a_{ij} x_j$, $i = 0, 1, \ldots, n$. Then we have $G(\varphi_A) = \mathbb{P}^n_k$, and φ_A determines an automorphism $f_A = {}^a\varphi_A : \mathbb{P}^n_k \to \mathbb{P}^n_k$. We call f_A a projective transformation.

(1) For $\alpha \in k^\times = k \setminus \{0\}$, prove that $f_{\alpha A} = f_A$. Prove also that for $A, B \in GL(n+1, k)$, we have $f_{AB} = f_B \circ f_A$.

(2) Let $A^{-1} = (b_{ij})$. For a point $(a_0 : a_1 : \cdots : a_n)$ of $\mathbb{P}^n_k$, show that

$$f_A((a_0 : a_1 : \cdots : a_n)) = \left(\sum_{j=0}^n b_{0j} a_j : \sum_{j=0}^n b_{1j} a_j : \cdots : \sum_{j=0}^n b_{nj} a_j \right).$$

(3) For $n = 1$, show that there exists a unique projective transformation f mapping distinct points P_1, P_2 and P_3 on $\mathbb{P}^1_k$ to distinct points Q_1, Q_2 and Q_3 in that order.

5.3. A linear form $a_0 x_0 + a_1 x_1 + a_2 x_2$ in the polynomial ring $k[x_0, x_1, \ldots, x_n]$ over a field k determines a closed subscheme of $\mathbb{P}^2_k$ which is called a line over k. When k is algebraically closed, the totality of k-valued points on the closed subscheme in $\mathbb{P}^2(k)$ is the line discussed in Chapter 1. As in Chapter 1, a line as a closed subscheme can be expressed as $a_0 x_0 + a_1 x_1 + a_2 x_2 = 0$.

(1) For lines l_1 and l_2, prove that there exists a projective transformation mapping l_1 and l_2 onto $x_0 = 0$ and $x_1 = 0$. Let points Q_1, Q_2, Q_3, Q_4 be in general position, i.e., no three points are on the same line. Prove that there exists a unique projective transformation mapping Q_1, Q_2, Q_3, Q_4 onto R_1, R_2, R_3, R_4 in general position in that order.

(2) For distinct points $(a_0 : a_1 : a_2)$ and $(b_0 : b_1 : b_2)$ in $\mathbb{P}^2(k)$, prove that there is a unique line going through these points whose equation is given by

$$\begin{vmatrix} a_1 & a_2 \\ b_1 & b_2 \end{vmatrix} x_0 + \begin{vmatrix} a_2 & a_0 \\ b_2 & b_0 \end{vmatrix} x_1 + \begin{vmatrix} a_0 & a_1 \\ b_0 & b_1 \end{vmatrix} x_2 = 0.$$

(3) Prove that two distinct lines $l_1 : a_0 x_0 + a_1 x_1 + a_2 x_2 = 0$ and $l_2 : b_0 x_0 + b_1 x_1 + b_2 x_2 = 0$ always intersect at a uniquely determined point, and the unique point of intersection is given by

$$\left(\begin{vmatrix} a_1 & a_2 \\ b_1 & b_2 \end{vmatrix} : \begin{vmatrix} a_2 & a_0 \\ b_2 & b_0 \end{vmatrix} : \begin{vmatrix} a_0 & a_1 \\ b_0 & b_1 \end{vmatrix} \right).$$

(4) Let l_1 and l_2 be distinct lines, and let A_1, A_2, A_3 be three distinct points on l_1 and B_1, B_2, B_3 be three distinct points on l_2. Let us denote the line going through A_i and B_j by $\overline{A_i B_j}$. Let P, Q, R be the points of intersection of $\overline{A_1 B_2}$ and $\overline{A_2 B_1}$, $\overline{A_1 B_3}$ and $\overline{A_3 B_1}$, $\overline{A_2 B_3}$ and $\overline{A_3 B_2}$, respectively. Prove that P, Q and R are on a line (Pappus' Theorem).

(5) The correspondence between a line over k, i.e., $a_0 x_0 + a_1 x_1 + a_2 x_2 = 0$, and a point $(a_0 : a_1 : a_2) \in \mathbb{P}^2(k)$ is one-to-one. Then 'a line over k going through two points' corresponds to 'a point of intersection of two lines'. This sort of correspondence between lines and points is called a *duality principle*. State the dual of Pappus' Theorem in (4).

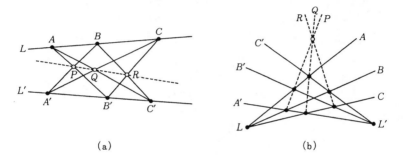

FIGURE 5.2. (a) Pappus' Theorem (b) The dual of Pappus' Theorem.

5.4. Let X be the closed subscheme in the n-dimensional projective space $P = \mathbb{P}_R^n$ over a Noetherian ring R defined by a coherent sheaf of ideals $\mathcal{J}$. Set $I = \Gamma_*(\mathcal{O}_X)$. Then I is a homogeneous ideal of $\Gamma_*(\mathcal{O}_P) = R[z_0, z_1, \ldots, z_n]$, and $S(X) = R[z_0, z_1, \ldots, z_n]/I$ is said to be the *homogeneous coordinate ring* of X. When $S(X)$ is an integral domain and is *integrally closed* in the quotient field (i.e., $S(X)$ is a normal ring), X is said to be *projectively normal*. For the scheme X, if $\mathcal{O}_{X,x}$ is a normal ring at every point $x \in X$, then X is said to be a *normal scheme*. In what will follow, let R be an algebraically closed field k. We consider the connected normal closed subscheme X of $\mathbb{P}_k^n$ defined by a coherent sheaf $\mathcal{J}$ of ideals.

(1) Show that $S = S(X)$ is an integral domain and that $S' = \bigoplus_{n \geq 0} \Gamma(X, \mathcal{O}_X(n))$ is a normal ring.

(2) Show that for a sufficiently large d, we have $S_d = S'_d$.

(3) Show that for a sufficiently large d, $S^{(d)}$ is a normal ring.

5.5. Construct a graded sheaf of $\mathcal{O}_X$-algebras $\mathcal{S} = \bigoplus_{n=0}^{\infty} \mathcal{J}^n$, where $\mathcal{J}$ is a coherent sheaf of ideals over a Noetherian scheme X. Define $\mathcal{S}^0 = \mathcal{O}_X$. Then $\widetilde{X} = \operatorname{Proj} \mathcal{S}$, or the natural morphism $\pi : \widetilde{X} \to X$ is said to be a *blowing up* or a *monoidal transformation* of X.

(1) Prove that the $\mathcal{O}_{\widetilde{X}}$-ideal sheaf $\widetilde{\mathcal{J}} = \pi^{-1}\mathcal{J} \cdot \mathcal{O}_{\widetilde{X}}$ generated by $\pi^{-1}\mathcal{J}$ is an invertible sheaf.

(2) Let Y be the closed scheme of X defined by $\mathcal{J}$ and let $U = X \setminus Y$. Prove that π induces an isomorphism from $\pi^{-1}(U)$ to U.

(3) Let $f : Z \to X$ be a morphism so that $f^{-1}\mathcal{J} \cdot \mathcal{O}_Z$ is an invertible sheaf. Then prove that there is induced a unique morphism $g : Z \to \widetilde{X}$ satisfying the following commutative diagram:

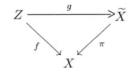

CHAPTER 6

Cohomology of Coherent Sheaves

We will develop the cohomology theory of quasicoherent and coherent sheaves over a scheme. Hence we will provide a general theory of cohomology which is suited also for the study of complex manifolds and functions of several complex variables. For the first reading the proofs may be skipped if one grasps the essence of the theorems. The explicit computation of the sheaf cohomology is more insightful than the formal proof.

However, the method of homological algebra plays a key role in various parts of mathematics. The reader is advised to read all the proofs during the second reading. Several examples are given with minimum preparation. We hope that the reader will enjoy the power of cohomology theory for the study of schemes in Chapter 7.

6.1. Cohomology of Sheaves

(a) Flabby Sheaves

In what will follow, we will consider sheaves of additive groups over a topological space X. We will define cohomology with a coefficient in a sheaf which is applicable to the general situation. First, we will define a flabby sheaf. A sheaf $\mathcal{F}$ over a topological space X is said to be a *flabby sheaf* if every section of $\mathcal{F}$ over an arbitrary open set U of X can be extended to a section over X. Namely, the restriction map $\rho_{U,X} : \mathcal{F}(X) \to \mathcal{F}(U)$ is surjective.

EXAMPLE 6.1. Let $\mathbb{F}$ be the sheafed space of a sheaf $\mathcal{F}$ over a topological space X (see Exercise **2.4**). For an open set U of X, define

(6.1)
$$C^0(\mathcal{F})(U) = \left\{ s : U \to \mathbb{F} \;\middle|\; \begin{array}{l} s \text{ is a map satisfying } p \circ s = \mathrm{id}_U, \\ \text{where } s \text{ need not be continuous} \end{array} \right\}.$$

For open sets $V \subset U$, define the restriction map $\rho_{V,U}$ as
$$\rho_{V,U} : C^0(\mathcal{F})(U) \to C^0(\mathcal{F})(V),$$
$$s \mapsto s|V,$$
where $s|V$ is the restriction of $s : U \to \mathbb{F}$ to V. For s and t in $C^0(\mathcal{F})(U)$, define $(s+t)(x) = s(x) + t(x)$, $x \in U$. Then $s + t \in C^0(\mathcal{F})(U)$, and $C^0(\mathcal{F})(U)$ becomes an additive group, so that $\rho_{V,U}$ is a group homomorphism. One can show that $C^0(\mathcal{F})$ is a sheaf over X.

We will prove that $C^0(\mathcal{F})$ is a flabby sheaf. Let U be an open set and let $s \in C^0(\mathcal{F})(U)$. Define
$$\tilde{s}(x) = \begin{cases} s(x) & \text{for } x \in U, \\ 0_x & \text{for } x \notin U, \end{cases}$$
where 0_x is the zero element of $\mathcal{F}_x$. Then $\tilde{s} \in C^0(\mathcal{F})(X)$, and $\tilde{s}|U = s$. That is, the restriction map $\rho_{U,X} : C^0(\mathcal{F})(X) \to C^0(\mathcal{F})(U)$ is surjective. Hence $C^0(\mathcal{F})$ is a flabby sheaf. Since we can regard
$$\mathcal{F}(U) = \{s : U \to \mathbb{F} | s \in C^0(\mathcal{F})(U) \text{ and } s \text{ is continuous}\}$$
(see Exercise **2.4**(2)), we have the natural map
(6.2) $$\iota_\mathcal{F} : \mathcal{F} \to C^0(\mathcal{F}). \quad \square$$

One reason why flabby sheaves are useful for defining the cohomology of a sheaf is the following.

PROPOSITION 6.2. (i) *For an exact sequence of sheaves of additive groups over a topological space X*
$$0 \to \mathcal{F} \xrightarrow{\varphi} \mathcal{G} \xrightarrow{\psi} \mathcal{H} \to 0,$$
if $\mathcal{F}$ is a flabby sheaf, then for an arbitrary open set U the sequence
$$0 \to \Gamma(U, \mathcal{F}) \xrightarrow{\varphi_U} \Gamma(U, \mathcal{G}) \xrightarrow{\psi_U} \Gamma(U, \mathcal{H}) \to 0$$
is an exact sequence of additive groups.

(ii) *For an exact sequence of sheaves of additive groups*
$$0 \to \mathcal{F} \to \mathcal{G} \to \mathcal{H} \to 0,$$
if $\mathcal{F}$ and $\mathcal{G}$ are flabby sheaves, then $\mathcal{H}$ is also a flabby sheaf.

PROOF. (i) It is enough to prove that $\Gamma(U, \mathcal{G}) \to \Gamma(U, \mathcal{H})$ is surjective. For $s \in \Gamma(U, \mathcal{H})$, let
$$\mathcal{M} = \left\{ (t, V) \; \middle| \; \begin{array}{l} V \text{ is an open set contained in } U \text{ such that} \\ t \in \mathcal{G}(V) \text{ satisfies } \psi_V(t) = \rho_{V,U}(s) \end{array} \right\}.$$

6.1. COHOMOLOGY OF SHEAVES

Since ψ is surjective, we have $\mathcal{M} \neq \varnothing$. Define $\langle t_1, V \rangle < \langle t_2, V_2 \rangle$ if $V_1 \subset V_2 \subset U$ and $t_1 = \rho_{V_1 V_2}(t_2)$. Then $\mathcal{M}$ is an ordered set. When $\{(t_\lambda, V_\lambda)\}_{\lambda \in \Lambda}$ is a totally ordered subset of $\mathcal{M}$ satisfying $(t_1, V_1) < (t_2, V_2) < \cdots$, let $V = \bigcup_{\lambda \in \Lambda} V_\lambda$. Since $V_\lambda \subset U$, we have $V \subset U$. Furthermore, for $\lambda < \mu$ we have $\rho_{V_\lambda V_\mu}(t_\mu) = t_\lambda$. Hence, there exists $t \in \mathcal{G}(U)$ to satisfy $\rho_{V_\lambda, V}(t) = t_\lambda$. Namely, we get $(t, V) \in \mathcal{M}$, and also for an arbitrary $\lambda \in \Lambda$ we have $(t_\lambda, V_\lambda) < (t, V)$. That is, $\mathcal{M}$ is an inductively ordered set. Therefore, Zorn's lemma implies that there is a maximal element $(\tilde{t}, \tilde{V})$ in $\mathcal{M}$. We will prove that $\tilde{V} = U$. Suppose $\tilde{V} \neq U$. Then for $x \in U \setminus \tilde{V}$, we can find an open neighborhood $V_x \subset U$ of x and $t_x \in \mathcal{G}(V_x)$ such that $\psi_{V_x}(t_x) = \rho_{V_x, U}(s)$. Namely, $(t_x, V_x) \in \mathcal{M}$. Then let

$$W = \tilde{V} \cap V_x \quad \text{and} \quad u = \rho_{W, V_x}(t_x) - \rho_{W, \tilde{V}}(\tilde{t}).$$

We get $\psi_W(u) = 0$. Hence, there exists $v \in \mathcal{F}(W)$ so that $\varphi_W(v) = u$. Since $\mathcal{F}$ is flabby, there is $\tilde{v} \in \mathcal{F}(V_x)$ satisfying $\rho_{W, V_x}(\tilde{v}) = v$. Then let $\tilde{t}_x = t_x - \varphi_{V_x}(\tilde{v})$. We have

$$\rho_{W, V_x}(\tilde{t}_x) = \rho_{W, V_x}(t_x) - \rho_{W, V_x}(\varphi_{V_x}(\tilde{v})) = \rho_{W, V_x}(t_x) - \varphi_W(\rho_{W, V_x}(\tilde{v}))$$
$$= \rho_{W, V_x}(t_x) - \varphi_W(v) = \rho_{W, V_x}(t_x) - u = \rho_{W, \tilde{V}}(\tilde{t}).$$

Therefore, there exists $\hat{t} \in \mathcal{G}(V_x \cup \tilde{V})$ so that

$$\rho_{V_x, V_x \cup \tilde{V}}(\hat{t}) = t_x \quad \text{and} \quad \rho_{\tilde{V}, V_x \cup \tilde{V}}(\hat{t}) = \hat{t}.$$

Consequently, $(\hat{t}, V_x \cup \tilde{V}) \in \mathcal{M}$, which contradicts the maximality of $(\tilde{t}, \tilde{V}) \in \mathcal{M}$. That is, $\tilde{V} = U$, i.e, ψ_U is a surjection.

(ii) For an open set U of X and for $s \in \mathcal{H}(U)$, the flabbiness of $\mathcal{F}$ in (i) implies that there is $t \in \mathcal{G}(U)$ satisfying $\psi_U(t) = s$. Since $\mathcal{G}$ is flabby, there exists $\tilde{t} \in \mathcal{G}(X)$ such that $\rho_{U,X}(\tilde{t}) = t$. Let $\tilde{s} = \psi_X(\tilde{t})$. Then we get

$$\rho_{U,X}(\tilde{s}) = \rho_{U,X}(\psi_X(\tilde{t})) = \psi_U(\rho_{U,X}(\tilde{t})) = \psi_U(t) = s.$$

Therefore, $\rho_{U,X} : \mathcal{H}(X) \to \mathcal{H}(U)$ is surjective, i.e., $\mathcal{H}$ is a flabby sheaf. $\square$

For a sheaf $\mathcal{F}$ over a topological space X, an exact sequence

(6.3) $\quad 0 \to \mathcal{F} \xrightarrow{\iota} \mathcal{G}^0 \xrightarrow{\delta^0} \mathcal{G}^1 \xrightarrow{\delta^1} \mathcal{G}^2 \xrightarrow{\delta^2} \mathcal{G}^3 \to \cdots$

is said to be a *flabby resolution* of $\mathcal{F}$ if all the $\mathcal{G}^j$, $j \geq 0$, are flabby sheaves.

EXAMPLE 6.3. For a sheaf $\mathcal{F}$ of additive groups over a topological space X, from Example 6.1, we have an exact sequence of sheaves of additive groups
$$0 \to \mathcal{F} \xrightarrow{\iota_\mathcal{F}} C^0(\mathcal{F}).$$
Recall that $C^0(\mathcal{F})$ is a flabby sheaf. Put
$$\mathcal{F}_1 = C^0(\mathcal{F})/\iota_\mathcal{F}(\mathcal{F}).$$
Then the sequence
(6.4) $$0 \to \mathcal{F} \xrightarrow{\iota_\mathcal{F}} C^0(\mathcal{F}) \xrightarrow{\eta_\mathcal{F}} \mathcal{F}_1 \to 0$$
is exact. Then, by repeating the same construction, we get the exact sequence
(6.5) $$0 \to \mathcal{F}_1 \xrightarrow{\iota_{\mathcal{F}_1}} C^0(\mathcal{F}_1) \xrightarrow{\eta_{\mathcal{F}_2}} \mathcal{F}_2 \to 0,$$
where $C^0(\mathcal{F}_1)$ is flabby and
$$\mathcal{F}_2 = C^0(\mathcal{F}_1)/\iota_{\mathcal{F}_1}(\mathcal{F}_1).$$
Define
$$\iota = \iota_\mathcal{F}, \quad \delta^* = \iota_{\mathcal{F}_1} \circ \eta_\mathcal{F}, \quad C^1(\mathcal{F}) = C^0(\mathcal{F}_1).$$
Then we get
(6.6) $$0 \to \mathcal{F} \xrightarrow{\iota} C^0(\mathcal{F}) \xrightarrow{\delta^0} C^1(\mathcal{F}).$$
Since $\iota_{\mathcal{F}_1}$ is injective, (6.4) implies
$$\operatorname{Ker} \delta^0 = \operatorname{Ker}(\iota_{\mathcal{F}_1} \circ \eta_\mathcal{F}) = \operatorname{Ker} \eta_\mathcal{F} = \operatorname{Im} \iota_\mathcal{F} = \operatorname{Im} \iota.$$
Namely, (6.6) is exact. Repeat the same process for $\mathcal{F}_2$ to obtain
(6.7) $$0 \to \mathcal{F}_2 \xrightarrow{\iota_{\mathcal{F}_2}} C^0(\mathcal{F}_2) \xrightarrow{\eta_{\mathcal{F}_2}} \mathcal{F}_3,$$
where $C^0(\mathcal{F}_2)$ is a flabby sheaf and $\mathcal{F}_3 = C^0(\mathcal{F}_2)/\iota_{\mathcal{F}_2}(\mathcal{F}_2)$. Then let
$$\delta^1 = \iota_{\mathcal{F}_2} \circ \eta_{\mathcal{F}_1} \quad \text{and} \quad C^2(\mathcal{F}) = C^0(\mathcal{F}_2).$$
We obtain
(6.8) $$0 \to \mathcal{F} \xrightarrow{\iota} C^0(\mathcal{F}) \xrightarrow{\delta^0} C^0(\mathcal{F}) \xrightarrow{\delta^1} C^2(\mathcal{F}).$$
Since $\iota_{\mathcal{F}_2}$ is injective, (6.5) implies that
$$\operatorname{Ker} \delta^1 = \operatorname{Ker} \eta_{\mathcal{F}_1} = \operatorname{Im} \iota_{\mathcal{F}_1}.$$
Furthermore, since $\eta_\mathcal{F}$ is surjective, (6.4) implies that
$$\operatorname{Im} \iota_{\mathcal{F}_1} = \operatorname{Im}(\iota_{\mathcal{F}_1} \circ \eta_\mathcal{F}) = \operatorname{Im} \delta^0.$$

6.1. COHOMOLOGY OF SHEAVES

That is, (6.8) is an exact sequence. Thus, we obtain a flabby resolution of $\mathcal{F}$ as follows:

$$(6.9) \quad 0 \to \mathcal{F} \xrightarrow{\iota} C^0(\mathcal{F}) \xrightarrow{\delta^0} C^1(\mathcal{F}) \xrightarrow{\delta^1} C^2(\mathcal{F}) \xrightarrow{\delta^2} \cdots$$
$$\cdots \xrightarrow{\delta^{n-1}} C^n(\mathcal{F}) \xrightarrow{\delta^n} C^{n+1}(\mathcal{F}) \to \cdots.$$

The resolution (6.9) is called the *canonical flabby resolution* of $\mathcal{F}$, sometimes written as $(C^\bullet(\mathcal{F}), \delta^\bullet)$. □

The canonical flabby resolution obtained in (6.9) has a desired property.

LEMMA 6.4. *For an exact sequence of additive groups*

$$0 \to \mathcal{F} \to \mathcal{G} \to \mathcal{H} \to 0,$$

the diagram induced by the canonical flabby resolutions of $\mathcal{F}, \mathcal{G}, \mathcal{H}$

$$\begin{array}{ccccccccc}
& & 0 & & 0 & & 0 & & 0 & & 0 \\
& & \downarrow & & \downarrow & & \downarrow & & \downarrow & & \downarrow \\
0 \to & \mathcal{F} & \to & C^0(\mathcal{F}) & \to & C^1(\mathcal{F}) & \to & C^2(\mathcal{F}) & \to & C^3(\mathcal{F}) & \to \cdots \\
& \downarrow & & \downarrow & & \downarrow & & \downarrow & & \downarrow \\
0 \to & \mathcal{G} & \to & C^0(\mathcal{G}) & \to & C^1(\mathcal{G}) & \to & C^2(\mathcal{G}) & \to & C^3(\mathcal{G}) & \to \cdots \\
& \downarrow & & \downarrow & & \downarrow & & \downarrow & & \downarrow \\
0 \to & \mathcal{H} & \to & C^0(\mathcal{H}) & \to & C^1(\mathcal{H}) & \to & C^2(\mathcal{H}) & \to & C^3(\mathcal{H}) & \to \cdots \\
& \downarrow & & \downarrow & & \downarrow & & \downarrow & & \downarrow \\
& & 0 & & 0 & & 0 & & 0 & & 0
\end{array}$$

is commutative, and also all the vertical and horizontal sequences are exact sequences.

PROOF. First we will show that

$$(6.10) \qquad 0 \to C^0(\mathcal{F}) \to C^0(\mathcal{G}) \to C^0(\mathcal{H}) \to 0$$

is an exact sequence. Since, for an arbitrary point $x \in X$,

$$0 \to \mathcal{F}_x \to \mathcal{G}_x \to \mathcal{H}_x \to 0$$

is exact, the definition (6.1) implies that the sequence

$$0 \to C^0(\mathcal{F})(U) \to C^0(\mathcal{G})(U) \to C^0(\mathcal{H})(U) \to 0$$

is an exact sequence for any open set U. Therefore, (6.10) is an exact sequence. Notice that the diagram

$$\begin{array}{ccccccc}
& & 0 & & 0 & & 0 \\
& & \downarrow & & \downarrow & & \downarrow \\
0 & \longrightarrow & \mathcal{F} & \longrightarrow & \mathcal{G} & \longrightarrow & \mathcal{H} & \longrightarrow & 0 \\
& & {\scriptstyle \iota_\mathcal{F}}\downarrow & & {\scriptstyle \iota_\mathcal{G}}\downarrow & & {\scriptstyle \iota_\mathcal{H}}\downarrow & & \\
0 & \longrightarrow & C^0(\mathcal{F}) & \longrightarrow & C^0(\mathcal{G}) & \longrightarrow & C^0(\mathcal{H}) & \longrightarrow & 0
\end{array}$$

is commutative. This commutative diagram implies the diagram

$$\begin{array}{ccccccc}
& & 0 & & 0 & & 0 \\
& & \downarrow & & \downarrow & & \downarrow \\
0 & \longrightarrow & \mathcal{F} & \longrightarrow & \mathcal{G} & \longrightarrow & \mathcal{H} & \longrightarrow & 0 \\
& & {\scriptstyle \iota_\mathcal{F}}\downarrow & & {\scriptstyle \iota_\mathcal{G}}\downarrow & & {\scriptstyle \iota_\mathcal{H}}\downarrow & & \\
0 & \longrightarrow & C^0(\mathcal{F}) & \longrightarrow & C^0(\mathcal{G}) & \longrightarrow & C^0(\mathcal{H}) & \longrightarrow & 0 \\
& & \downarrow & & \downarrow & & \downarrow \\
0 & \longrightarrow & \mathcal{F}_1 & \longrightarrow & \mathcal{G}_1 & \longrightarrow & \mathcal{H}_1 & \longrightarrow & 0 \\
& & \downarrow & & \downarrow & & \downarrow \\
& & 0 & & 0 & & 0
\end{array}$$

where all the vertical and horizontal sequences are exact. For the bottom horizontal exact sequence, as before we also get the exact sequence

$$0 \to C^0(\mathcal{F}_1) \to C^0(\mathcal{G}_1) \to C^0(\mathcal{H}_1) \to 0.$$

By repeating this construction, we obtain Lemma 6.4. □

EXERCISE 6.5. Let

(6.11) $\qquad 0 \to \mathcal{F} \xrightarrow{f} \mathcal{G}_0 \xrightarrow{g_0} \mathcal{G}_1 \xrightarrow{g_1} \mathcal{G}_2 \xrightarrow{g_2} \mathcal{G}_3 \to \cdots$

be a flabby resolution of a flabby sheaf, i.e., all the sheaves in (6.11) are flabby, and the sequence (6.11) is exact. Then for any open subset U, the sequence

$$0 \to \Gamma(U, \mathcal{F}) \to \Gamma(U, \mathcal{G}_0) \to \Gamma(U, \mathcal{G}_1) \to \Gamma(U, \mathcal{G}_2) \to \cdots$$

is exact.

PROOF. The exact sequence (6.11) gives an exact sequence

$$0 \to \mathcal{F} \to \mathcal{G}_0 \to \operatorname{Im} g_0 \to 0.$$

Since $\mathcal{F}$ and $\mathcal{G}$ are flabby, Proposition 6.2(ii) implies that $\mathrm{Im}\, g_0$ is a flabby sheaf. By Proposition 6.2(i), the sequence

(6.12) $\qquad 0 \to \Gamma(U, \mathcal{F}) \to \Gamma(U, \mathcal{G}) \to \Gamma(U, \mathrm{Im}\, g_0) \to 0$

is exact. The sequence (6.11) also implies that

$$0 \to \mathrm{Im}\, g_0 \to \mathcal{G}_1 \to \mathrm{Coker}\, g_0 \to 0$$

is exact. Since $\mathrm{Im}\, g_0$ and $\mathcal{G}_1$ are flabby, $\mathrm{Coker}\, g_0$ is a flabby sheaf. Once again, we get the exact sequence

(6.13) $\qquad 0 \to \Gamma(U, \mathrm{Im}\, g_0) \to \Gamma(U, \mathcal{G}_1) \to \Gamma(U, \mathrm{Coker}\, g_0) \to 0.$

Then (6.12) and (6.13) imply that we get

(6.14) $\qquad 0 \to \Gamma(U, \mathcal{F}) \to \Gamma(U, \mathcal{G}_0) \to \Gamma(U, \mathcal{G}_1).$

The exactness of (6.12) and (6.13) shows that (6.14) is an exact sequence.

Moreover, since $\mathrm{Im}\, g_0 = \mathrm{Ker}\, g_1$, we have $\mathrm{Coker}\, g_0 = \mathcal{G}_1/\mathrm{Im}\, g_0 = \mathcal{G}_1/\mathrm{Ker}\, g_1$. Therefore, by (6.11), we get the exactness of

$$0 \to \mathrm{Coker}\, g_0 \to \mathcal{G}_2 \to \mathrm{Im}\, g_2 \to 0.$$

Since $\mathrm{Coker}\, g_0$ and $\mathcal{G}_2$ are flabby, $\mathrm{Im}\, g_2$ is also a flabby sheaf. Hence we have the exact sequence

(6.15) $\qquad 0 \to \Gamma(U, \mathrm{Coker}\, g_2) \to \Gamma(U, \mathcal{G}_2) \to \Gamma(U, \mathrm{Im}\, g_2) \to 0.$

By (6.13), (6.14) and (6.15), we get

(6.16) $\qquad 0 \to \Gamma(U, \mathcal{F}) \to \Gamma(U, \mathcal{G}_0) \to \Gamma(U, \mathcal{G}_1) \to \Gamma(U, \mathcal{G}_2).$

The exactness of (6.14) together with the exactness of (6.13) and (6.15) implies that (6.16) is exact. By repeating this construction, the exactness in Exercise 6.5 will follow. $\square$

PROBLEM 1. For an exact sequence of sheaves

$$0 \to \mathcal{F}_1 \xrightarrow{f_1} \mathcal{F}_2 \xrightarrow{f_2} \mathcal{F}_3 \xrightarrow{f_3} \mathcal{F}_4 \to \cdots,$$

the canonical flabby resolutions $(C^\bullet(\mathcal{F}_j), f^\bullet)$ of $\mathcal{F}_j$, $j = 1, 2, 3, \ldots$, give the following commutative diagram:

$$\begin{array}{ccccccccc}
& & 0 & & 0 & & 0 & & 0 & & 0 \\
& & \downarrow & & \downarrow & & \downarrow & & \downarrow & & \downarrow \\
0 \to & \mathcal{F}_1 & \to & C^0(\mathcal{F}_1) & \to & C^1(\mathcal{F}_1) & \to & C^2(\mathcal{F}_1) & \to & C^3(\mathcal{F}_1) & \to \cdots \\
& \downarrow & & \downarrow & & \downarrow & & \downarrow & & \downarrow \\
0 \to & \mathcal{F}_2 & \to & C^0(\mathcal{F}_2) & \to & C^1(\mathcal{F}_2) & \to & C^2(\mathcal{F}_2) & \to & C^3(\mathcal{F}_2) & \to \cdots \\
& \downarrow & & \downarrow & & \downarrow & & \downarrow & & \downarrow \\
0 \to & \mathcal{F}_3 & \to & C^0(\mathcal{F}_3) & \to & C^1(\mathcal{F}_3) & \to & C^2(\mathcal{F}_3) & \to & C^3(\mathcal{F}_3) & \to \cdots \\
& \downarrow & & \downarrow & & \downarrow & & \downarrow & & \downarrow \\
0 \to & \mathcal{F}_4 & \to & C^0(\mathcal{F}_4) & \to & C^1(\mathcal{F}_4) & \to & C^2(\mathcal{F}_4) & \to & C^3(\mathcal{F}_4) & \to \cdots \\
& \downarrow & & \downarrow & & \downarrow & & \downarrow & & \downarrow \\
& \vdots & & \vdots & & \vdots & & \vdots & & \vdots
\end{array}$$

where all the vertical and horizontal sequences are exact.

(b) **Cohomology Group**

As we saw in Example 6.3, for a sheaf $\mathcal{F}$ of additive groups over a topological space X, there exists a flabby resolution of $\mathcal{F}$

(6.17) $\qquad 0 \to \mathcal{F} \xrightarrow{\iota} \mathcal{G}^0 \xrightarrow{\delta^0} \mathcal{G}^1 \xrightarrow{\delta^1} \mathcal{G}^2 \xrightarrow{\delta^2} \mathcal{G}^3 \to \cdots.$

From this exact sequence we get a sequence of additive groups

(6.18)
$$0 \to \Gamma(X, \mathcal{G}^0) \xrightarrow{\delta_X^0} \Gamma(X, \mathcal{G}^1) \xrightarrow{\delta_X^1} \Gamma(X, \mathcal{G}^2) \xrightarrow{\delta_X^2} \Gamma(X, \mathcal{G}^3) \to \cdots.$$

Since (6.17) is exact, we have at least $\delta^{n+1} \circ \delta^n = 0$, $n = 0, 1, 2, \ldots$. Hence we get

$$\delta_X^{n+1} \circ \delta_X^n = 0, \quad n = 0, 1, 2, \ldots.$$

Namely, $(A^n = \Gamma(X, \mathcal{G}^n), \delta_X^n)$ is a complex of additive groups. We will define cohomology groups measuring how far the sequence (6.18) is from being exact. That is, we define

(6.19) $\qquad H^n(X, \mathcal{F}) = \operatorname{Ker} \delta_X^n / \operatorname{Im} \delta_X^{n-1}, \quad n = 0, 1, 2, \ldots,$

where we let $\delta_X^{-1} = 0$. The left-hand side of (6.19) may appear unreasonable since for the complex $\mathcal{D} = (A^n = \Gamma(X, \mathcal{G}^n), \delta_X^n)$, the cohomology group should be written as $H^n(\mathcal{D})$. However, we will show that the right-hand side of (6.19) is independent of the choice of any flabby resolution of $\mathcal{F}$. This fact justifies the notation of the

6.1. COHOMOLOGY OF SHEAVES

cohomology group, and $H^n(X, \mathcal{F})$ is said to be the *n-th cohomology group of the sheaf* $\mathcal{F}$.

From the exactness of (6.17), we have the exact sequence
$$0 \to \Gamma(X, \mathcal{F}) \xrightarrow{\delta_X} \Gamma(X, \mathcal{G}^0) \xrightarrow{\delta_X^0} \Gamma(X, \mathcal{G}^1).$$
This is because the exact sequences of sheaves
$$\begin{array}{ccccccccc} 0 & \to & \mathcal{F} & \to & \mathcal{G}^0 & \to & \operatorname{Im} \delta^0 & \to & 0, \\ 0 & \to & \operatorname{Im} \delta^0 & \to & \mathcal{G}^1 & \to & \operatorname{Coker} \delta^0 & \to & 0, \end{array}$$
induced by (6.17), imply the following exact sequences of additive groups:
$$\begin{array}{ccccc} 0 & \to & \Gamma(X, \mathcal{F}) & \to & \mathcal{F}(X, \mathcal{G}^0) & \to & \Gamma(X, \operatorname{Im} \delta^0), \\ 0 & \to & \Gamma(X, \operatorname{Im} \delta^0) & \to & \Gamma(X, \mathcal{G}^1) & \to & \Gamma(X, \operatorname{Coker} \delta^0). \end{array}$$
Then the composition of maps
$$\Gamma(X, \mathcal{G}^0) \to \Gamma(X, \operatorname{Im} \delta^0) \to \Gamma(X, \mathcal{G}^1)$$
is exactly δ_X^0, i.e.,
$$\operatorname{Ker} \delta_X^0 = \operatorname{Im} \iota.$$
Hence (6.19) implies that
$$H^0(X, \mathcal{F}) = \operatorname{Ker} \delta_X^0 \xrightarrow{\sim} \Gamma(X, \mathcal{F}).$$
By identifying the isomorphic groups, we obtain the following.

LEMMA 6.6.
$$H^0(X, \mathcal{F}) = \Gamma(X, \mathcal{F}). \quad \square$$

LEMMA 6.7. *For a flabby sheaf* $\mathcal{F}$, *we have*
$$H^n(X, \mathcal{F}) = 0, \qquad n = 1, 2, 3, \ldots.$$

PROOF. For a flabby resolution of $\mathcal{F}$
$$0 \to \mathcal{F} \xrightarrow{\iota} \mathcal{G}^0 \xrightarrow{\delta^0} \mathcal{G}^1 \xrightarrow{\delta^1} \mathcal{G}^2 \xrightarrow{\delta^2} \cdots,$$
by Exercise 6.5 we get the exact sequence
$$0 \to \Gamma(X, \mathcal{F}) \to \Gamma(X, \mathcal{G}^0) \xrightarrow{\delta_X^0} \Gamma(X, \mathcal{G}^1) \xrightarrow{\delta_X^1} \Gamma(X, \mathcal{G}^2) \xrightarrow{\delta_X^2} \cdots.$$
Therefore, for $n \geq 1$ we have
$$\operatorname{Ker} \delta_X^n = \operatorname{Im} \delta_X^{n-1}.$$
Namely, $H^n(X, \mathcal{F}) = 0$ for $n \geq 1$. $\quad \square$

120   6. COHOMOLOGY OF COHERENT SHEAVES

We are ready to prove that the cohomology group does not depend upon the choice of a flabby resolution.

THEOREM 6.8. *The cohomology group $H^n(X, \mathcal{F})$ is uniquely determined by $\mathcal{F}$ and X, independently of the particular choice of a flabby resolution of $\mathcal{F}$.*

PROOF. Let

$$0 \to \mathcal{F} \to \mathcal{G}^0 \to \mathcal{G}^1 \to \mathcal{G}^2 \to \mathcal{G}^3 \to \cdots$$

be a flabby resolution of $\mathcal{F}$ and let

$$0 \to \mathcal{F} \to C^0(\mathcal{F}) \to C^1(\mathcal{F}) \to C^2(\mathcal{F}) \to C^3(\mathcal{F}) \to \cdots$$

be the canonical flabby resolution of $\mathcal{F}$. We will show that the cohomology groups defined by the flabby resolution and by the canonical flabby resolution are isomorphic. From Problem 1, we have the following commutative diagram consisting of exact sequences:

$$\begin{array}{ccccccccc}
& & 0 & & 0 & & 0 & & 0 \\
& & \downarrow & & \downarrow & & \downarrow & & \downarrow \\
0 & \to & \mathcal{F} & \to & \mathcal{G}^0 & \to & \mathcal{G}^1 & \to & \mathcal{G}^2 & \to \cdots \\
& & \downarrow & & \downarrow & & \downarrow & & \downarrow \\
0 & \to & C^0(\mathcal{F}) & \to & C^0(\mathcal{G}^0) & \to & C^0(\mathcal{G}^1) & \to & C^0(\mathcal{G}^2) & \to \cdots \\
& & \downarrow & & \downarrow & & \downarrow & & \downarrow \\
0 & \to & C^1(\mathcal{F}) & \to & C^1(\mathcal{G}^0) & \to & C^1(\mathcal{G}^1) & \to & C^1(\mathcal{G}^2) & \to \cdots \\
& & \downarrow & & \downarrow & & \downarrow & & \downarrow \\
0 & \to & C^2(\mathcal{F}) & \to & C^2(\mathcal{G}^0) & \to & C^2(\mathcal{G}^1) & \to & C^2(\mathcal{G}^2) & \to \cdots \\
& & \downarrow & & \downarrow & & \downarrow & & \downarrow \\
& & \vdots & & \vdots & & \vdots & & \vdots
\end{array}$$

The above diagram induces the commutative diagram of additive groups

$$\begin{array}{ccccccccc}
& & 0 & & 0 & & 0 & & \\
& & \downarrow & & \downarrow & & \downarrow & & \\
0 & \longrightarrow & \Gamma(X, \mathcal{G}^0) & \xrightarrow{\delta_X^0} & \Gamma(X, \mathcal{G}^1) & \xrightarrow{\delta_X^1} & \Gamma(X, \mathcal{G}^2) & \longrightarrow & \cdots \\
& & \downarrow & & \downarrow & & \downarrow & & \\
0 \longrightarrow \Gamma(X, C^0(\mathcal{F})) & \longrightarrow & \Gamma(X, C^0(\mathcal{G}^0)) & \longrightarrow & \Gamma(X, C^0(\mathcal{G}^1)) & \longrightarrow & \Gamma(X, C^0(\mathcal{G}^2)) & \longrightarrow & \cdots \\
d_X^0 \downarrow & & \downarrow & & \downarrow & & \downarrow & & \\
0 \longrightarrow \Gamma(X, C^1(\mathcal{F})) & \longrightarrow & \Gamma(X, C^1(\mathcal{G}^0)) & \longrightarrow & \Gamma(X, C^1(\mathcal{G}^1)) & \longrightarrow & \Gamma(X, C^1(\mathcal{G}^2)) & \longrightarrow & \cdots \\
d_X^1 \downarrow & & \downarrow & & \downarrow & & \downarrow & & \\
0 \longrightarrow \Gamma(X, C^2(\mathcal{F})) & \longrightarrow & \Gamma(X, C^2(\mathcal{G}^0)) & \longrightarrow & \Gamma(X, C^2(\mathcal{G}^1)) & \longrightarrow & \Gamma(X, C^2(\mathcal{G}^2)) & \longrightarrow & \cdots \\
& & \downarrow & & \downarrow & & \downarrow & & \\
& & \vdots & & \vdots & & \vdots & &
\end{array}$$

Then, by Exercise 6.5, all the rows and columns are exact sequences except the first row and the first column. For simplicity, let $A^n = \Gamma(X, \mathcal{G}^n)$, $B^n = \Gamma(X, C^n(\mathcal{F}))$ and $C^{i,j} = \Gamma(X, C^i(\mathcal{G}^j))$. Consider the commutative diagram

(6.20)

$$\begin{array}{ccccccccccc}
& & 0 & & 0 & & 0 & & 0 & & \\
& & \downarrow & & \downarrow & & \downarrow & & \downarrow & & \\
& 0 \longrightarrow & A^0 & \xrightarrow{\delta^0} & A^1 & \xrightarrow{\delta^1} & A^2 & \xrightarrow{\delta^2} & A^3 & \longrightarrow & \cdots \\
& & e_0 \downarrow & & e^1 \downarrow & & e^2 \downarrow & & e^3 \downarrow & & \\
0 \longrightarrow & B^0 \xrightarrow{\varepsilon^0} & C^{0,0} & \xrightarrow{\delta^{0,0}} & C^{0,1} & \xrightarrow{\delta^{0,1}} & C^{0,2} & \xrightarrow{d^{0,2}} & C^{0,3} & \longrightarrow & \cdots \\
& d^0 \downarrow & d^{0,0} \downarrow & & d^{0,1} \downarrow & & d^{0,2} \downarrow & & d^{0,3} \downarrow & & \\
0 \longrightarrow & B^1 \xrightarrow{\varepsilon^1} & C^{1,0} & \xrightarrow{\delta^{1,0}} & C^{1,1} & \xrightarrow{\delta^{1,1}} & C^{1,2} & \xrightarrow{d^{1,2}} & C^{1,3} & \longrightarrow & \cdots \\
& d^1 \downarrow & d^{1,0} \downarrow & & d^{1,1} \downarrow & & d^{1,2} \downarrow & & d^{1,3} \downarrow & & \\
0 \longrightarrow & B^2 \xrightarrow{\varepsilon^2} & C^{2,0} & \xrightarrow{\delta^{2,0}} & C^{2,1} & \xrightarrow{\delta^{2,1}} & C^{2,2} & \xrightarrow{d^{2,2}} & C^{2,3} & \longrightarrow & \cdots \\
& d^2 \downarrow & d^{2,0} \downarrow & & d^{2,1} \downarrow & & d^{2,2} \downarrow & & d^{2,3} \downarrow & & \\
0 \longrightarrow & B^3 \xrightarrow{\varepsilon^3} & C^{3,0} & \xrightarrow{\delta^{3,0}} & C^{3,1} & \xrightarrow{\delta^{3,1}} & C^{3,2} & \xrightarrow{d^{3,2}} & C^{3,3} & \longrightarrow & \cdots \\
& \downarrow & \downarrow & & \downarrow & & \downarrow & & \downarrow & & \\
& \vdots & \vdots & & \vdots & & \vdots & & \vdots & &
\end{array}$$

where, except the first row $0 \to A^0 \to A^1 \to \cdots$ and the first column $0 \to B^0 \to B^1 \to B^2 \to \cdots$, all the row and column sequences are exact. What we need to prove is the existence of an isomorphism

(6.21) $\quad f^n : \operatorname{Ker} d^n / \operatorname{Ker} d^{n-1} \overset{\sim}{\to} \operatorname{Ker} \delta^n / \operatorname{Ker} \delta^{n-1}, \quad n \geq 0,$

where $d^{-1} = 0$ and $\delta^{-1} = 0$. The isomorphism (6.21) can be proved easily by the double complex method. Note that in order to regard (6.20) as a double complex, the maps $d^{p,q}$ need to be redefined as $d'^{p,q} = (-1)^p d^{p,q}$ so that one may get $d'^{p,q} \circ \delta^{p,q} + \delta^{p+1,q} \circ d'^{p,q-1} = 0$ instead of $d^{p,q} \circ \delta^{p,q-1} = \delta^{p+1,q-1} \circ d^{p,q-1}$. We will give an elementary proof for the isomorphism in (6.21). One can observe from Lemma 6.6 that there is an isomorphism $\operatorname{Ker} d^0 \overset{\sim}{\to} \operatorname{Ker} \delta^0$ ($\overset{\sim}{\to} \Gamma(X, \mathcal{F})$). We will construct the isomorphism using diagram (6.20):

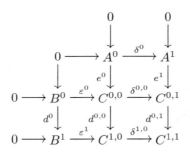

Let $b \in B^0$ satisfy $d^0(b) = 0$, and let $\varepsilon^0(b) = c$. Then

$$d^{0,0}(c) = d^{0,0}(\varepsilon^0(b)) = (\varepsilon^1 \circ d^0)(b) = \varepsilon^1(d^0(b)) = 0.$$

Since $\operatorname{Ker} d^{0,0} = \operatorname{Im} e^0$, there exists $a \in A^0$ to satisfy $e^0(a) = c$. Then we have $e^1(\delta^0(a)) = \delta^{0,0}(e^0(a)) = \delta^{0,0}(c) = \delta^{0,0}(\varepsilon^0(b)) = 0$. Since e^1 is injective, we get $\delta^0(a) = 0$, i.e., $a \in \operatorname{Ker} \delta^0$. Next we will show that for b, a is uniquely determined. Notice that from the above construction c is uniquely determined. Since e^0 is injective, an a satisfying $e^0(a) = c$ is uniquely determined. That is, we obtain a map $f^0 : \operatorname{Ker} d^0 \to \operatorname{Ker} \delta^0$. Since $\varepsilon^j, e^j, \delta^{i,j}, d^{i,j}$ are all homomorphisms of additive groups, f^0 is a homomorphism. As in the above, for $a \in \operatorname{Ker} \delta^0$, we get $b \in \operatorname{Ker} d^0$, obtaining a homomorphism $g^0 : \operatorname{Ker} \delta^0 \to \operatorname{Ker} d^0$. One can easily verify that $g^0 \circ f^0 = \operatorname{id}$ and $f^0 \circ g^0 = \operatorname{id}$. Hence

6.1. COHOMOLOGY OF SHEAVES

f^0 is an isomorphism. Now consider the diagram

$$\begin{array}{ccccccc}
 & & A^0 & \xrightarrow{\delta^0} & A^1 & \xrightarrow{\delta^1} & A^2 \\
 & & \downarrow{e^0} & & \downarrow{e^1} & & \downarrow{e^2} \\
B^0 & \xrightarrow{\varepsilon^0} & C^{0,0} & \xrightarrow{\delta^{0,0}} & C^{0,1} & \longrightarrow & C^{0,2} \\
\downarrow{d^0} & & \downarrow{d^{0,0}} & & \downarrow{d^{0,1}} & & \\
B^1 & \xrightarrow{\varepsilon^1} & C^{1,0} & \xrightarrow{\delta^{1,0}} & C^{1,1} & & \\
\downarrow{d^1} & & \downarrow{d^{1,0}} & & & & \\
B^2 & \xrightarrow{\varepsilon^2} & C_{2,0} & & & &
\end{array}$$

Let $b \in \operatorname{Ker} d^1$ and let $c^{1,0} = \varepsilon^1(b)$. Since we have
$$d^{1,0}(c^{1,0}) = d^{1,0}(\varepsilon^1(b)) = \varepsilon^2(d^1(b)) = 0$$
and
$$\operatorname{Ker} d^{1,0} = \operatorname{Im} d^{0,0},$$
there exists $c^{0,0} \in C^{0,0}$ satisfying $d^{0,0}(c^{0,0}) = c^{1,0}$. Let $d^{0,0}(\tilde{c}^{0,0}) = c^{1,0}$. We have $c^{0,0} - \tilde{c}^{0,0} \in \operatorname{Im} e^0$. Hence we can write
$$\tilde{e}^{0,0} = c^{0,0} + e^0(a^0), \qquad a^0 \in A^0.$$
Let
$$c^{0,1} = \delta^{0,0}(c^{0,0}) \quad \text{and} \quad \tilde{c}^{0,1} = \delta^{0,0}(\tilde{c}^{0,0}).$$
The commutativity of the above diagram implies
$$d^{0,1}(c^{0,1}) = d^{0,1}(\delta^{0,0}(c^{0,0})) = \delta^{1,0}(d^{0,0}(c^{0,0}))$$
$$= \delta^{1,0}(c^{1,0}) = \delta^{1,0}(\varepsilon^1(b)) = 0.$$
Hence there is an element $a \in A^1$ to satisfy $e^1(a) = c^{0,1}$. Similarly, we get $d^{0,1}(\tilde{c}^{0,1}) = 0$. We have an element $\tilde{a} \in A^1$ to satisfy $e^1(\tilde{a}) = c^{0,1}$. Since e^1 is injective, we get
$$\tilde{a} = a + \delta^0(a^0).$$
Furthermore, we have
$$e^2(\delta^1(a)) = \delta^{0,1}(e^1(a)) = \delta^{0,1}(c^{0,1}) = \delta^{0,1}(\delta^{0,0}(c^{0,0})) = 0.$$
Since e^2 is injective, we get $\delta^1(a) = 0$, i.e., $a \in \operatorname{Ker} \delta^1$. Similarly, we get $\tilde{a} \in \operatorname{Ker} \delta^1$. Note that $b \in \operatorname{Ker} d^1$ does not uniquely determine $a \in \operatorname{Ker} \delta^1$, and $a + a'$, $a' \in \operatorname{Im} \delta^0$, also satisfies the condition. However, the homomorphism of additive groups
$$\operatorname{Ker} d^1 \to \operatorname{Ker} \delta^1 / \operatorname{Im} \delta^0$$

is uniquely determined. As in the above, for $b \in \operatorname{Im} d^0$, we have $a \in \operatorname{Im} \delta^0$. Namely, we obtain the homomorphism

$$f' : \operatorname{Ker} d^1 / \operatorname{Im} d^0 \to \operatorname{Ker} \delta^1 / \operatorname{Im} \delta^0.$$

Conversely, for a given $a \in \operatorname{Ker} \delta^1$, we obtain

$$g' : \operatorname{Ker} \delta^1 / \operatorname{Im} \delta^0 \to \operatorname{Ker} d^1 / \operatorname{Im} d^0.$$

Then one can confirm that

$$g' \circ f' = \operatorname{id} \quad \text{and} \quad f' \circ g' = \operatorname{id},$$

i.e., f' is an isomorphism.

For the case $n \geq 2$, a similar argument as above implies that f^n is an isomorphism. In diagram (6.20), for $b \in \operatorname{Ker} d^0$, consider

$$c^{n,0} = \varepsilon^n(b) \in C^{n,0}, \quad c^{n-1,1} \in C^{n-1,1}, \ldots.$$

The reader is expected to complete the proof. $\square$

This theorem indicates that the cohomology groups $H^n(X, \mathcal{F})$ can be computed by any flabby resolution of $\mathcal{F}$. For example, if $\mathcal{F}$ is a flabby sheaf, then the following flabby resolution can be used:

$$0 \to \mathcal{F} \to \mathcal{F} \to 0 \to 0 \to \cdots.$$

From this resolution, we get

$$H^n(X, \mathcal{F}) = \begin{cases} \Gamma(X, \mathcal{F}), & n = 0, \\ 0, & n \geq 1. \end{cases}$$

PROBLEM 2. Let Y be a closed subspace of a topological space X. Let $\iota : Y \to X$ be the natural map induced by $Y \subset X$. For a flabby sheaf $\mathcal{G}$ over Y, prove that $\iota_* \mathcal{G}$ is a flabby sheaf over X.

PROBLEM 3. Using the same notation as in Problem 2, prove that there is an isomorphism

$$H^n(Y, \mathcal{F}) \xrightarrow{\sim} H^n(X, \iota_* \mathcal{F}),$$

where $\mathcal{F}$ is a sheaf over Y.

Theorem 6.8 implies a very useful property of the sheaf cohomology theory.

THEOREM 6.9. *For an exact sequence of sheaves of additive groups over a topological space X*

$$0 \to \mathcal{F} \xrightarrow{\varphi} \mathcal{G} \xrightarrow{\psi} \mathcal{H} \to 0,$$

6.1. COHOMOLOGY OF SHEAVES

we have the induced long exact sequence of cohomology groups

$$0 \to H^0(X,\mathcal{F}) \to H^0(X,\mathcal{G}) \to H^0(X,\mathcal{H}) \to H^1(X,\mathcal{F}) \to$$
$$\to H^1(X,\mathcal{G}) \to H^1(X,\mathcal{H}) \to H^2(X,\mathcal{F}) \to H^2(X,\mathcal{G}) \to$$
$$\to H^2(X,\mathcal{H}) \to H^3(X,\mathcal{F}) \to H^3(X,\mathcal{G}) \to \cdots.$$

PROOF. By Lemma 6.4, the canonical flabby resolutions of $\mathcal{F}, \mathcal{G}$ and $\mathcal{H}$ give the following commutative diagram of exact sequences:

(6.22)
$$\begin{array}{ccccccc}
& & 0 & & 0 & & 0 \\
& & \downarrow & & \downarrow & & \downarrow \\
0 & \to & \mathcal{F} & \to & \mathcal{G} & \to & \mathcal{H} & \to 0 \\
& & \downarrow & & \downarrow & & \downarrow \\
0 & \to & C^0(\mathcal{F}) & \to & C^0(\mathcal{G}) & \to & C^0(\mathcal{H}) & \to 0 \\
& & \downarrow & & \downarrow & & \downarrow \\
0 & \to & C^1(\mathcal{F}) & \to & C^1(\mathcal{G}) & \to & C^1(\mathcal{H}) & \to 0 \\
& & \downarrow & & \downarrow & & \downarrow \\
0 & \to & C^2(\mathcal{F}) & \to & C^2(\mathcal{G}) & \to & C^2(\mathcal{H}) & \to 0 \\
& & \downarrow & & \downarrow & & \downarrow \\
& & \vdots & & \vdots & & \vdots
\end{array}$$

For the sake of simplicity, put

$$F^n = \Gamma(X, C^n(\mathcal{F})), \quad G^n = \Gamma(X, C^n(\mathcal{G})), \quad H^n = \Gamma(X, C^n(\mathcal{H})).$$

From (6.22) we get the commutative diagram

(6.23)
$$\begin{array}{ccccccc}
& & 0 & & 0 & & 0 \\
& & \downarrow & & \downarrow & & \downarrow \\
0 & \to & F^0 & \xrightarrow{\varphi_0} & G^0 & \xrightarrow{\psi_0} & H^0 & \to 0 \\
& & d_F^0 \downarrow & & d_G^0 \downarrow & & d_H^0 \downarrow \\
0 & \to & F^1 & \xrightarrow{\varphi_1} & G^1 & \xrightarrow{\psi_1} & H^1 & \to 0 \\
& & d_F^1 \downarrow & & d_G^1 \downarrow & & d_H^1 \downarrow \\
0 & \to & F^2 & \xrightarrow{\varphi_2} & G^2 & \xrightarrow{\psi_2} & H^2 & \to 0 \\
& & d_F^2 \downarrow & & d_G^2 \downarrow & & d_H^2 \downarrow \\
0 & \to & F^3 & \xrightarrow{\varphi_3} & G^3 & \xrightarrow{\psi_3} & H^3 & \to 0 \\
& & \downarrow & & \downarrow & & \downarrow \\
& & \vdots & & \vdots & & \vdots
\end{array}$$

All the rows of (6.23) are exact sequences. For $f \in \operatorname{Ker} d_F^n$, we have
$$d_G^n(\varphi_n(f)) = \varphi_{n+1}(d_F^n(f)) = 0.$$
Namely, $\varphi_n(f) \in \operatorname{Ker} d_G^n$. Since we have
$$\varphi_n(d_F^{n-1}(a)) = d_G^{n-1}(\varphi_{n-1}(a)),$$
we also get
$$\varphi_n(\operatorname{Im} d_F^{n-1}) \subset \operatorname{Im} d_G^{n-1}.$$
Hence one can define a homomorphism
$$\overline{\varphi}_n : H^n(X, \mathcal{F}) = \operatorname{Ker} d_F^n / \operatorname{Im} d_F^{n-1} \to H^n(X, \mathcal{G}) = \operatorname{Ker} d_G^n / \operatorname{Im} d_G^{n-1}.$$
Similarly,
$$\overline{\psi}_n : H^n(X, \mathcal{G}) = \operatorname{Ker} d_G^n / \operatorname{Im} d_G^{n-1} \to H^n(X, \mathcal{H}) = \operatorname{Ker} d_H^n / \operatorname{Im} d_H^{n-1}.$$
Since $\psi_n \circ \varphi_n = 0$, we get $\overline{\psi}_n \circ \overline{\varphi}_n = 0$. Let $[g]$ be the element of $H^n(X, \mathcal{G})$ determined by $g \in \operatorname{Ker} d_G^n$. Suppose $\overline{\psi}_n([g]) = 0$. Namely, $\psi_n(g) \in \operatorname{Im} d_H^{n-1}$. Hence there exists $h \in H^{n-1}$ satisfying $\psi_n(g) = d_H^{n-1}(h)$. Since all the horizontal sequences in (6.23)

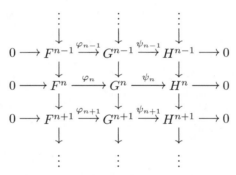

are exact, there exists $g_{n-1} \in G^{n-1}$ to satisfy $h = \psi_{n-1}(g_{n-1})$. Let $\overline{g} = g - d_G^{n-1}(g_{n-1})$. Then $\tilde{g} \in \operatorname{Ker} d_G^n$, and $[\tilde{g}] = [g]$. On the other hand, we have
$$\psi_n(\tilde{g}) = \psi_n(g) - \psi_n(d_G^{n-1}(g_{n-1})) = \psi_n(g) - d_H^{n-1}(\psi_{n-1}(g_{n-1}))$$
$$= \psi_n(g) - d_H^{n-1}(h) = 0.$$
Hence, there exists $f \in F^n$ such that $\varphi_n(f) = \tilde{g}$. Since $\varphi_{n+1}(d_F^n(f)) = d_G^n(\varphi_n(f)) = d_G^n(\tilde{g}) = 0$ and φ_{n+1} is injective, we get $f \in \operatorname{Ker} d_F^n$. Therefore, we have $\overline{\varphi}_n([f]) = [g]$. Consequently, for the sequence
$$H^n(X, \mathcal{F}) \xrightarrow{\overline{\varphi}_n} H^n(X, \mathcal{G}) \xrightarrow{\overline{\psi}_n} H^n(X, \mathcal{H})$$

6.1. COHOMOLOGY OF SHEAVES

we have proved that

(6.24) $$\operatorname{Ker} \overline{\psi}_n = \operatorname{Im} \overline{\varphi}_n.$$

In particular, $\overline{\varphi}_0$ coincides with $\Gamma(X, \mathcal{F}) \xrightarrow{\varphi_X} \Gamma(X, \mathcal{G})$, and $\overline{\varphi}_0$ is injective.

Next we will show that there exists a homomorphism

(6.25) $$\delta^n : H^n(X, \mathcal{H}) \to H^{n+1}(X, \mathcal{F}).$$

For $h \in \operatorname{Ker} d_H^n$ there exists $g \in G^n$ satisfying $\psi_n(g) = h$. Since $\psi_{n+1}(d_G^n(g)) = d_H^n(\psi_n(g)) = d_H^n(h) = 0$, there is an element $f \in F^{n+1}$ to satisfy $\varphi_{n+1}(f) = d_G^n(g)$. Since we have

$$\varphi_{n+2}(d_F^{n+1}(f)) = d_G^{n+1}(\varphi_{n+1}(f)) = d_G^{n+1}(d_G^n(g)) = 0,$$

we get $f \in \operatorname{Ker} d_F^{n+1}$. For $h \in \operatorname{Ker} d_H^n$, let $\psi_n(\tilde{g}) = h$. Then $g - \tilde{g} \in \operatorname{Im} \varphi_n$. Then we can write $\tilde{g} = g + \varphi_n(a)$. For $\tilde{f}$ satisfying $\varphi_{n+1}(\tilde{f}) = d_G^n(\tilde{g})$, we have $\tilde{f} = f + d_F^n(a)$. Hence $[f]$ is uniquely determined by h:

(6.26)

$$\begin{array}{ccccccccc}
0 & \to & F^n & \to & G^n & \to & H^n & \to & 0 \\
& & & & g & \mapsto & h & & \\
& & \downarrow & & \downarrow & \downarrow & \downarrow & & \\
& & f \mapsto & & d_G^n(g) & & 0 & & \\
0 & \to & F^{n+1} & \to & G^{n+1} & \to & H^{n+1} & \to & 0 \\
& & \downarrow & & \downarrow & & \downarrow & &
\end{array}$$

Moreover, let $\hat{h} = h + d_H^{n-1}(b)$, $b \in H^{n-1}$. Going through the same construction as above, we get $\hat{f} \in F^{n+1}$. Then one can easily show that $[\hat{f}] = [f]$. Therefore, the homomorphism in (6.25) can be obtained by defining $\delta_n([h]) = [f]$. Namely, we have the sequence of homorphisms

(6.27)
$$H^n(X, \mathcal{G}) \xrightarrow{\overline{\psi}_n} H^n(X, \mathcal{H}) \xrightarrow{\delta_n} H^{n+1}(X, \mathcal{F}) \xrightarrow{\overline{\varphi}_{n+1}} H^{n+1}(X, \mathcal{G}).$$

For $h \in \operatorname{Ker} d_H^n$, suppose $\delta_n([h]) = 0$. In other words, there exists $c \in F^n$ satisfying $f = d_F^n(c)$ in diagram (6.26). Then we have $d_G^n(g) = \varphi_n(f) = \varphi_n(d_F^n(c)) = d_G^n(\varphi_n(c))$. Hence $g - \varphi_n(c) \in \operatorname{Ker} d_G^n$. Since $g \in G^n$ can be chosen arbitrarily as long as $\psi_n(g) = h$, we redefine

g as $g - \varphi_n(c)$, i.e., $g \in \operatorname{Ker} d_G^n$. That is, $[h] = \overline{\psi}_n([g])$, and so $\operatorname{Ker} \delta_n \subset \operatorname{Im} \overline{\psi}_n$. On the other hand, if $[h] = \overline{\psi}_n([g])$, the above argument implies that the $f \in F^{n+1}$ corresponding to h has the form $f = d_F^n(c)$. Namely, $[f] = 0$. Consequently, $\operatorname{Im} \overline{\psi}_n \subset \operatorname{Ker} \delta_n$. We obtain

(6.28) $$\operatorname{Ker} \delta_n = \operatorname{Im} \overline{\psi}_n.$$

Next, we will consider $\operatorname{Ker} \overline{\varphi}_{n+1}$ in (6.27). Suppose that $f \in \operatorname{Ker} d_F^{n+1}$ satisfies $\overline{\varphi}_{n+1}([f]) = 0$. Namely, there exists $g \in G^n$ such that $\varphi_{n+1}(f) = d_G^n(g)$. Let $h = \psi_n(g)$. Then we get

$$d_H^n(h) = d_H^n(\psi_n(g)) = \psi_{n+1}(d_G^n(g)) = \psi_{n+1}(\varphi_{n+1}(f)) = 0.$$

Hence $h \in \operatorname{Ker} d_H^n$. From (6.26), we have $\delta_n([h]) = [f]$. We conclude that $\operatorname{Ker} \overline{\varphi}_{n+1} \subset \operatorname{Im} \delta_n$. Conversely, consider $h \in \operatorname{Ker} d_H^n$ such that $\delta_n([h]) = [f]$, $f \in \operatorname{Ker} d_F^{n+1}$. Then, since $\varphi_{n+1}(f) = d_G^n(g)$ in (6.26), we get $\overline{\varphi}_{n+1}([f]) = 0$. Therefore $\operatorname{Im} \delta_n \subset \operatorname{Ker} \overline{\varphi}_{n+1}$. That is, we obtain

(6.29) $$\operatorname{Im} \delta_n = \operatorname{Ker} \overline{\varphi}_{n+1}.$$

We have proved the theorem by obtaining (6.24), (6.28), (6.29). □

Thus, for a sheaf $\mathcal{F}$ over a topological space X, the cohomology groups $H^n(X, \mathcal{F})$ can be defined to satisfy Theorem 6.9. Practically speaking, a flabby resolution is seldom used to compute the sheaf cohomology. We will show how the sheaf cohomology group can be computed for a scheme.

(c) **Cohomology of Affine Schemes**

Our first goal is to prove the following theorem.

THEOREM 6.10. *For a quasicoherent $\mathcal{O}_X$-module $\mathcal{F}$ over an affine scheme X, one has*

$$H^n(X, \mathcal{F}) = 0, \qquad n \geq 1.$$

Since a quasicoherent $\mathcal{F}$ over an affine scheme $X = \operatorname{Spec} R$ can be written as $\widetilde{M}$ for the R-module $M = \Gamma(X, \mathcal{F})$, we will construct a flabby resolution of $\widetilde{M}$. The flabby sheaves $C^n(\mathcal{F})$ in the canonical flabby resolution (6.9) need not be quasicoherent. Hence, one cannot use Corollary 4.22 and Problem 12 in Chapter 4. We will define the notion of an injective R-module. Let $f : M \to N$ be an arbitrary

6.1. COHOMOLOGY OF SHEAVES

injective R-homomorphism. For any R-homomorphism $g : M \to I$, if there exists an R-homomorphism $h : N \to I$ making the diagram

$$\begin{array}{ccc} 0 \to M & \xrightarrow{f} & N \\ {\scriptstyle g}\downarrow & \swarrow_{h} & \\ I & & \end{array}$$

commutative, then the R-module I is said to be an *injective R-module*. That is, for any short exact sequence

$$0 \to M \to N,$$

if the induced sequence

$$\mathrm{Hom}_R(N, I) \to \mathrm{Hom}_R(M, I) \to 0$$

is exact, then I is an injective R-module.

It is well known that an arbitrary R-module M is an R-submodule of an injective R-module. Namely, one can find an injective R-module I^0 and an injective R-homomorphism ι so that

$$0 \to M \xrightarrow{\iota} I^0$$

is exact. Similarly, there exists an injective R-module I^1 so that

$$0 \to I^0/\iota(M) \to I^1$$

is exact. By composing those exact sequences, we get an exact sequence of R-modules

$$0 \to M \xrightarrow{\iota} I^0 \xrightarrow{\eta^0} I^1.$$

Thus, we obtain the exact sequence of injective R-modules I^j,

(6.30) $\qquad 0 \to M \xrightarrow{\iota} I^0 \xrightarrow{\eta^0} I^1 \xrightarrow{\eta^1} I^2 \xrightarrow{\eta^2} I^3 \to \cdots .$

The exact sequence (6.30) is said to be an *injective resolution* of M.

For simplicity, in what follows we assume that R is Noetherian.

LEMMA 6.11. *For a Noetherian ring R, the $\mathcal{O}_X$-module $\tilde{I}$ determined by an injective R-module I is a flabby sheaf over the affine scheme $X = \mathrm{Spec}\, R$.*

We need the following lemma to prove Lemma 6.11.

LEMMA 6.12. *Let $\mathfrak{a}$ be an ideal of a Noetherian ring R, and let I be an injective R-module. Then*

$$J = \{a \in I | \mathfrak{a}^n a = 0 \text{ for some positive integer } n\}$$

is also an injective R-module.

PROOF. For an ideal $\mathfrak{b}$ of R, the inclusion $\mathfrak{b} \subset R$ induces an R-homomorphism
$$(6.31) \qquad \operatorname{Hom}_R(R, J) \to \operatorname{Hom}_R(\mathfrak{b}, J).$$
We prove that (6.31) is surjective. For a given $\varphi \in \operatorname{Hom}_R(\mathfrak{b}, J)$ there exists a positive integer n to satisfy $\mathfrak{a}^n \varphi(b) = 0$, $b \in \mathfrak{b}$. Since $\mathfrak{b}$ is finitely generated, there exists a positive integer N so that $\mathfrak{a}^N \varphi(\mathfrak{b}) = (0)$. The Artin-Rees Lemma (see Problem 4 below) implies that for some positive integer r, we get $\mathfrak{a}^m \cap \mathfrak{b} = \mathfrak{a}^{m-r}(\mathfrak{a}^r \cap \mathfrak{b})$ for $m \geq r$. Therefore, for $n \geq N + r$,
$$\varphi(\mathfrak{a}^n \cap \mathfrak{b}) = \varphi(\mathfrak{a}^{n-r}(\mathfrak{a}^r \cap \mathfrak{b})) = \mathfrak{a}^{n-r}\varphi(\mathfrak{a}^r \cap \mathfrak{b}) \subset \mathfrak{a}^{n-r}\varphi(\mathfrak{b}) = (0).$$
Namely, $\varphi : \mathfrak{b} \to J$ induces $\overline{\varphi} : \mathfrak{b}/\mathfrak{b} \cap \mathfrak{a}^n \to J \subset I$, and for an injective R-module I, there exists an R-homomorphism $\overline{\psi} : R/\mathfrak{a}^n \to I$ so that the composition $\mathfrak{b}/\mathfrak{b} \cap \mathfrak{a}^n \subset R/\mathfrak{a}^n \to I$ coincides with $\overline{\varphi}$. Let ψ be the composition $\overline{\psi} \circ p$, where $p : R \to R/\mathfrak{a}^n$ is the natural homomorphism. Then, for $a \in R$, we have $\mathfrak{a}^n \psi(a) = \psi(\mathfrak{a}^n a) \subset \psi(\mathfrak{a}^n) = \overline{\psi}(\overline{0}) = 0$, i.e., $\psi(a) \in J$. Therefore, $\psi \in \operatorname{Hom}_R(R, J)$. Then $\mathfrak{b} \subset R \xrightarrow{\psi} J$ coincides with φ. Consequently, (6.31) is surjective.

Next, we will prove that if, for an arbitrary ideal $\mathfrak{b}$ of R, (6.31) is surjective, then J is an injective R-module.

Consider the exact sequence
$$0 \to M \to N,$$
i.e., M may be regarded as an R-submodule of N. Then the map
$$(6.32) \qquad \operatorname{Hom}_R(N, J) \to \operatorname{Hom}_R(M, J)$$
is defined by the restriction $f|M$ of $f \in \operatorname{Hom}_R(N, J)$. We will show the surjectiveness of (6.32). For $g \in \operatorname{Hom}_R(M, J)$, let
$$\mathcal{M} = \left\{ (g', M') \,\middle|\, \begin{array}{l} g' \in \operatorname{Hom}_R(M', J), g'|M = g, \text{ where } M' \text{ is an} \\ R\text{-submodule of } N \text{ satisfying } M \subset M' \subset N \end{array} \right\}.$$
Since $(g, M) \in \mathcal{M}$, $\mathcal{M} \neq \varnothing$. Define an order $(g', M') < (g'', M'')$ by $M' \subset M''$ and $g''|M' = g'$. We will show that $\mathcal{M}$ is an inductively ordered set. Let $\{(g_\lambda, M_\lambda)_{\lambda \in \Lambda}\}$ be a totally ordered subset of $\mathcal{M}$. Namely, Λ is a totally ordered set such that $(g_\lambda, M_\lambda) < (g_\mu, M_\mu)$ for $\lambda < \mu$. Since $M_\lambda \subset N$, if you let $\widetilde{M} = \bigcup_{\lambda \in \Lambda} M_\lambda$, then $\widetilde{M}$ is also an R-submodule of M. For $m \in \widetilde{M}$, choose $\mu \in \Lambda$ to satisfy $m \in M_\mu$. Then define $\tilde{g} \in \operatorname{Hom}_R(\widetilde{M}, J)$ by $\tilde{g}(m) = g_\mu(m)$. If $m \in M_{\mu'}$, then either $\mu < \mu'$ or $\mu' < \mu$. If $\mu < \mu'$, then $g_{\mu'}|M_\mu = g_\mu$ implies $g_{\mu'}(m) = g_\mu(m)$. If $\mu' < \mu$, then $g_\mu|M_{\mu'} = g_{\mu'}$ implies $g_{\mu'}(m) =$

6.1. COHOMOLOGY OF SHEAVES   131

$g_\mu(m)$. Namely, $\tilde{g}$ is properly defined as a map. Since $M_\lambda \subset \widetilde{M}$ and $\tilde{g}|M_\lambda = g_\lambda$, we get $\tilde{g}|M = g$. Hence $(\tilde{g}, \widetilde{M}) \in \mathcal{M}$ and $(g_\lambda, M_\lambda) < (\tilde{g}, \widetilde{M})$ for all $\lambda \in \Lambda$. Therefore, $\mathcal{M}$ is an inductively ordered set. Let $(\hat{g}, \widehat{M})$ be a maximal element. If $\widehat{M} = N$, we have $\hat{g}|M = g$. That is, g is the image of $\hat{g} \in \mathrm{Hom}_R(M, J)$ under the map (6.32). Suppose $\widehat{M} \neq N$. Let $n \in N \setminus \widehat{M}$ and let $\mathfrak{b} = \{a \in R | an \in \widehat{M}\}$. Then $\mathfrak{b}$ is an ideal of R. As we showed,

$$\mathrm{Hom}_R(R, J) \to \mathrm{Hom}_R(\mathfrak{b}, J)$$

is surjective. Define $\varphi \in \mathrm{Hom}_R(\mathfrak{b}, J)$ by $\varphi(a) = \hat{g}(an)$. Then, for $a \in \mathfrak{b}$, there exists $\psi \in \mathrm{Hom}_R(R, J)$ to satisfy $\psi(a) = \varphi(a)$. Then let $\widetilde{M} = \widehat{M} + R^n \subset N$ and let $\tilde{g}(\hat{m} + rm) = \hat{g}(\hat{m}) + \psi(r)$, $\hat{m} \in \widehat{M}$, $r \in R$. For $\hat{m} \in \widehat{M} \cap R^n$, write $\hat{m} = bn$. Then we have $\tilde{g}(\hat{m}) = \tilde{g}(bn) = \varphi(b) = \psi(b)$. That is, the above map $\hat{g}$ is properly defined as a map. Since $\widehat{M} \subsetneq \widetilde{M} \subset N$, $\tilde{g} \in \mathrm{Hom}_R(\widetilde{M}, J)$ and $\tilde{g}|M = g$, we have $(\tilde{g}, \widetilde{M}) \in \mathcal{M}$ and $(\hat{g}, \widehat{M}) < (\tilde{g}, \widetilde{M})$. This contracts the maximality of $(\hat{g}, \widehat{M})$ in $\mathcal{M}$. Namely, $\widehat{M} = N$, and so (6.32) is a surjection. Hence, J is an injective R-module. This proves Lemma 6.12. □

PROOF OF LEMMA 6.11. For $f \in R$, we will prove that the restriction map

(6.33) $\qquad I = \Gamma(X, \tilde{I}) \to \Gamma(D(f), \tilde{I}) = I_f$

is surjective. Note that (6.33) is the natural homomorphism induced by localization. If one can show that an arbitrary element $\frac{a}{f^m}$ of I_f, $a \in I$, can be written as $\frac{a}{f^m} = \frac{b}{1}$, $b \in I$, then (6.33) is a surjection. Let

$$\mathfrak{a} = \{r \in R | f^n r = 0 \text{ for some positive integer } n\}.$$

Then $\mathfrak{a}$ is an ideal of R. If $\mathfrak{a} = R$, there exists a positive integer n satisfying $f^n \cdot 1 = 0$. Namely, f is a nilpotent element, and $D(f) = \varnothing$ and $I_f = 0$. Hence, (6.33) is trivially surjective. Therefore, let us assume $\mathfrak{a} \neq R$. Since R is Noetherian, $\mathfrak{a}$ is finitely generated. By choosing a finite set of generators, one can find a positive integer n to satisfy $f^n \mathfrak{a} = (0)$. Hence, the R-homomorphism

$$\alpha : R/\mathfrak{a} \to R,$$
$$[r] \mapsto f^{n+m}r,$$

is injective, where $[r]$ is the element of $R/\mathfrak{a}$ determined by $r \in R$. Define $\varphi \in \mathrm{Hom}_R(R/\mathfrak{a}, I)$ by $\varphi([r]) = rf^n a$. Since I is an injective

R-module, and α is injective, there exists $\psi \in \operatorname{Hom}_R(R, I)$ satisfying $\psi \circ \alpha = \varphi$. Let $b = \psi(1)$. Then
$$f^n a = \varphi([1]) = \psi(\alpha([1])) = \psi(f^{n+m}) = f^{n+m} b.$$
That is,
$$\frac{a}{f^m} = \frac{f^n a}{f^{n+m}} = \frac{b}{1}.$$
Therefore, (6.33) is a surjection.

Next, we will show that for an arbitrary open set U, the restriction map
$$\rho_{V,X} : \Gamma(X, \widetilde{I}) = I \to \Gamma(U, \widetilde{I})$$
is surjective. Let $Y_0 = \overline{\operatorname{supp} \widetilde{I}}$, i.e., the closure of $\operatorname{supp} \widetilde{I}$. (Note that for a finitely generated R-module I, $\operatorname{supp} \widetilde{I}$ is a closed set (see Exercise 4.1). But, for a general R-module, $\operatorname{supp} \widetilde{I}$ need not be a closed set.) Let $\xi = \Gamma(U, \widetilde{I})$. Choose an open set $D(f)$ to satisfy $D(f) \subset U$ and $D(f) \cap Y_0 \neq \varnothing$. Then, since (6.33) is surjective, one can find $\eta_1 \in \Gamma(X, \widetilde{I}) = I$ so that $\rho_{D(f),U}(\xi - \rho_{U,X}(\eta_1)) = 0$. If $\xi = \rho_{U,X}(\eta_1)$, then nothing is left to be proved. Let $\xi - \rho_{U,X}(\eta_1) \neq 0$. Since X is Noetherian (Proposition 2.36), U is also Noetherian. Hence U is quasicompact. Therefore, there exists a positive integer n satisfying $f^n(\xi - \rho_{U,X}(\eta_1)) = 0$. Define an R-submodule I_1 of I as follows:
$$I_1 = \{a \in I \mid f^n a = 0 \text{ for some positive integer } n\}.$$
Then Lemma 6.12 implies that I_1 is also an injective R-module. Put $Y_1 = \overline{\operatorname{supp} \widetilde{I_1}}$. Then we have $Y_1 \subset Y_0$, $Y_1 \cap D(f) = \varnothing$ and $Y_0 \cap D(f) \neq \varnothing$. Namely, $Y_1 \subsetneq Y_0$. Choose an open set $D(f_1)$ to satisfy $D(f_1) \subset U$ and $D(f_1) \cap Y_1 \neq \varnothing$. By applying the above idea to $\xi - \rho_{U,X}(\eta_1)$, one can take $\eta_2 \in \Gamma(X, \widetilde{I_1}) = I_1$ so that
$$\rho_{D(f_1),U}(\xi - \rho_{U,X}(\eta_1) - \rho_{U,X}(\eta_2)) = 0.$$
If $\xi = \rho_{U,X}(\eta_1 + \eta_2)$, there is nothing to prove. If $\xi - \rho_{U,X}(\eta_1 + \eta_2) \neq 0$, then define
$$I_2 = \{b \in I_1 \mid f_1^n b = 0 \text{ for some positive integer } n\}.$$
Once again, from Lemma 6.12, I_2 is an injective R-module. Let $Y_2 = \overline{\operatorname{supp} \widetilde{I_2}}$. We have $Y_2 \subsetneq Y_1$. Choose $D(f_2)$ to satisfy $D(f_2) \subset U$ and $D(f_2) \cap Y_2 \neq \varnothing$. Then one can choose $\eta_3 \in \Gamma(X, \widetilde{I_2}) = I_2$ to satisfy
$$\rho_{D(f_2),U}(\xi - \rho_{U,X}(\eta_1 + \eta_2) - \rho_{U,X}(\eta_3)) = 0.$$

By continuing this process, we get the descending chain of closed sets $X \supset Y_0 \supsetneq Y_1 \supsetneq Y_2 \supsetneq \cdots$. Since X is a Noetherian topological space (Chapter 2, Problem 20), there exists a finite number l so that $\eta = \eta_1 + \eta_2 + \cdots + \eta_l \in I = \Gamma(X, \widetilde{I})$ satisfies $\mu = \rho_{U,X}(\eta)$. Namely, (6.33) is surjective, i.e., $\widetilde{I}$ is a flabby sheaf. □

Finally, we can finish the proof of Theorem 6.10. For a module M over a Noetherian ring R, there exists an injective resolution

$$0 \to M \to I^0 \to I^1 \to I^2 \to \cdots.$$

Then we obtain an exact sequence of quasicoherent sheaves over $X = \operatorname{Spec} R$,

$$0 \to \widetilde{M} \to \widetilde{I}^0 \to \widetilde{I}^1 \to \widetilde{I}^2 \to \cdots$$

(see Example 4.19). Since $\widetilde{I}^j$ is a flabby sheaf, the above sequence is a flabby resolution of $\widetilde{M}$. Furthermore, since $\Gamma(X, \widetilde{M}) = M$ and $\Gamma(X, \widetilde{I}^j) = I^j$, we get the exact sequence

$$0 \to \Gamma(X, \widetilde{M}) \to \Gamma(X, \widetilde{I}^0) \to \Gamma(X, \widetilde{I}^1) \to \Gamma(X, \widetilde{I}^2) \to \cdots.$$

Thus $H^n(X, \widetilde{M}) = 0$, $n \geq 1$, and Theorem 6.10 is proved. □

The assumption that R is Noetherian was used in the proof above. However, Theorem 6.10 still holds without that assumption (see EGA, III, 1.3.1 in the bibliography to *Algebraic Geometry* 3).

PROBLEM 4. Let M be a finite module over a Noetherian ring R, let M' be an R-submodule of M, and let $\mathfrak{a}$ be an ideal of R. Prove that there exists a positive integer m so that

$$(\mathfrak{a}^n M) \cap M' = \mathfrak{a}^{n-m}(\mathfrak{a}^m M \cap M')$$

for all $n \geq m$. This assertion is called the Artin-Rees Lemma.

(d) **Čech Cohomology Groups**

For the sake of computation, we will define Čech cohomology groups (E. Čech, 1893–1960), since the definition via a flabby resolution is not so practical for computation. We will focus on sheaves of additive groups over a scheme. Let $\mathcal{U} = \{U_i\}_{i \in I}$ be an open covering of a scheme X, and let $U_{i_0 i_1 \cdots i_n} = U_{i_0} \cap U_{i_1} \cap \cdots \cap U_{i_n}$. We always assume $U_{i_0 i_1 \cdots i_n} \neq \varnothing$. Let $\mathcal{F}$ be a sheaf of additive groups over X. Then the p-th alternating cochains $\{f_{i_0 i_1 \cdots i_p}\}$ with values in $\mathcal{F}$ with respect to $\mathcal{U}$ are defined as

(6.34) $f_{i_0 i_1 \cdots i_p} \in \Gamma(U_{i_0 i_1 \cdots i_p}, \mathcal{F})$, $f_{i_0 \cdots i_k i_{k+1} \cdots i_p} = -f_{i_0 \cdots i_{k+1} i_k \cdots i_p}$.

The second condition of (6.34) means that for a permutation σ

(6.35) $$f_{i_{\sigma(0)}i_{\sigma(1)}\cdots i_{\sigma(p)}} = \operatorname{sgn}\sigma f_{i_0 i_1 \cdots i_p},$$

where $\operatorname{sgn}\sigma$ indicates the signature of σ. When there are two or more identical integers among $\{i_0, i_1, \ldots, i_p\}$, from (6.34) and (6.35) we get $f_{i_0 i_1 \cdots i_p} = 0$. We will only consider cochains of mutually distinct indices $i_0, i_1, \ldots, i_p$. Let $C^p(\mathcal{U}, \mathcal{F})$ be the totality of p-th alternating cochains $\{f_{i_0 i_1 \cdots i_p}\}$ with values in $\mathcal{F}$ with respect to $\mathcal{U}$. For $\{f_{i_0 i_1 \cdots i_p}\}, \{g_{i_0 i_1 \cdots i_p}\} \in C^p(\mathcal{U}, \mathcal{F})$, define, for each component,

$$\{f_{i_0 i_1 \cdots i_p}\} \pm \{g_{i_0 i_1 \cdots i_p}\} = \{f_{i_0 i_1 \cdots i_p} \pm g_{i_0 i_1 \cdots i_p}\}.$$

Then $C^0(\mathcal{U}, \mathcal{F})$ becomes an additive group. Moreover, if $\mathcal{F}$ is a sheaf of R-modules, then $C^p(\mathcal{U}, \mathcal{F})$ becomes an R-module. Define an additive group homomorphism $\delta^p : C^p(\mathcal{U}, \mathcal{F}) \to C^{p+1}(\mathcal{U}, \mathcal{F})$ as follows:

(6.36)
$$\delta^p \{f_{i_0 i_1 \cdots i_p}\} = \{g_{i_0 i_1 \cdots i_{p+1}}\}, \text{ where}$$
$$g_{i_0 i_1 \cdots i_{p+1}} = \sum (-1)^k f_{i_0 i_1 \cdots \check{i}_k \cdots i_{p+1}}.$$

Here $\check{i}_k$ indicates the exclusion of i_k. The right-hand side of the second equation of (6.36) is actually the restriction of $f_{i_0 i_1 \cdots \check{i}_k \cdots i_p}$ to $U_{i_0 i_1 \cdots i_{p+1}}$. We will use this notation to avoid unnecessary complex notation. We have the following important property of δ^p.

LEMMA 6.13.
$$\delta^{p+1} \circ \delta^p = 0, \quad p = 0, 1, 2, \ldots.$$

PROBLEM 5. Prove Lemma 6.13.

From Lemma 6.13, one defines

(6.37) $$\check{H}^p(\mathcal{U}, \mathcal{F}) = \operatorname{Ker} \delta^p / \operatorname{Im} \delta^{p-1}, \quad p = 0, 1, 2, \ldots.$$

(6.37) is said to be the p-th Čech cohomology group with respect to the open covering $\mathcal{U}$. Note that we set $\delta^{-1} = 0$. $\check{H}(\mathcal{U}, \mathcal{F})$ depends upon the choice of an open covering $\mathcal{U}$. We will consider the direct limit of $\check{H}(\mathcal{U}, \mathcal{F})$ over $\mathcal{U}$ by taking refinements of $\mathcal{U}$. Let $\mathcal{U} = \{U_i\}_{i \in I}$ and $\mathcal{V} = \{V_\lambda\}_{\lambda \in \Lambda}$ be open coverings of X. Then $\mathcal{V}$ is said to be a *refinement* of $\mathcal{U}$ if for each V_λ there exists $U_{\alpha(\lambda)}$ in $\mathcal{U}$ to satisfy $V_\lambda \subset U_{\alpha(\lambda)}$. We denote the refinement by $\mathcal{V} > \mathcal{U}$. Then define $\beta : I \to \Lambda$ to satisfy $V_{\beta(i)} \subset U_i$. In general, β is not determined uniquely. For $\beta : I \to \Lambda$, one can define a homomorphism

$$f_\beta^p : C^p(\mathcal{U}, \mathcal{F}) \to C^p(\mathcal{V}, \mathcal{F}),$$
$$\{f_{i_0 i_1 \cdots i_p}\} \mapsto \{f_{\beta(i_0)\beta(i_1)\cdots\beta(i_p)}\},$$

6.1. COHOMOLOGY OF SHEAVES

where $f_{\beta(i_0)\beta(i_1)\cdots\beta(i_p)}$ is the restriction of $f_{i_0 i_1 \cdots i_p}$ to $V_{\beta(i_0)\beta(i_1)\cdots\beta_{i_p}}$. Let $\delta^p_{\mathcal{U}}$ be δ^p for $C^p(\mathcal{U},\mathcal{F})$ and let $\delta^p_{\mathcal{V}}$ be δ^p for $C^p(\mathcal{V},\mathcal{F})$. Then we have

$$f^{p+1}_\beta \circ \delta^p_{\mathcal{U}} = \delta^p_{\mathcal{V}} \circ f^p_\beta.$$

Therefore, $\{f^p_\beta\}$ induces an additive group homomorphism

$$f^p_{\beta*} : \check{H}^p(\mathcal{U},\mathcal{F}) \to \check{H}^p(\mathcal{V},\mathcal{F}).$$

Notice that $f^p_{\beta*}$ is determined by $\mathcal{U}$ and $\mathcal{V}$ only, i.e., independently of $\beta : I \to \Lambda$. Hence, one can take the inductive limit over all the open coverings $\mathcal{U}$ with respect to the ordering $>$ for refinements. We define the p-th Čech cohomology group as

(6.38) $$\check{H}^p(X,\mathcal{F}) = \varinjlim_{\mathcal{U}} \check{H}^p(\mathcal{U},\mathcal{F}).$$

Note that the above definition is given for a topological space X, and X need not be a scheme. Notice also that

(6.39) $$\check{H}^0(\mathcal{U},\mathcal{F}) = \Gamma(X,\mathcal{F}),$$

and therefore

(6.40) $$\check{H}^0(X,\mathcal{F}) = \Gamma(X,\mathcal{F}).$$

(6.39) may be explained as follows. For $\xi = \{f_i\} \in \operatorname{Ker} \delta^0$, if $U_i \cap U_j \neq \varnothing$, then the restrictions of f_i and f_j onto $U_i \cap U_j$ coincide. Then, by the definition of a sheaf, there exists a unique $f \in \Gamma(X,\mathcal{F})$ to satisfy $\rho_{U_i,X}(f) = f_i$, $i \in I$. Namely, $\operatorname{Ker} \delta = \Gamma(X,\mathcal{F})$.

It is important for the Čech cohomology groups $\check{H}^p(X,\mathcal{F})$ to satisfy Theorem 6.9. However, for a general topological space X, $\check{H}^p(X,\mathcal{F})$ does not satisfy Theorem 6.9. For the case of a scheme, we have the following.

THEOREM 6.14. *If X is a separated scheme (i.e., $X \to \operatorname{Spec} \mathbb{Z}$ is separated), and if $\mathcal{U} = \{U_i\}_{i \in I}$ is an affine covering of X, then for an exact sequence of quasicoherent sheaves*

$$0 \to \mathcal{F} \to \mathcal{G} \to \mathcal{H} \to 0,$$

we have the exact sequence of Čech cohomology groups

(6.41) $0 \to \Gamma(X,\mathcal{F}) \to \Gamma(X,\mathcal{G}) \to \Gamma(X,\mathcal{H}) \to \check{H}^1(\mathcal{U},\mathcal{F})$
$\to \check{H}^1(\mathcal{U},\mathcal{G}) \to \check{H}^1(\mathcal{U},\mathcal{H}) \to \check{H}^2(\mathcal{U},\mathcal{F}) \to \check{H}^2(\mathcal{U},\mathcal{G}) \to \cdots.$

PROOF. Since $\pi : X \to \operatorname{Spec} \mathbb{Z}$ is a separated morphism, the intersection $U \cap V$ of affine open sets U and V is also an affine open set (see Exercise 4.39(3)). Therefore, for an affine covering $\mathcal{U} = \{U_i\}_{i \in I}$, $U_{i_0 i_1 \cdots i_p}$ is an affine open set. Hence, by Corollary 4.22, we have the exact sequence
$$0 \to \mathcal{F}(U_{i_0 i_1 \cdots i_p}) \to \mathcal{G}(U_{i_0 i_1 \cdots i_p}) \to \mathcal{H}(U_{i_0 i_1 \cdots i_p}) \to 0.$$
We obtain the following diagram with exact rows and exact columns:

(6.42)
$$\begin{array}{ccccccccc}
& & 0 & & 0 & & 0 & & \\
& & \downarrow & & \downarrow & & \downarrow & & \\
0 & \to & C^0(\mathcal{U}, \mathcal{F}) & \xrightarrow{\varphi^0} & C^0(\mathcal{U}, \mathcal{G}) & \xrightarrow{\psi^0} & C^0(\mathcal{U}, \mathcal{H}) & \to & 0 \\
& & \downarrow \delta_{\mathcal{F}}^0 & & \downarrow \delta_{\mathcal{G}}^0 & & \downarrow \delta_{\mathcal{H}}^0 & & \\
0 & \to & C^1(\mathcal{U}, \mathcal{F}) & \xrightarrow{\varphi^1} & C^1(\mathcal{U}, \mathcal{G}) & \xrightarrow{\psi^1} & C^1(\mathcal{U}, \mathcal{H}) & \to & 0 \\
& & \downarrow \delta_{\mathcal{F}}^1 & & \downarrow \delta_{\mathcal{G}}^1 & & \downarrow \delta_{\mathcal{H}}^1 & & \\
0 & \to & C^2(\mathcal{U}, \mathcal{F}) & \xrightarrow{\varphi^2} & C^2(\mathcal{U}, \mathcal{G}) & \xrightarrow{\psi^2} & C^1(\mathcal{U}, \mathcal{H}) & \to & 0 \\
& & \downarrow \delta_{\mathcal{F}}^2 & & \downarrow \delta_{\mathcal{G}}^2 & & \downarrow \delta_{\mathcal{H}}^1 & & \\
0 & \to & C^3(\mathcal{U}, \mathcal{F}) & \xrightarrow{\varphi^3} & C^3(\mathcal{U}, \mathcal{G}) & \xrightarrow{\psi^3} & C^3(\mathcal{U}, \mathcal{H}) & \to & 0 \\
& & \downarrow & & \downarrow & & \downarrow & & \\
& & \vdots & & \vdots & & \vdots & &
\end{array}$$

As in the proof of Theorem 6.9, we get the exact sequence (6.41) from the diagram (6.42). $\square$

If a scheme is separated, then, without taking the inductive limit as in (6.38), the Čech cohomology groups for an affine covering have the desired property. The next theorem tells how a cohomology group can be computed without taking an inductive limit.

THEOREM 6.15 (Leray). *Let $\mathcal{U} = \{U_i\}_{i \in I}$ be an open covering of a topological space X and let $\mathcal{F}$ be a sheaf of additive groups over X. If, for an arbitrary finite intersection $U_{i_0 i_1 \cdots i_p} = U_{i_0} \cap U_{i_1} \cap \cdots \cap U_{i_p} \neq \varnothing$,*
$$H^n(U_{i_0 i_1 \cdots i_p}, \mathcal{F}) = 0, \quad n \geq 1,$$
then we have an isomorphism
$$\check{H}^n(\mathcal{U}, \mathcal{F}) \xrightarrow{\sim} H^n(X, \mathcal{F}), \quad n \geq 1. \quad \square$$

That is, there always exists a natural homomorphism from the Čech cohomology group $\check{H}^n(\mathcal{U}, \mathcal{F})$ to $H^n(X, \mathcal{F})$, and Leray's theorem asserts that this homomorphism is actually an isomorphism. We will

6.1. COHOMOLOGY OF SHEAVES

not provide a proof for the above theorem. From this theorem, we obtain the following.

COROLLARY 6.16. *For an affine covering $\mathcal{U} = \{U_i\}_{i \in I}$ of a separated scheme X, we have*

$$\check{H}^n(\mathcal{U}, \mathcal{F}) \overset{\sim}{\to} \check{H}^n(X, \mathcal{F}) \overset{\sim}{\to} H^n(X, \mathcal{F}),$$

where $\mathcal{F}$ is an arbitrary quasicoherent sheaf.

This corollary is helpful for computing cohomology groups.

EXAMPLE 6.17. Let $\mathcal{O}_{\mathbb{P}^1_k}$ be the structure sheaf of 1-dimensional projective space $\mathbb{P}^1_k = \operatorname{Proj} k[x_0, x_1]$ over a field k. Let us compute $H^n(\mathbb{P}^1_k, \mathcal{O}_{\mathbb{P}^1_k})$. Consider an affine covering $\mathcal{U} = \{U_0 = D_+(x_0), U_1 = D_+(x_1)\}$. We have $C^0(\mathcal{U}, \mathcal{O}_X) = \mathcal{O}_X(U_0) \oplus \mathcal{O}_X(U_1)$ and $C^1(\mathcal{U}, \mathcal{O}_X) = \mathcal{O}_X(U_{01})$, and $C^l(\mathcal{U}, \mathcal{O}_X) = 0$, $l \geq 2$. Therefore, we obtain

$$\mathcal{O}_X(U_0) = k\left[\frac{x_1}{x_0}\right], \quad \mathcal{O}_X(U_1) = k\left[\frac{x_0}{x_1}\right], \quad \mathcal{O}_X(U_{01}) = k\left[\frac{x_1}{x_0}, \frac{x_0}{x_1}\right].$$

For the sake of simplicity, let $x = \frac{x_1}{x_0}$ and $y = \frac{x_0}{x_1}$. Over U_{01}, we get $y = \frac{1}{x}$. For $\xi = \{f_0, f_1\} \in C^0(\mathcal{U}, \mathcal{O}_X)$, we can write $f_0 = f_0(x) \in k[x]$, $f_1 = f_1(y) \in k[y]$, and $(\delta^0 \xi)_{01} = f_1 - f_0 = f_1(\frac{1}{x}) - f_0(x) \in k[x, \frac{1}{x}]$. If $\delta^0 \xi = 0$, then $f_1(\frac{1}{x}) = f_0(x)$, i.e., $f_1 = f_0 = \alpha \in k$. Namely, $\operatorname{Ker} \delta^0 = k$. That is, $H^0(\mathbb{P}^1_k, \mathcal{O}_{\mathbb{P}^1_k}) = k$, which has been shown in Example 4.4.

For $f_{01} \in k[x, \frac{1}{x}]$, one can choose $f(x)$ and $g(x)$ in $k[x]$ to satisfy $f_{01} = f(\frac{1}{x}) - g(x)$. Then let $f_0 = g(x)$ and $f_1(y) = f(y)$, and let $\xi = \{f_0, f_1\} \in C^0(\mathcal{U}, \mathcal{O}_{\mathbb{P}^1_k})$. We have $(\delta^0 \xi)_{01} = f_{01}$ and $\operatorname{Im} \delta^0 = C^1(\mathcal{U}, \mathcal{O}_{\mathbb{P}^1_k})$. That is, $H^1(\mathbb{P}^1_k, \mathcal{O}_{\mathbb{P}^1_k}) = 0$. □

We also obtain the following from Corollary 6.16.

LEMMA 6.18. *For an n-dimensional projective space $X = \mathbb{P}^n_R$ over a Noetherian ring R and a quasicoherent sheaf $\mathcal{F}$ over $\mathbb{P}^n_k$, we have*

$$H^p(X, \mathcal{F}) = 0, \quad p \geq n+1.$$

PROOF. Since $S = R[x_0, x_1, \ldots, x_n]$ and $X = \operatorname{Proj} S$, we obtain an affine covering $\mathcal{U} = \{U_j\}_{0 \leq j \leq n}$ of X, where $U_j = D_+(x_j)$, $j = 0, 1, \ldots, n$. Then $C^p(\mathcal{U}, \mathcal{F}) = 0$ for $p \geq n+1$. The conclusion follows. □

Even though we have used alternating cochains to define Čech cohomology groups, the cochains need not be alternating to get the same cohomology groups.

PROBLEM 6. Let $\mathcal{F}$ and $\mathcal{G}$ be quasicoherent sheaves over a separated scheme X and let $\mathcal{U} = \{U_j\}_{j \in J}$ be an open covering of X. Then define a pairing for Čech complexes

$$C^p(\mathcal{U}, \mathcal{F}) \times C^q(\mathcal{U}, \mathcal{G}) \to \widetilde{C}^{p+q}(\mathcal{U}, \mathcal{F} \otimes \mathcal{G}),$$
$$(\{f_{i_0 \cdots i_p}\}, \{g_{i_0 \cdots i_q}\}) \mapsto \{h_{i_0 i_1 \cdots i_{p+q}}\},$$

where $h_{i_0 \cdots i_{p+q}} = f_{i_0 \cdots i_p} \otimes g_{i_p i_{p+1} \cdots i_{p+q}} | U_{i_0 i_1 \cdots i_{p+q}}$. Note that $\widetilde{C}^n(\mathcal{F})$ is the totality of cochains, not necessarily alternating. Then prove that on cohomology groups one can define the pairing

$$H^p(X, \mathcal{F}) \to H^q(X, \mathcal{G}) \to H^{p+q}(X, \mathcal{F} \otimes \mathcal{G}).$$

6.2. Cohomology of a Projective Scheme

(a) Cohomology of a Projective Space

We are going to study the cohomology of the invertible sheaf $\mathcal{O}_X(m)$ of n-dimensional projective space $\mathbb{P}^n_R = \operatorname{Proj} R[x_0, \ldots, x_n]$ over a Noetherian ring R. It is not an overstatement to say that this cohomology is the foundation for the cohomology theory of a projective scheme.

THEOREM 6.19. *For an n-dimensional projective space $X = \mathbb{P}^n_R = \operatorname{Proj} S$, $S = R[x_0, x_1, \ldots, x_n]$, over a Noetherian ring R, we have the following.*

(i) *The natural homomorphism*

$$S \to \Gamma_*(\mathcal{O}_X) = \bigoplus_{m \in \mathbb{Z}} H^0(X, \mathcal{O}_X(m))$$

is an isomorphism of graded S-modules.

(ii) *For all integers m and for $0 < i < n$, we have $H^i(X, \mathcal{O}_X(m)) = 0$.*

(iii) $H^n(X, \mathcal{O}_X(-n-1)) \xrightarrow{\sim} R$.

(iv) *For all integers m, a pairing*

$$H^0(X, \mathcal{O}_X(m)) \times H^n(X, \mathcal{O}_X(-m-n-1)) \to H^n(X, \mathcal{O}_X(-n-1))$$

can be defined, and this pairing is a perfect pairing between the free R-modules. That is, if $\alpha \in H^0(X, \mathcal{O}_X(m))$ satisfies $\alpha\beta = 0$ for all $\beta \in H^n(X, \mathcal{O}_X(-m-n-1))$, then $\alpha = 0$, and if β satisfies $\alpha\beta = 0$ for all $\alpha \in H^0(X, \mathcal{O}_X(m))$, then $\beta = 0$.

6.2. COHOMOLOGY OF A PROJECTIVE SCHEME

PROOF. We have already proved (i) in Theorem 5.21. We will use Čech cohomology to prove (ii), (iii) and (iv). Let $U_j = D_+(x_j)$, $j = 0, 1, \ldots, n$. Then $\mathcal{U} = \{U_j\}_{0 \leq j \leq n}$ is an affine covering of X. Notice that we get $U_{j_0 j_1 \cdots j_p} = D_+(x_{j_0} x_{j_1} \cdots x_{j_p})$. Furthermore, we can write

$$\Gamma(U_{j_0 j_1 \cdots j_p}, \mathcal{O}_X(m))$$
$$= \left\{ \frac{f}{(x_{j_0} x_{j_1} \cdots x_{j_p})^l} \;\middle|\; \begin{array}{l} f \text{ is a homogeneous element} \\ \text{of degree } l(p+1) + m \end{array} \right\}.$$

Therefore, we can identify $C^n(\mathcal{U}, \mathcal{O}_X(m))$ with $S(m)_{(x_0 x_1 \cdots x_n)}$ and $C^{n-1}(\mathcal{U}, \mathcal{O}_X(m))$ with $\bigoplus_{i=0}^n S(m)_{(x_0 x_1 \cdots \check{x}_i \cdots x_n)}$. Then

$$\delta^{n-1} : C^{n-1}(\mathcal{U}, \mathcal{O}_X(m)) \to C^n(\mathcal{U}, \mathcal{O}_X(m))$$

is the R-homomorphism

(6.43)
$$d^{t-1} : \bigoplus_{i=0}^n S(m)_{(x_0 x_1 \cdots \check{x}_i \cdots x_n)} \to S(m)_{(x_0 x_1 \cdots x_n)},$$
$$\left(\frac{f_0}{(x_1 \cdots x_n)^{l_0}}, \frac{f_1}{(x_0 x_2 \cdots x_n)^{l_1}}, \ldots, \frac{f_n}{(x_0 \cdots x_{n-1})^{l_n}} \right)$$
$$\mapsto \frac{\sum (-1)^i x_i^i (x_0 \cdots x_{i-1} x_{i+1} \cdots x_n)^{l - l_i} f_i}{(x_0 x_1 \cdots x_n)^l},$$

where $l = \max_{0 \leq i \leq n}\{l_i\}$. Note that an element of $S(m)_{(x_0 x_1 \cdots x_n)}$ can be written as $\frac{f}{(x_0 x_1 \cdots x_n)^l}$, where f is of homogeneous degree $l(n+1) + m$. Since f is a linear combination of monomials with coefficients in R, as an R-module base, one can choose

$$\frac{x_0^{m_0} x_1^{m_1} \cdots x_n^{m_n}}{(x_0 x_1 \cdots x_n)^l} = x_0^{m_0 - l} x_1^{m_1 - l} \cdots x_n^{m_n - l},$$

where $m_0 + m_1 + \cdots + m_n = l(n+1) + m$. Therefore, as a base of a free R-module for $S(m)_{(x_0 x_1 \cdots x_n)}$, one can choose the totality of

(6.44)
$$x_0^{l_0} x_1^{l_1} \cdots x_n^{l_n}, \quad l_0 + l_1 + \cdots + l_n = m, \quad l_j \in \mathbb{Z}, \quad j = 0, 1, \ldots, n.$$

(We regard, e.g.,

$$x_0^{-m+1} x_1^{-1} = \frac{x_0^{l-m+1} x_1^{l-1} x_2^l \cdots x_n^l}{(x_0 x_1 \cdots x_n)^l}, \quad l \geq -m + 1.)$$

In (6.44), if $l_j \geq 0$, then those elements in (6.44) form a base for $S(m)_{(x_0 x_1 \cdots \check{x}_i \cdots x_n)}$ over R. Then the homomorphism (6.43) is induced by the natural injection $S(m)_{(x_0 x_1 \cdots \check{x}_i \cdots x_n)} \subset S(m)_{(x_0 x_1 \cdots x_n)}$, and $\operatorname{Im} d^{n-1}$ is a linear combination of the elements in (6.44), satisfying l_j for at least one j. In particular, for $m = -n-1$, the only term with all $l_j < 0$ is

$$x_0^{-1} x^{-1} \cdots x_n^{-1} = \frac{1}{x_0 x_1 \cdots x_n}.$$

Therefore, we obtain

$$H^n(X, \mathcal{O}_X(-n-1)) \xrightarrow{\sim} R \cdot \frac{1}{x_0 x_1 \cdots x_n},$$

completing the proof of (iii).

We will prove (iv). From (i), for $m < 0$, we have $H^0(X, \mathcal{O}_X(m)) = 0$. Using the above method, for $m \geq -n$ there is at least one l_j that is zero. Then we have $H^n(X, \mathcal{O}_X(m)) = 0$. Therefore, for $m < 0$, (iv) holds. Next let $m \geq 0$. As a base for the R-free module $H^0(X, \mathcal{O}_X(m))$, one can take $x_0^{m_0} x_1^{m_1} \cdots x_n^{m_n}$, $m_0 + m_1 + \cdots + m_n = m$, $m_j \geq 0$, $j = 0, 1, 2, \ldots, n$. We can take $x_1^{l_0} x_2^{l_1} \cdots x_n^{l_n}$, $l_0 + l_1 \cdots + l_n = -m - n - 1$, $l_j < 0$, $j = 0, 1, 2, \ldots, n$, as a base for the R-free module $H^n(X, \mathcal{O}_X(-m-n-1))$. The pairing in (iv) for those bases is given by

(6.45) $(x_0^{m_0} x_1^{m_1} \cdots x_n^{m_n}, x_0^{l_0} x_1^{l_1} \cdots x_n^{l_n}) \mapsto x_0^{l_0+m_0} x_1^{l_1+m_1} \cdots x_n^{l_n+m_n}.$

When $l_j + m_j \geq 0$ for a certain j, as an element of $H^n(X, \mathcal{O}_X(-n-1))$, the right-hand side of (6.45) is 0. Fix $x_0^{m_0} x_1^{m_1} \cdots x_n^{m_n}$, $m_0 + m_1 + \cdots + m_n = m$, $m_j \geq 0$, $j = 0, 1, 2, \ldots, n$. Let $l_j = -m_j - 1$. Then $x_0^{l_0} x_1^{l_1} \cdots x_n^{l_n}$ is a non-zero element of $H^n(X, \mathcal{O}_X(-m-n-1))$, and the image of the map in (6.45) becomes $x_0^{-1} x_1^{-1} \cdots x_n^{-1}$. For a non-zero element $x_0^{l_0} x_1^{l_1} \cdots x_n^{l_n}$ of $H^n(X, \mathcal{O}_X(-m-n-1))$, let $m_j = -l_j - 1$. Then $x_0^{m_0} x_1^{m_1} \cdots x_n^{m_n}$ is a non-zero element of $H^0(X, \mathcal{O}_X(m))$, and the image of the map (6.45) is $x_0^{-1} x_1^{-1} \cdots x_n^{-1}$. Therefore, the pairing in (iv) is a perfect pairing.

Finally, we will prove (ii). We will prove it by induction on the dimension n of the projective space. For $n = 1$, by (iii) and (iv), we have

$$H^1(\mathbb{P}_R^1, \mathcal{O}_{\mathbb{P}_R^1}(m)) = 0, \qquad m \geq -1.$$

For $n = 1$, (ii) asserts nothing. Consider the following exact sequence of homomorphisms of graded S-modules:

(6.46) $\qquad 0 \to S(-1) \xrightarrow{\times x_n} S \to S/(x_n) \to 0.$

6.2. COHOMOLOGY OF A PROJECTIVE SCHEME

Note that $S/(x_n) \xrightarrow{\sim} R[x_0, x_1, \ldots, x_{n-1}]$. If you let $H = V(x_n)$, then $H \xrightarrow{\sim} \mathbb{P}_R^{n-1}$. From (6.46), we obtain the exact sequence of $\mathcal{O}_X$-modules

$$0 \to \mathcal{O}_X(-1) \to \mathcal{O}_X \to \mathcal{O}_H \to 0.$$

By (6.46), the induced long exact sequence from

(6.47) $\qquad 0 \to \mathcal{O}_X(m-1) \to \mathcal{O}_X(m) \to \mathcal{O}_H(m) \to 0$

(obtained by tensoring the above sequence by $\mathcal{O}_X(m)$) becomes

$$0 \to H^0(X, \mathcal{O}_X(m-1)) \to H^0(X, \mathcal{O}_X(m)) \to H^0(H, \mathcal{O}_H(m)) \to 0.$$

Therefore, we obtain the following exact sequence of cohomology groups:

$$0 \to H^1(X, \mathcal{O}_X(m-1)) \to H^1(X, \mathcal{O}_X(m)) \to H^1(H, \mathcal{O}_H(m))$$
$$\to H^2(X, \mathcal{O}_X(m-1)) \to H^2(X, \mathcal{O}_X(m)) \to H^2(H, \mathcal{O}_H(m))$$
$$\to \cdots$$
$$\to H^{n-1}(X, \mathcal{O}_X(m-1)) \to H^{n-1}(X, \mathcal{O}_X(m))$$
$$\to H^{n-1}(H, \mathcal{O}_H(m))$$
$$\to H^n(X, \mathcal{O}_X(m-1)) \to H^n(X, \mathcal{O}_X(m)) \to 0.$$

By the inductive assumption, for $0 < i < n-1$, we get $H^i(H, \mathcal{O}_H(m)) = 0$. Therefore,

$$H^i(X, \mathcal{O}_X(m-1)) \xrightarrow{\times x_n} H^i(X, \mathcal{O}_X(m))$$

is an isomorphism, $0 < i < n - 1$. Next, consider

$$H^{n-1}(H, \mathcal{O}_H(m)) \xrightarrow{\delta_{n-1}} H^n(X, \mathcal{O}_X(m-1)).$$

For $m \geq -n+1$, we have

$$H^{n-1}(H, \mathcal{O}_H(m)) = 0.$$

For $m \leq -n$, a base for the free R-module $H^{n-1}(H, \mathcal{O}_H(m))$ can be given by $x_0^{l_0} x_1^{l_1} \cdots x_{n-1}^{l_{n-1}}$, $l_0 + l_1 + \cdots + l_{n-1} = m$, $l_j < 0$, $j = 0, 1, 2, \ldots, n-1$. The map δ_{n-1} can be obtained from diagram (6.42), but in this case it is given by $x_0^{l_0} x_1^{l_1} \cdots x_{n-1}^{l_{n-1}} \mapsto x_0^{l_0} x_1^{l_1} \cdots x_{n-1}^{l_{n-1}} x_n^{-1}$. Namely, δ_{n-1} is an injection. Therefore,

$$H^{n-1}(X, \mathcal{O}_X(m-1)) \xrightarrow{\times x_n} H^{n-1}(X, \mathcal{O}_X(m))$$

is also an isomorphism. Hence, we need to show that $H^p(X, \mathcal{O}_X) = 0$, $0 < p < n$. An element of $C^p(\mathcal{U}, \mathcal{O}_X)$ is uniquely determined by an

element of
$$\bigoplus_{i_0<i_1<\cdots<i_p} \mathcal{O}_X(U_{i_0i_1\cdots i_p}) = \bigoplus_{i_0<i_1<\cdots<i_p} S_{(x_{i_0}x_{i_1}\cdots x_{i_p})}.$$

Such an element can be written as
$$\xi = \left(\frac{f_{i_0i_1\cdots i_p}}{(x_{i_0}x_{i_1}\cdots x_{i_p})^r}\right), \quad i_0<i_1<\cdots<i_p, \quad f_{i_0i_1\cdots i_p} \in S_{(r(p+1))}.$$

For a permutation σ on $\{0,1,\ldots,p\}$, let
$$f_{i_{\sigma(0)}i_{\sigma(1)}\cdots i_{\sigma(p)}} = \operatorname{sgn}(\sigma)f_{i_0i_1\cdots i_p}.$$

Then ξ determines a unique element $\tilde{\xi}$ in $C^p(\mathcal{U},\mathcal{O}_X)$. Since
$$(\delta^p\tilde{\xi})_{i_0i_1\cdots i_p} = \sum_{k=0}^{p+1}\frac{(-1)^k x_{i_k}^r f_{i_0i_1\cdots\check{i}_k\cdots i_{p+1}}}{(x_{i_0}x_{i_1}\cdots x_{i_{p+1}})^r},$$

for $\delta^p\tilde{\xi} = 0$, we get
$$\sum_{k=0}^{p+1}(-1)^k x_{i_k}^r f_{i_0i_1\cdots\check{i}_k\cdots i_{p+1}} = 0.$$

We can rewrite the above as

(6.48) $$x_{i_0}^p f_{i_1\cdots i_{p+1}} = \sum_{k=1}^{p+1}(-1)^{k-1}x_{i_k}^r f_{i_0i_1\cdots\check{i}_k\cdots i_{p+1}}.$$

Then let
$$f_{i_0i_1\cdots\check{i}_k\cdots i_{p+1}} = x_{i_0}^r g_{i_0i_1\cdots\check{i}_k\cdots i_{p+1}} + h_{i_0i_1\cdots\check{i}_k\cdots i_{p+1}},$$

where $h_{i_0i_1\cdots\check{i}_k\cdots i_{p+1}}$ is of degree less than or equal to $r-1$ in x_{i_0}. We obtain the following from (6.48):
$$\sum_{i=1}^{p+1}(-1)^{k-1}x_{i_k}^r h_{i_0i_1\cdots\check{i}_k\cdots i_{p+1}} = 0.$$

From (6.48), we also get

(6.49) $$f_{i_1\cdots i_{p+1}} = \sum_{k=1}^{p+1}(-1)^{k-1}x_{i_k}^r g_{i_0i_1\cdots\check{i}_k\cdots i_{p+1}}.$$

For a fixed i_0, define
$$\eta_{j_0j_1\cdots j_{p-1}} = \frac{g_{i_0j_0j_1\cdots j_{p-1}}}{(x_{j_0}x_{j_1}\cdots x_{p-1})^r}.$$

6.2. COHOMOLOGY OF A PROJECTIVE SCHEME

Then $\eta = (\eta_{j_0 j_1 \cdots j_{p-1}}) \in C^{p-1}(\mathcal{U}, \mathcal{O}_X)$. By (6.49), we obtain the following:

$$(\delta^{p-1}\eta)_{j_0 j_1 \cdots j_p} = \sum_{k=0}^{p} (-1)^k \eta_{j_0 j_1 \cdots \check{j}_k \cdots j_p}$$

$$= \sum_{k=0}^{p} (-1)^k \frac{x_{j_k}^r g_{i_0 j_0 \cdots \check{j}_k \cdots j_p}}{(x_{j_0} x_{j_1} \cdots x_{j_k} \cdots x_{j_p})^r}$$

$$= \frac{f_{j_0 j_1 \cdots j_p}}{(x_{j_0} x_{j_1} \cdots x_{j_p})^r} = (\tilde{\xi})_{j_0 j_1 \cdots j_p}.$$

Namely, $\tilde{\xi} = \delta^{p-1}\eta$. Therefore, $H^p(X, \mathcal{O}_X) = 0$ for $0 < p < n$. □

PROBLEM 7. By using the same method as above, prove also that for any integer m, $H^p(X, \mathcal{O}_X(m)) = 0$, $0 < p < n$.

If the Noetherian ring R is a field k, then we can rewrite Theorem 6.19 as follows.

COROLLARY 6.20. *For an n-dimensional projective space $X = \mathbb{P}^n_k$ over a field k and the invertible sheaf $\mathcal{O}_X(m)$, $H^p(X, \mathcal{O}_X(m))$ does not equal 0 if and only if either $p = 0$, $m \geq 0$ or $p = n$, $m \leq -n - 1$. Then*

$$\dim_k H^0(X, \mathcal{O}_X(m)) = \binom{m+n}{n}, \qquad m \geq 0,$$
$$\dim_k H^n(X, \mathcal{O}_X(-m-n-1)) = \binom{m+n}{n}, \qquad m \geq 0.$$

Furthermore, for $m \geq 0$, we have a perfect pairing:

$$H^0(X, \mathcal{O}_X(m)) \times H^n(X, \mathcal{O}_X(-m-n-1)) \to H^n(X, \mathcal{O}_X(-n-1)).$$

Theorem 6.19 and Corollary 6.20 are fundamental results for the theory of Noetherian projective schemes.

(b) **Finiteness of Cohomology of Projective Schemes**

We will consider a finitely generated graded ring $S = \bigoplus_{m=0}^{\infty} S_n$ over a Noetherian ring R, i.e., a finitely generated graded ring satisfying $S_0 = R$. Furthermore, we assume that as an $S_0 = R$-algebra, S is generated by elements $f_0, f, \ldots, f_n$ in S_1. Then we have a surjective homomorphism of R-graded rings

(6.50)
$$\varphi : R[x_0, x_1, \ldots, x_n] \to S,$$
$$F(x_0, x_1, \ldots, x_n) \mapsto F(f_0, f_1, \ldots, f_n).$$

Then $I = \operatorname{Ker} \varphi$ is a graded ideal of $R[x_0, x_1, \ldots, x_n]$. The surjectivity implies the closed immersion

(6.51) $$\iota : X = \operatorname{Proj} S \to \mathbb{P}_R^n.$$

Conversely, if X is a closed subscheme $V(I)$ of $\mathbb{P}_R^n$, we can regard $X = \operatorname{Proj} S$. In what will follow, we consider the cohomology of a coherent sheaf over a projective scheme $X = \operatorname{Proj} S$. The following theorem is fundamental.

THEOREM 6.21. *Let $S = \bigoplus_{m=0}^{\infty} S_m$ be a graded ring that is generated by finitely many elements of S_1 as an algebra over a Noetherian ring $S_0 = R$. Then, for a coherent sheaf $\mathcal{F}$ over $X = \operatorname{Proj} S$, $H^p(X, \mathcal{F})$ is a finite R-module (here p is arbitrary). Furthermore, there exists an integer m_0 so that for any $m \geq m_0$,*

$$H^p(X, \mathcal{F}(m)) = 0, \qquad p = 0, 1, 2, \ldots.$$

For an arbitrary coherent sheaf $\mathcal{F}$ over X, there exists a positive integer n_0 such that for $p \geq n_0$ we have

$$H^p(X, \mathcal{F}) = 0.$$

PROOF. Suppose that elements $f_1, f_2, \ldots, f_n$ in S_1 generate S as an $S_0 = R$-algebra. Consider the surjective homomorphism φ in (6.50) and the closed immersion ι in (6.51). From Problem 3, we have an isomorphism

$$H^p(X, \mathcal{F}) \xrightarrow{\sim} H^p(\mathbb{P}_R^n, \iota_* \mathcal{F}).$$

By Theorem 5.24(iii), there exists a finite graded S-module M satisfying $\mathcal{F} = \widetilde{M}$. Let M' be the $R[x_0, x_1, \ldots, x_n]$-module determined by homomorphism φ and M. Then M' is a finite graded $R[x_0, x_1, \ldots, x_n]$-module satisfying $\iota_* \mathcal{F} = \widetilde{M'}$, where $\iota_* \mathcal{F}$ is a coherent sheaf over $\mathbb{P}_R^n$. Furthermore, we have $(\iota_* \mathcal{F})(m) = \iota_*(\mathcal{F}(m))$. Therefore, it is enough to prove the assertion for the case where

$X = \mathbb{P}_R^n$ and $\mathcal{F}$ is a coherent sheaf over X. Then, for an arbitrary coherent sheaf $\mathcal{F}$, we have
$$H^p(X, \mathcal{F}) = 0$$
for $p \geq n+1$. This proves the last part of the theorem.

Put $S = R[x_0, x_1, \ldots, x_n]$. For a finite graded S-module M satisfying $\mathcal{F} = \widetilde{M'}$, there is an exact sequence of graded S-modules
$$\bigoplus_{i=1}^{l} S(m_i) \to M \to 0,$$
so that we may obtain the exact sequence of $\mathcal{O}_X$-modules
$$\bigoplus_{i=1}^{l} \mathcal{O}_X(m_i) \xrightarrow{\eta_0} \mathcal{F} \to 0.$$
Let $\mathcal{F}_1 = \operatorname{Ker} \eta_0$. By Corollary 4.29, $\mathcal{F}_1$ is a coherent sheaf. Then we get the exact sequence

(6.52) $$0 \to \mathcal{F}_1 \to \bigoplus_{i=1}^{l} \mathcal{O}_X(m_i) \xrightarrow{\eta_0} \mathcal{F} \to 0.$$

By repeating this process, we have

(6.53) $$0 \to \mathcal{F}_2 \to \bigoplus_{i=1}^{l_1} \mathcal{O}_X(m_i^{(1)}) \xrightarrow{\eta_1} \mathcal{F}_1 \to 0,$$

where $\mathcal{F}_2$ is coherent. Namely, for $j = 1, 2, \ldots$,

(6.54) $$0 \to \mathcal{F}_{j+1} \to \bigoplus_{i=1}^{l_j} \mathcal{O}_X(m_i^{(j)}) \xrightarrow{\eta_j} \mathcal{F}_j \to 0$$

is an exact sequence of $\mathcal{O}_X$-modules. From (6.52), there is induced the exact sequence of cohomology groups
$$\bigoplus_{i=1}^{l} H^p(X, \mathcal{O}_X(m_i)) \to H^p(X, \mathcal{F}) \to H^{p+1}(X, \mathcal{F}_1)$$
$$\to \bigoplus_{i=1}^{l} H^{p+1}(X, \mathcal{O}_X(m_i)) \to \cdots.$$

By Theorem 6.19, $H^i(X, \mathcal{O}_X(m))$ is a finite R-module. Therefore, $H^p(X, \mathcal{F})$ is a finite R-module if and only if $H^{p+1}(X, \mathcal{F}_1)$ is a finite R-module. From the exact sequence of cohomology groups obtained by (6.53), one can also assert the following statement: $H^{p+1}(X, \mathcal{F}_1)$

is a finite R-module if and only if $H^{p+2}(X, \mathcal{F}_2)$ is a finite R-module. Similarly, by exact sequence (6.54), $H^p(X, \mathcal{F})$ is a finite R-module if and only if for a positive integer j (hence for every positive integer j), $H^{p+j}(X, \mathcal{F}_j)$ is a finite R-module. On the other hand, from Lemma 6.18, $H^{p+j}(X, \mathcal{F}_j) = 0$ for $p + j \geq n + 1$. Hence $H^p(X, \mathcal{F})$ is a finite R-module.

For an arbitrary integer M, by tensoring $\mathcal{O}_X(m)$ to (6.52), we get

$$(6.55) \qquad 0 \to \mathcal{F}_1(m) \to \bigoplus_{i=1}^{l} \mathcal{O}_X(m + m_i) \to \mathcal{F}(m) \to 0,$$

and also (6.54) provides

$$(6.56) \quad 0 \to \mathcal{F}_{j+1}(m) \to \bigoplus_{i=1}^{l} \mathcal{O}_X(m + m_i^{(j)}) \to \mathcal{F}_j(m) \to 0,$$

$$j = 1, 2, 3, \ldots.$$

For $m + m_i \geq -n$, $i = 1, 2, \ldots, l$, we have $H^q(X, \mathcal{O}_X(m + m_i)) = 0$, $q \geq 1$. From the induced exact sequence of cohomology groups associated with (6.55), we get $H^q(X, \mathcal{F}(m)) \overset{\sim}{\to} H^{q+1}(X, \mathcal{F}_1(m))$, $q = 1, 2, \ldots$. Therefore, by (6.56), we obtain

$$H^q(X, \mathcal{F}_j(m)) \overset{\sim}{\to} H^{q+1}(X, \mathcal{F}_{j+1}(m)),$$

$q = 1, 2, \ldots$, for $m + m_i^{(j)} \geq -n$, $j = 1, 2, \ldots, l_j$. Namely, choose m so that $m + m_i \geq -n$, $i = 1, 2, \ldots, l$, and $m + m_i^{(j)} \geq -n$, $i = 1, 2, \ldots, l_j$, $j = 1, 2, \ldots, n$. Then we get

$$H^q(X, \mathcal{F}(m)) \overset{\sim}{\to} H^{q+1}(X, \mathcal{F}_1(m)) \overset{\sim}{\to} H^{q+2}(X, \mathcal{F}_2(m))$$
$$\overset{\sim}{\to} \cdots \overset{\sim}{\to} H^{q+n}(X, \mathcal{F}_n(m)) = 0, \qquad q = 1, 2, \ldots.$$

□

From (6.52) and (6.53), we get the exact sequence of $\mathcal{O}_X$-modules

$$(6.57) \quad \to \bigoplus_{i=1}^{l_j} \mathcal{O}_X(m_i^{(j)}) \to \bigoplus_{i=1}^{l_j-1} \mathcal{O}_X(m_i^{(j-1)}) \to \cdots \to \bigoplus_{i=1}^{l_j} \mathcal{O}_X(m_i^{(j)})$$

$$\to \bigoplus_{i=1}^{l} \mathcal{O}_X(m_i) \to \mathcal{F} \to 0.$$

Since $\bigoplus_{i=1}^{l_j} \mathcal{O}_X(m_i^{(j)})$ is a locally free sheaf, (6.57) is said to be a *locally free resolution* of $\mathcal{F}$.

6.2. COHOMOLOGY OF A PROJECTIVE SCHEME

Let us rewrite Theorem 6.21 when R is a field k.

COROLLARY 6.22. *Suppose that a graded ring $S = \bigoplus_{m=0}^{\infty} S_m$ over a field k is generated by finitely many elements in S_1 as a k-algebra. Then, for an arbitrary coherent sheaf $\mathcal{F}$ over $X = \operatorname{Proj} S$, we have*

$$\dim_k H^p(X, \mathcal{F}) < \infty.$$

Furthermore, one can choose n_0 so that for $p \geq n_0$

$$H^p(X, \mathcal{F}) = 0.$$

One can also find m_0 so that for $m \geq m_0$

$$H^p(X, \mathcal{F}(m)) = 0, \qquad p = 1, 2, 3, \ldots. \quad \square$$

In what will follow, we assume that the graded ring $S = \bigoplus_{m=0}^{\infty} S_m$ is generated by finitely many elements of S_1 as a k-algebra. Hence, we can consider $X = \operatorname{Proj} S$ as a closed subscheme of $\mathbb{P}_k^n$. As shown in the proof of Theorem 6.21, one chooses n that equals the number of generators $f_0, f_1, \ldots, f_n \in S_1$ as a k-algebra minus one. Therefore, we focus on closed subschemes of a projective space $\mathbb{P}_k^n$ over a field k. From Corollary 6.22, for a coherent sheaf $\mathcal{F}$ over X, we can consider

$$(6.58) \qquad \chi(X, \mathcal{F}) = \sum_{q=0}^{\infty} (-1)^q \dim_k H^q(X, \mathcal{F}).$$

This is because the terms on the right have finite values and, for a large q, $H^q(X, \mathcal{F}) = 0$. Then $\chi(X, \mathcal{F})$ is said to be the *Euler characteristic* or *Euler-Poincaré characteristic* of the coherent sheaf $\mathcal{F}$.

EXERCISE 6.23. For an exact sequence of coherent sheaves

$$0 \to \mathcal{F} \to \mathcal{G} \to \mathcal{H} \to 0$$

over a closed subscheme X of $\mathbb{P}_k^n$, we have

$$(6.59) \qquad \chi(X, \mathcal{G}) = \chi(X, \mathcal{F}) + \chi(X, \mathcal{H}).$$

PROOF. The above exact sequence induces the long exact sequence of coherent groups

$$0 \to H^0(X, \mathcal{F}) \to H^0(X, \mathcal{G}) \to H^0(X, \mathcal{H})$$
$$\to H^1(X, \mathcal{F}) \to H^1(X, \mathcal{G}) \to H^1(X, \mathcal{H})$$
$$\to \cdots.$$

All the cohomology groups are finite-dimensional vector spaces over the field k, and for $q \geq n+1$ we have $H^q(X, \mathcal{F}) = H^q(X, \mathcal{G}) = H^q(X, \mathcal{H}) = 0$. From the equation relating dimensions associated with the exact sequence of vector spaces (see Problem 8 below), we get (6.59). □

PROBLEM 8. For an exact sequence of finite-dimensional vector spaces over a field k and linear maps
$$0 \to V_0 \xrightarrow{f_0} V_1 \xrightarrow{f_1} V_2 \xrightarrow{f_2} \cdots \xrightarrow{f_{m-2}} V_{m-1} \xrightarrow{f_{m-1}} V_m \to 0,$$
prove that
$$\sum_{j=0}^{n} (-1)^j \dim_k V_j = 0.$$

In algebraic geometry, it is important to determine the dimension of $H^0(X, \mathcal{F})$. Even though it is a difficult problem, for a non-singular X (see *Algebraic Geometry* 3) we can compute the Euler characteristic $\chi(X, \mathcal{F})$ via an invariant called the Chern class associated with X and $\mathcal{F}$. That is, the computation can be done by the Riemann-Roch Theorem. See *Algebraic Geometry* 3 for the curve case.

PROBLEM 9. Let $X = \mathbb{P}_k^n$ be an n-dimensional projective space over a field k. For an integer m, prove that
$$\chi(X, \mathcal{O}_X(m)) = \begin{cases} \binom{m+n}{n}, & m \geq 0, \\ (-1)\binom{-m-1}{n}, & m \leq -n-1, \\ 0, & \text{otherwise.} \end{cases}$$
In particular, $\chi(X, \mathcal{O}_X) = 1$.

EXAMPLE 6.24. Let $X = X_d$ be a d-dimensional hypersurface in an n-dimensional projective space $\mathbb{P}_k^n$ over a field k. Namely, X_d is the closed subscheme determined by a homogeneous polynomial of degree d, i.e., $X_d = V(F)$. The sheaf $\mathcal{J}_X$ of ideals of X_d is isomorphic to $\mathcal{O}_\mathbb{P}(-d)$ as an $\mathcal{O}_\mathbb{P}$-module. Then

(6.60) $$0 \to \mathcal{O}_\mathbb{P}(-d) \to \mathcal{O}_\mathbb{P} \to \mathcal{O}_X \to 0$$

is exact. From Example 6.23, we get
$$\chi(X, \mathcal{O}_X) = \chi(\mathbb{P}_k^n, \mathcal{O}_\mathbb{P}) - \chi(\mathbb{P}_k^n, \mathcal{O}_\mathbb{P}(-d)).$$

Therefore, for $d \leq n$ we have $\chi(X, \mathcal{O}_X) = 1$. On the other hand, for $d \geq n+1$, we have $\chi(X, \mathcal{O}_X) = 1 + (-1)^{n-1}\binom{d-1}{n}$. In fact, Corollary

6.20 implies that
$$\dim_k H^0(X, \mathcal{O}_X) = 1,$$
$$\dim_k H^{n-1}(X, \mathcal{O}_X) = \begin{cases} 0, & d \leq n, \\ \binom{d-1}{n}, & d \geq n+1, \end{cases}$$

and also $H^q(X, \mathcal{O}_X) = 0$, $1 \leq q \leq n-2$.

Furthermore, for a positive integer e, by tensoring (6.60) with $\mathcal{O}_\mathbb{P}(-e)$, we get the exact sequence
$$0 \to \mathcal{O}_\mathbb{P}(-e-d) \to \mathcal{O}_\mathbb{P}(-e) \to \mathcal{O}_X(-e) \to 0.$$

From this, we get
$$\chi(X, \mathcal{O}_X(-e)) = \chi(X, \mathcal{O}_\mathbb{P}(-e)) - \chi(X, \mathcal{O}_\mathbb{P}(-e-d)).$$

Consequently, we obtain
$$\chi(X, \mathcal{O}_X(-e)) = \begin{cases} (-1)^{n-1}\left\{\binom{e-1}{n} - \binom{e+d-1}{n}\right\}, & \text{for } e \geq n+1, \\ (-1)^n \binom{e+d-1}{n}, & \text{for } e+d \geq n+1,\ e \leq n, \\ 0, & \text{for } e+d \leq n. \end{cases}$$

(c) **Bézout's Theorem**

We will give a proof of Bézout's Theorem (Theorem 1.32) as an application of sheaf cohomology. Let $C_1 = V(F)$ and $C_2 = V(G)$ be the closed subschemes in a projective plane $X = \mathbb{P}^2_k$ determined by homogeneous polynomials $F, G \in k[x_0, x_1, x_2]$ with coefficients in an algebraically closed field k. Let m and n be the degrees of F and G, respectively. Assume that C_1 and C_2 do not have a common component, intersecting only at points. Let $\mathcal{O}_X(-C_1)$ and $\mathcal{O}_X(-C_2)$ be the defining ideals of C_1 and C_2. Then we have the exact sequences

(6.61) $\quad\quad 0 \to \mathcal{O}_X(-C_1) \to \mathcal{O}_X \to \mathcal{O}_{C_1} \to 0,$

(6.62) $\quad\quad 0 \to \mathcal{O}_X(-C_2) \to \mathcal{O}_X \to \mathcal{O}_{C_2} \to 0.$

As $\mathcal{O}_X$-modules, we can identify $\mathcal{O}_X(-C_1)$ and $\mathcal{O}_X(-C_2)$ with $\mathcal{O}_X(-m)$ and $\mathcal{O}_X(-n)$. Let x be a point of intersection of C_1 and C_2. Let f_x and g_x be the defining equations (i.e., generators as ideals of $\mathcal{O}_{X,x}$ for $\mathcal{O}_X(-C_1)_x$ and $\mathcal{O}_X(-C_2)_x$). Define the local intersection multiplicity of C_1 and C_2 at x as
$$I_x(C_1, C_2) = \dim_k \mathcal{O}_{X,x}/(f_x, g_x).$$

The intersection multiplicity $C_1 \cdot C_2$ is defined by

$$C_1 \cdot C_2 = \sum_{x \in C_1 \cap C_2} \dim_k \mathcal{O}_{X,x}/(f_x, g_x).$$

That is, for the closed subscheme $C_1 \cap C_2 = V((F, G))$, we can write

(6.63) $\quad C_1 \cdot C_2 = \dim_k H^0(C_1 \cap C_2, \mathcal{O}_{C_1 \cap C_2}) = \chi(C_1 \cap C_2, \mathcal{O}_{C_1 \cap C_2}).$

Here is a proof. Let $V_j = D_+(x_j)$, $j = 0, 1, 2$, be an affine covering of $\mathbb{P}_k^2$. Over V_0, let $x = \frac{x_1}{x_0}, y = \frac{x_2}{x_0}$, $f(x, y) = x_0^{-m} F(x_0, x_1, x_2)$, $g(x, y) = x_0^{-n} G(x_0, x_1, x_2)$. Over V_0, $\mathcal{O}_X(-C_1)$ is the ideal sheaf (f) generated by f, and $\mathcal{O}_X(-C_2)$ is the ideal sheaf (g) generated by g, and the defining ideal of $C_1 \cap C_2$ over V_0 is the ideal sheaf (f, g) generated by f and g. Therefore we have $\mathcal{O}_{C_1 \cap C_2 | V_0} = \mathcal{O}_{V_0}/(f, g)$, and $\mathrm{supp}(\mathcal{O}_{C_1 \cap C_2 | V_0})$ is the totality of all the intersection points in V_0 of C_1 and C_2. For $x \in \mathrm{supp}(\mathcal{O}_{C_1 \cap C_2 | V_0})$, we have $\mathcal{O}_{C_1 \cap C_2, x} = \mathcal{O}_{V_0}/(f_x, g_x)$, where f_x and g_x are corresponding elements in $\mathcal{O}_{V_0, x}$ of f and g, respectively. Similar results will be obtained for other affine open sets V_1 and V_2. Each point of $C_1 \cap C_2$ is isolated. We can cover the space by mutually disjoint affine open sets. (In fact, each point of $C_1 \cap C_2$ is a closed and open set.) Therefore, we obtain $H^q(C_1 \cap C_2, \mathcal{O}_{C_1 \cap C_2}) = 0$, $q \geq 1$, which proves (6.63). One can observe by the restriction to each V_j that the sequence

(6.64) $\quad 0 \to \mathcal{O}_X(-C_2) \otimes_{\mathcal{O}_X} \mathcal{O}_{C_1} \to \mathcal{O}_{C_1} \to \mathcal{O}_{C_1 \cap C_2} \to 0$

is exact, considering $\mathcal{O}_X(-C_2) \otimes_{\mathcal{O}_X} \mathcal{O}_{C_1} \subset \mathcal{O}_{C_1}$. From this, we obtain

(6.65)
$\chi(C_1 \cap C_2, \mathcal{O}_{C_1 \cap C_2}) = \chi(C_1, \mathcal{O}_{C_1}) - \chi(C_1, \mathcal{O}_X(-C_2) \otimes_{\mathcal{O}_X} \mathcal{O}_{C_1}).$

Since

$$\mathcal{O}_X(-C_1) = \mathcal{O}_X(-m),$$
$$\mathcal{O}_X(-C_2) = \mathcal{O}_X(-n),$$
$$\mathcal{O}_X(-C_2) \otimes_{\mathcal{O}_X} \mathcal{O}_{C_1} = \mathcal{O}_{C_1}(-n),$$

by tensoring (6.61) and (6.62) with $\mathcal{O}_X(-n)$, we get

$$\chi(C_1, X_2) = \chi(X, \mathcal{O}_X) - \chi(X, \mathcal{O}_X(-m))$$
$$= 1 - \chi(X, \mathcal{O}_X(-m)),$$
$$\chi(C_1, \mathcal{O}_{C_1}(-n)) = \chi(X, \mathcal{O}_X(-n)) - \chi(X, \mathcal{O}_X(-m-n)).$$

6.2. COHOMOLOGY OF A PROJECTIVE SCHEME

Namely, we obtain
(6.66)
$$\chi(C_1 \cap C_2, \mathcal{O}_{C_1 \cap C_2}) = 1 - \chi(X, \mathcal{O}_X(-m)) - \chi(C_1, \mathcal{O}_{C_1}(-n))$$
$$+ \chi(X, \mathcal{O}_X(-m-n)).$$

If $m \geq 3$ and $n \geq 3$, then Problem 7 implies
$$\chi(C_1 \cap C_2, \mathcal{O}_{C_1 \cap C_2}) = 1 - \binom{m-1}{2} - \binom{n-1}{2} + \binom{m+n-1}{2}$$
$$= mn.$$

For $m = 1, 2$ or $n = 1, 2$, we also get the above equality from Problem 7. This completes the proof of Bézout's Theorem.

Note that in (6.64), C_1 can be replaced by C_2. In fact, (6.66) is symmetric with respect to C_1 and C_2.

Bézout's Theorem computes the intersection multiplicity of curves in $\mathbb{P}^2_k$. We used $\mathbb{P}^2_k$ only at the end of the above proof for the explicit computation. That is, (6.65) is still valid for non-singular algebraic surfaces (see Chapter 7 for the definition).

We began with the non-symmetric situation for C_1 and C_2, but the result was symmetric for C_1 and C_2. A proof can be given symmetrically. That is, since
$$\mathcal{O}_{C_1} \otimes_{\mathcal{O}_X} \mathcal{O}_{C_2|V_0} = \mathcal{O}_{V_0}/(f) \otimes_{\mathcal{O}_{V_0}} \mathcal{O}_{V_0}/(g) = \mathcal{O}_{V_0}/(f,g),$$
one can show that $\mathcal{O}_{C_1} \otimes_{\mathcal{O}_X} \mathcal{O}_{C_2} = \mathcal{O}_{C_1 \cap C_2}$. From (6.61) and (6.62), we get an $\mathcal{O}_X$-module sequence
$$\mathcal{O}_X(-C_1 - C_2) \to \mathcal{O}_X(-C_1) \otimes \mathcal{O}_X(-C_2)$$
$$\to \mathcal{O}_X \to \mathcal{O}_{C_1} \otimes_{\mathcal{O}_X} \mathcal{O}_{C_2} \to 0.$$

When this sequence is restricted to V_0, we get
$$(fg) \to (f) \oplus (g) \to \mathcal{O}_{V_0} \to \mathcal{O}_{V_0}/(f,g) \to 0,$$
$$\alpha fg \mapsto (\alpha gf, \alpha fg),$$
$$(\beta f, \gamma g) \mapsto \beta f - \gamma g.$$

Namely, the above sequence is exact, and the homomorphism on the far left is injective. This argument holds for V_1 and V_2. We have the following exact sequence:
(6.67)
$$0 \to \mathcal{O}_X(-C_1 - C_2) \to \mathcal{O}_X(-C_1) \oplus \mathcal{O}_X(-C_2)$$
$$\to \mathcal{O}_X \to \mathcal{O}_{C_1 \cap C_2} \to 0.$$

This exact sequence is symmetric with respect to C_1 and C_2. We get
(6.68)
$$\chi(C_1 \cap C_2, \mathcal{O}_{C_1 \cap C_2}) = \chi(X, \mathcal{O}_X) - \chi(X, \mathcal{O}_X(-C_1))$$
$$- \chi(X, \mathcal{O}_X(-C_2)) + \chi(X, \mathcal{O}_X(-C_1 - C_2)),$$
which is exactly (6.66) itself. This is the equation for the case where X is a general non-singular algebraic surface. Namely, we have obtained the following theorem.

THEOREM 6.25. *Let X be a non-singular algebraic surface defined over an algebraically closed field k. If algebraic curves C_1 and C_2 on X do not have a common component, then the intersection multiplicity $C_1 \cdot C_2$ defined by (6.63) can be computed as*
$$C_1 \cdot C_2 = \chi(X, \mathcal{O}_X) - \chi(X, \mathcal{O}_X(-C_1))$$
$$- \chi(X, \mathcal{O}_X(-C_2)) + \chi(X, \mathcal{O}_X(-C_1 - C_2)). \quad \Box$$

This theorem is the generalization of Bézout's Theorem..

(d) **Criterion for Ampleness**

We will use cohomology to determine whether an invertible sheaf is ample or not.

THEOREM 6.26. *Let X be a proper scheme over a Noetherian ring R, i.e., $X \to Y = \operatorname{Spec} R$ is a proper morphism, and let $\mathcal{L}$ be an invertible sheaf over X. Then the following (i) and (ii) are equivalent.*

(i) $\mathcal{L}$ *is an ample invertible sheaf.*

(ii) *For any coherent sheaf $\mathcal{F}$ over X, there exists a positive integer n_0 so that for $n \geq n_0$ we have $H^q(X, \mathcal{F} \otimes \mathcal{L}^{\otimes n}) = 0$, $q \geq 1$.*

PROOF. From Theorem 5.41, one can choose M so that $\mathcal{L}^{\otimes m}$ is ample with respect to f. Let $\mathcal{F} \otimes \mathcal{L}^{\otimes l} = \mathcal{F}(l)$. Apply Theorem 6.21 to $\mathcal{F}, \mathcal{F}(1), \ldots, \mathcal{F}(m-1)$ to obtain
$$H^q(X, \mathcal{F}(km)) = 0, H^q(X, \mathcal{F}(1 + km)) = 0,$$
$$\ldots, H^q(X, \mathcal{F}(m - 1 + km)) = 0,$$
$q \geq 1$ and $k \geq k_0$ for some k_0. Then for $n_0 = k_0 m$, we obtain (ii).

We will prove that (ii) implies (i). It is enough to prove that for any coherent sheaf $\mathcal{F}$, there exists a positive integer n_0 with the following property: for $n \geq n_0$, $\mathcal{F} \otimes \mathcal{L}^{\otimes n}$ is generated as an $\mathcal{O}_X$-module by $\Gamma(X, \mathcal{F} \otimes \mathcal{L}^{\otimes n})$. Let $\mathfrak{m}_x$ be the defining ideal of a closed point x of X and let $k(x) = \mathcal{O}_{X,x}/\mathfrak{m}_x$. Then we have the exact sequence
$$0 \to \mathfrak{m}_x \mathcal{F} \to \mathcal{F} \to \mathcal{F} \otimes_{\mathcal{O}_X} k(x) \to 0.$$

By tensoring with $\mathcal{L}^{\otimes n}$, the exact sequence

$$0 \to (\mathfrak{m}_x \mathcal{F}) \otimes \mathcal{L}^{\otimes n} \to \mathcal{F} \otimes \mathcal{L}^{\otimes n} \to (\mathcal{F} \otimes \mathcal{L}^{\otimes n}) \otimes k(x) \to 0$$

is induced. By the assumption, there is an n_0 to satisfy

$$H^1(X, (\mathfrak{m}_x \mathcal{F}) \otimes \mathcal{L}^{\otimes n}) = 0 \quad \text{for} n \geq n_0.$$

By the exact sequence of the cohomology groups,

$$\Gamma(X, \mathcal{F} \otimes \mathcal{L}^{\otimes n}) \to \Gamma(X, (\mathcal{F} \otimes \mathcal{L}^{\otimes n}) \otimes k(x))$$

is surjective. Nakayama's lemma implies that the stalk of $\mathcal{F} \otimes \mathcal{L}^{\otimes n}$ at x is generated by $\Gamma(X, \mathcal{F} \otimes \mathcal{L}^{\otimes n})$. Then, for $n \geq n_0$, one can choose an open neighborhood U of x so that the stalk of $\mathcal{F} \otimes \mathcal{L}^{\otimes n}$ at each point x in U is generated by $\Gamma(X, \mathcal{F} \otimes \mathcal{L}^{\otimes n})$. (This U may vary for each n.)

In particular, for $\mathcal{F} = \mathcal{O}_X$, one can choose n_1 and an open neighborhood V of x so that the stalk of $\mathcal{L}^{\otimes n_1}$ at each point in V is generated by $\Gamma(X, \mathcal{L}^{\otimes n_1})$. Then, for a coherent sheaf and for $m = 0, 1, 2, \ldots, n_1 - 1$, one can choose an open neighborhood U_m of x so that the stalk of $\mathcal{F} \otimes \mathcal{L}^{\otimes (n_0+m)}$ at each point in U_m is generated by $\Gamma(X, \mathcal{F} \otimes \mathcal{L}^{\otimes (n_0+m)})$. For $n \geq n_0$, let $n = n_0 + rn_1 + m$, $0 \leq m \leq n_1 - 1$. Since we have $\mathcal{F} \otimes \mathcal{L}^{\otimes n} = (\mathcal{F} \otimes \mathcal{L}^{\otimes (n_0+m)}) \otimes (\mathcal{L}^{\otimes n_1})^{\otimes r}$, the stalk of $\mathcal{F} \otimes \mathcal{L}^{\otimes n}$ at each point in $U_x = V \cap U_0 \cap U_1 \cap \cdots \cap U_{n_1-1}$ is generated by $\Gamma(X, \mathcal{F} \otimes \mathcal{L}^{\otimes n})$.

By considering the above argument at each closed point x in X, we get an open covering $\{U_x\}$ of X. Since X is Noetherian, there exists a finite covering $\{U_{x_i}\}_{i=1}^s$. Again let n_0 be the largest n_0 determined at each point x. Then, for $n \geq n_0$, $\mathcal{F} \otimes \mathcal{L}^{\otimes n}$ is generated by the global sections $H^0(X, \mathcal{F} \otimes \mathcal{L}^{\otimes n})$. □

6.3. Higher Direct Image

(a) **Higher Direct Image**

For a continuous map of topological spaces $f : X \to Y$ and for an exact sequence of sheaves of additive groups

$$0 \to \mathcal{F}_1 \to \mathcal{F}_2 \to \mathcal{F}_3 \to 0,$$

we have the exact sequence induced by the direct image f_*,

$$0 \to f_* \mathcal{F}_1 \to f_* \mathcal{F}_2 \to f_* \mathcal{F}_3.$$

As we did for the cohomology groups, we will consider extending the above exact sequence to a long exact sequence. For an open set U of

Y, we have the following exact sequence of cohomology groups over $f^{-1}(U)$:

(6.69) $\quad 0 \to \Gamma(f^{-1}(U), \mathcal{F}_1) \to \Gamma(f^{-1}(U), \mathcal{F}_2) \to \Gamma(f^{-1}(U), \mathcal{F}_3)$
$\to H^1(f^{-1}(U), \mathcal{F}_1) \to H^1(f^{-1}(U), \mathcal{F}_2) \to H^1(f^{-1}(U), \mathcal{F}_3) \to \cdots$.

To a sheaf $\mathcal{F}$ of additive groups over X and an open set U of X, one associates $H^p(f^{-1}(U), \mathcal{F})$. For open sets $V \subset U$, we have the natural homomorphism

$$\rho_{V,U} : H^p(f^{-1}(U), \mathcal{F}) \to H^p(f^{-1}(V), \mathcal{F}).$$

For open sets $W \subset V \subset U$, we also have $\rho_{W,U} = \rho_{W,V} \circ \rho_{V,U}$. That is, we have a presheaf $U \mapsto H^p(f^{-1}(U), \mathcal{F})$. The sheafification of this presheaf is denoted as $R^p f_* \mathcal{F}$ and is said to be the p-th higher direct image of $\mathcal{F}$ under f. We call $R^p f_* \mathcal{F}$, $p \geq 1$, the higher direct images of $\mathcal{F}$. We sometimes write $f_* \mathcal{F}$ as $R^0 f_* \mathcal{F}$. Since direct limit preserves the exactness, from (6.69) we get the following proposition.

PROPOSITION 6.27. *For an exact sequence of sheaves of additive groups over a topological space X*

$$0 \to \mathcal{F}_1 \to \mathcal{F}_2 \to \mathcal{F}_3 \to 0,$$

a continuous map $f : X \to Y$ induces the long exact sequence of direct images and higher direct images over Y

$$0 \to f_* \mathcal{F}_1 \to f_* \mathcal{F}_2 \to f_* \mathcal{F}_3 \to R^1 f_* \mathcal{F}_1$$
$$\to R^1 f_* \mathcal{F}_2 \to R^1 f_* \mathcal{F}_3 \to R^2 f_* \mathcal{F}_1 \to \cdots . \quad \square$$

In algebraic geometry, for a quasicoherent (or coherent) sheaf $\mathcal{F}$ over X, it is important to consider the case where a scheme morphism $f : X \to Y$ induces the quasicoherent (or coherent) higher direct image over Y. For a quasicoherent sheaf, things are simple. When either f is a quasicompact separated morphism or X is a Noetherian scheme, we showed that, for a quasicoherent $\mathcal{O}_X$-module $\mathcal{F}$, the direct image $f_* \mathcal{F}$ is quasicoherent (i.e., Proposition 4.36). Using this fact, we will show that the higher direct images $R^p f_* \mathcal{F}$ are also quasicoherent.

First consider the case when $f : X \to Y$ is a quasicompact separated morphism. The quasicoherency of $R^p f_* \mathcal{F}$ is of a local nature with respect to Y. Hence, one may assume $Y = \operatorname{Spec} R$. Since f is quasicompact, there exists a finite affine open covering $\mathcal{U} = \{U_i\}_{i \in I}$. Since f is separated, $U_{ij} = U_i \cap U_j$, $U_{i_1 i_2 \cdots i_n} = U_{i_1} \cap U_{i_2} \cap \cdots \cap U_{i_n}$ are affine open sets. Let $U_i = \operatorname{Spec} A_i$. Then A_i is an R-algebra.

6.3. HIGHER DIRECT IMAGE

By Theorem 6.10 and Leray's Theorem (Theorem 6.15), the cohomology groups with coefficients in $\mathcal{F}$ can be computed by the Čech complex

(6.70) $\quad 0 \to C^0(\mathcal{U}, \mathcal{F}) \xrightarrow{\delta^0} C^1(\mathcal{U}, \mathcal{F}) \xrightarrow{\delta^1} C^2(\mathcal{U}, \mathcal{F}) \to \cdots,$

where $C^p(\mathcal{U}, \mathcal{F})$ is the totality of alternating cochains $\{f_{i_0 i_1 \cdots i_p}\}$. Since $\mathcal{F}$ is an $\mathcal{O}_X$-module, for $a \in R$, one can define $\{af_{i_0 i_1 \cdots i_p}\}$. Namely, $C^p(\mathcal{U}, \mathcal{F})$ is an R-module.

Since $\mathcal{F}$ is a quasicoherent $\mathcal{O}_X$-module, there exists an A_i-module M_i satisfying $\mathcal{F}|U_i = \widetilde{M_i}$. Then M_i is an R-module. For $a \in R$, consider the open set $D(a)$ of $\operatorname{Spec} R$. We have $f^{-1}(D(a)) \cap U_i = \operatorname{Spec}(A_i)_a$ and

$$\mathcal{F}|f^{-1}(D(a)) \cap U_i = (\widetilde{M_i})_a.$$

We also get $M_i = \Gamma(U_i, \mathcal{F})$ and

$$(M_i)_a = \Gamma(f^{-1}(D(a)) \cap U_i, \mathcal{F}) = M_i \otimes_R R_a.$$

Since localization takes an exact sequence of R-modules to an exact sequence of R_a-modules, the Čech complex $\{C^\bullet(\mathcal{U}_a, \mathcal{F}), \delta_a^\bullet\}$ for the covering $\mathcal{U}_a = \{f^{-1}(D(a)) \cap U_i\}_{i \in I}$ of $f^{-1}(D(a))$ is the localization of the complex (6.70) at a. That is, one can obtain the above complex by tensoring (6.70) with R_a as R-modules. Next, consider the relationship between localization and cohomology. We have

$$H^p(X, \mathcal{F}) = \ker \delta^p / \operatorname{Im} \delta^{p-1}, \quad H^p(f^{-1}(D(a)), \mathcal{F}) = \operatorname{Ker} \delta_a^p / \operatorname{Im} \delta_a^{p-1}.$$

By tensoring the exact sequence of R-modules

$$0 \to \operatorname{Ker} \delta^p \to C^p(\mathcal{U}, \mathcal{F}) \xrightarrow{\delta^p} C^{p+1}(\mathcal{U}, \mathcal{F})$$

with R_a over R, we get the exact sequence of R_a-modules

$$0 \to \operatorname{Ker} \delta^p \otimes_R R_a \to C^p(\mathcal{U}_a, \mathcal{F}) \xrightarrow{\delta_a^p} C^{p+1}(\mathcal{U}_a, \mathcal{F}).$$

Since $\operatorname{Ker}^{p-1} \otimes_R R_a = \operatorname{Ker} \delta_a^{p-1}$, we have $\operatorname{Im} \delta_a^{p-1} = \operatorname{Im} \delta^{p-1} \otimes_R R_a$. Namely,

(6.71) $\quad H^p(X, \mathcal{F}) \otimes_R R_a = H^p(f^{-1}(D(a)), \mathcal{F}).$

As noted, $R^p f_* \mathcal{F}$ is the sheafification of the presheaf $H^p(f^{-1}(U), \mathcal{F})$. Equation (6.71) indicates that this presheaf is obtained by the localization of $H^p(X, \mathcal{F})$ over $D(a)$, $a \in R$, as an R-module. That is, $R^p f_* \mathcal{F}$ is a sheaf of additive groups over Y constructed from the R-module $H^p(X, \mathcal{F})$. Namely, $R^p f_* \mathcal{F}$ is a quasicoherent sheaf over Y.

We have proved half of the following theorem.

THEOREM 6.28. *If a scheme morphism $f : X \to Y$ is quasi-compact and separated, or if X is Noetherian, then, for an arbitrary quasicoherent sheaf $\mathcal{F}$ over X, the sheaves $R^p f_* \mathcal{F}$, $p \geq 0$, are quasi-coherent sheaves over Y.*

In the case where X is a Noetherian scheme, the above proof fails. This is because for an affine open covering $\mathcal{U} = \{U_i\}_{i \in I}$, $U_{ij} = U_i \cap U_j$ need not be an affine open set. We will generalize the method of proof for a Noetherian affine scheme in Theorem 6.10.

LEMMA 6.29. *A quasicoherent sheaf $\mathcal{F}$ over a Noetherian scheme X can be embedded into a quasicoherent flabby sheaf as an $\mathcal{O}_X$-submodule.*

PROOF. Let $\{U_i\}_{i=1}^m$ be a finite affine open covering of X, and let $U_i = \operatorname{Spec} A_i$, $\mathcal{F}|U_i = \widetilde{M_i}$, where M_i is an A_i-module. Then there exists an injective A_i-module I_i containing M_i as an A_i-module. Lemma 6.11 implies that $\widetilde{I_i}$ is a flabby sheaf over U_i. Let $\iota^{(i)} : U_i \to X$ be an open immersion. Then $\iota_I^{(i)} \widetilde{I_i}$ is a flabby sheaf over X. Therefore, $\mathcal{J} = \bigoplus_{i=1}^m \iota_I^{(i)} \widetilde{I_i}$ is a flabby sheaf. On the other hand, we have the exact sequence of $\mathcal{O}_{U_i}$-modules

$$0 \to \mathcal{F}|U_i \to \widetilde{I_i},$$

which implies the homomorphism of $\mathcal{O}_X$-modules $\mathcal{F} \to \iota_I^{(i)} \widetilde{I_i}$. This homomorphism is injective over U_i. Consequently, we obtain the exact sequence of $\mathcal{O}_X$-modules

$$0 \to \mathcal{F} \to \mathcal{J} = \bigoplus_{i=1}^m \iota_*^{(i)} \widetilde{I_i}.$$

Since $\widetilde{I_i}$ is a quasicoherent sheaf and an open immersion is a quasicompact separated morphism, from Proposition 4.36 we conclude that $\iota_*^{(i)} \widetilde{I_i}$ is quasicoherent and that $\mathcal{J}$ is also a quasicoherent flabby sheaf. □

This lemma implies that for a quasicoherent sheaf $\mathcal{F}$ over X, there exists a resolution by quasicoherent flabby sheaves

$$0 \to \mathcal{F} \to \mathcal{G}^0 \to \mathcal{G}^1 \to \mathcal{G}^2 \to \cdots.$$

Therefore, for $f : X \to Y$ and for an open set U of Y, $H^p(f^{-1}(U), \mathcal{F})$ can be computed as a cohomology of the complex

$$0 \to \Gamma(f^{-1}(U), \mathcal{G}^0) \xrightarrow{\delta_U^0} \Gamma(f^{-1}(U), \mathcal{G}^1) \xrightarrow{\delta_U^1} \Gamma(f^{-1}(U), \mathcal{G}^2) \to \cdots.$$

In order to show the quasicoherency of $R^p f_* \mathcal{F}$, one may assume $Y = \operatorname{Spec} R$. For an open set U of Y, we can also write

$$\Gamma(f^{-1}(U), \mathcal{G}^i) = \Gamma(U, f_* \mathcal{G}^i).$$

From Proposition 4.36, $f_* \mathcal{G}^i$ is a quasicoherent sheaf over Y. Therefore, for the affine open set $D(a)$ determined by $a \in R$, we have

$$\Gamma(f^{-1}(D(a)), \mathcal{G}^i) = \Gamma(D(a), f_* \mathcal{G}^i) = \Gamma(Y, f_* \mathcal{G}^i) \otimes_R R_a.$$

As before, we get

$$\operatorname{Ker} \delta_{D(a)}^p = \operatorname{Ker} \delta_Y^p \otimes_R R_a, \quad \operatorname{Im} \delta_{D(a)}^{p-1} = \operatorname{Im} \delta_Y^{p-1} \otimes_R R_a.$$

By definition, we have

$$H^p(X, \mathcal{F}) = \operatorname{Ker} \delta_Y^p / \operatorname{Im} \delta_Y^{p-1},$$
$$H^p(f^{-1}(D(a)), \mathcal{F}) = \operatorname{Ker} \delta_{D(a)}^p / \operatorname{Im} \delta_{D(a)}^{p-1}.$$

Namely, we obtain

$$H^p(f^{-1}(D(a)), \mathcal{F}) = H^p(X, \mathcal{F}) \otimes_R R_a.$$

That is, $R^p f_* \mathcal{F}$ is nothing but the $\mathcal{O}_Y$-module constructed from the R-module $H^p(X, \mathcal{F})$. This proves that $R^p f_* \mathcal{F}$ is a quasicoherent sheaf, completing the proof of Theorem 6.28. □

PROBLEM 10. For a closed immersion $\iota : X \to Y$, we have $R^p f_* \mathcal{F} = 0$, $p \geq 1$, where $\mathcal{F}$ is an arbitrary quasicoherent sheaf over X.

(b) **Projective Morphisms**

One of the most important devices in algebraic geometry is the notion of a coherent sheaf. The proper mapping theorem states that, for a proper morphism $f : X \to Y$ and a coherent sheaf over X, if Y is a Noetherian scheme, then the direct image and the higher direct images of the coherent sheaf are coherent sheaves over Y. This theorem is crucial in algebraic geometry. We will prove this theorem in Chapter 7. Next, we will treat the case when $f : X \to Y$ is a projective morphism.

THEOREM 6.30. *For a projective morphism $f : X \to Y$ and for a coherent sheaf $\mathcal{F}$ over X, if Y is a locally Noetherian scheme, then, for all p, $R^p f_* \mathcal{F}$ is a coherent sheaf over Y.*

PROOF. It is sufficient to prove the case when Y is an affine scheme $\operatorname{Spec} R$. By the assumption, R is a Noetherian ring. Then there exists a graded ring $S = \bigoplus_{n=0}^{\infty} S_n$ such that $S_0 = R$. As an R-algebra, S is generated by the elements of S_1, and S_1 is a finite R-module satisfying $X = \operatorname{Proj} S$, where $f : X \to Y$ coincides with the structure morphism $\operatorname{Proj} S \to \operatorname{Spec} R$. Then, by Theorem 6.21, for all p, $H^p(X, \mathcal{F})$ is a finite R-module. The sheaf associated with it is $R^p f_* \mathcal{F}$. Namely, $R^p f_* \mathcal{F}$ is a coherent sheaf. □

PROBLEM 11. For a projective morphism $f : X \to Y$, let $\mathcal{L}$ be an ample invertible sheaf with respect to f. Assume that Y is a Noetherian scheme. For an arbitrary coherent sheaf $\mathcal{F}$ over X, let $\mathcal{F}(m) = \mathcal{F} \otimes \mathcal{L}^{\otimes m}$. Prove that one can choose a positive integer m_0 so that for $m \geq m_0$ we have $R^p f_* \mathcal{F}(m) = 0$ for $p \geq 1$.

Summary

6.1. Cohomology of a sheaf over a topological space X is defined by a flabby resolution of the sheaf. For a short exact sequence of sheaves of additive groups, there is induced a long exact sequence of cohomology groups.

6.2. Čech cohomology is defined for an open covering. If a scheme is separated, then the Čech cohomology coincides with the sheaf cohomology.

6.3. For an affine scheme, all the higher cohomologies of a quasicoherent sheaf vanish.

6.4. The cohomology of the invertible sheaf $\mathcal{O}_P(m)$ of a projective space $P = \mathbb{P}^n_R$ over a Noetherian ring R can be computed explicitly.

6.5. For an invertible sheaf over a proper scheme over a Noetherian ring, one can determine whether the sheaf is an ample invertible sheaf or not by a cohomological method.

6.6. For curves on a projective plane over an algebraically closed field, the intersection multiplicity theorem can be formulated in terms of cohomology, and from it one can prove Bézout's Theorem.

6.7. Let $f : X \to Y$ be a continuous map of topological spaces. For a sheaf $\mathcal{F}$ of additive groups, the higher direct image $R^p f_* \mathcal{F}$ can be defined. We have $R^0 f_* \mathcal{F} = f_* \mathcal{F}$. For a quasicompact and separated morphism of schemes $f : X \to Y$, if $\mathcal{F}$ is a quasicoherent sheaf over X, then their higher direct images $R^p f_* \mathcal{F}$ are quasicoherent.

Exercises

6.1. For a Noetherian scheme X, prove that the following conditions are equivalent.

(1) X is an affine scheme.

(2) For an arbitrary coherent sheaf $\mathcal{F}$, we have $H^n(X, \mathcal{F}) = 0$ for all $n \geq 1$.

(3) For an arbitrary coherent sheaf $\mathcal{J}$ of ideals, we have $H^n(X, \mathcal{J}) = 0$ for all $n \geq 1$.

6.2. Let $\mathbb{P}_k^2$ be a projective plane over an algebraically closed field k. Let $C = V(F)$ be a plane curve of degree n on $\mathbb{P}_k^2$, where F is a homogeneous polynomial of degree n. Prove that
$$\dim_k H^1(C, \mathcal{O}_C) = \frac{(n-1)(n-2)}{2}.$$

6.3. Prove that there is an isomorphism of commutative groups between isomorphic classes $\operatorname{Pic} X$ of invertible sheaves over a scheme X and $\check{H}^1(X, \mathcal{O}_X^*)$. Note that the group structure of the cohomology group $\check{H}^1(\mathcal{U}, \mathcal{O}_X^*)$ induced by an open covering $\mathcal{U} = \{U_i\}_{i \in I}$ of X is defined by $\{f_{ij}\} \cdot \{g_{ij}\} = \{f_{ij} \cdot g_{ij}\}$ for cochains $\{f_{ij}\}$ and $\{g_{ij}\}$. Then the inductive limit induces the group structure on $\check{H}^1(X, \mathcal{O}_X^*)$.

Solutions to Problems

Chapter 4

Problem 1. Since $^a\mathcal{G}$ is obviously a presheaf, we will show that $^a\mathcal{G}$ satisfies (F1) and (F2). For an open covering $\{V_\lambda\}_{\lambda \in \Lambda}$ of U, if $s = \{s(x)\}_{x \in U} \in {}^a\mathcal{G}(U)$ satisfies $\rho_{V_\lambda, U}(s) = 0$, $\lambda \in \Lambda$, then the definition of a restriction map implies $s(x) = 0_x$ for each $x \in U$. Note that 0_x is a zero element of $\mathcal{G}_x$. Therefore, $s = 0$ holds, proving (F1). Suppose that $s_\lambda = \{s_\lambda(y)\}_{y \in V_\lambda} \in {}^a\mathcal{G}(V_\lambda)$, $\lambda \in \Lambda$, satisfy $\rho_{V_{\lambda\mu}, V_\lambda}(s_\lambda) = \rho_{V_{\lambda\mu}, V_\mu}(s_\mu)$ for $V_{\lambda\mu} = V_\lambda \cap V_\mu \neq \varnothing$. Namely, for $z \in V_{\lambda\mu}$, we have $s_\lambda(z) = s_\mu(z) \in \mathcal{G}_z$. Then, for $x \in U$, choose one V_λ satisfying $x \in V_\lambda$ and define a map s from U to $\bigcup_{x \in U} \mathcal{G}_x$ by $s(x) = s_\lambda(x)$. Since for $x \in V_\mu$ we have $s_\lambda(x) = s_\mu(x)$, the definition of s is correct. Since $s_\lambda \in {}^a\mathcal{G}(V_\lambda)$, one can find an open set $W \subset V_\lambda$ containing x and $t \in \mathcal{G}(W)$ so that we always have $s_\lambda(y) = t_y$, $y \in W$. Since $W \subset U$, this means $s \in {}^a\mathcal{G}(U)$. From the construction of s, we have $\rho_{V_\lambda, U}(s) = s_\lambda$.

Problem 2. The topology on $\widetilde{\mathcal{G}} = \prod_{x \in X} \mathcal{G}_x$ is defined as follows. For an open set U of X, $\{t_x\}_{x \in U}$, where $t \in \mathcal{G}(U)$, is a base for open sets. Thus one identifies $^a\mathcal{G}(U)$ with $\widetilde{\mathcal{G}}(U)$.

Problem 4. From Example 2.32, for

$$R_j = k\left[\frac{x_0}{x_j}, \ldots, \frac{x_{j-1}}{x_j}, \frac{x_{j+1}}{x_j}, \ldots, \frac{x_m}{x_j}\right], \quad U_j = \operatorname{Spec} R_j,$$

we have $\mathbb{P}^n_k = \bigcup_{j=0}^n U_j$. Let F_j be the restriction of $f \in \Gamma(\mathbb{P}^n_k, \mathcal{O}_{\mathbb{P}^n_k})$ to U_j. Then, F_j is an element of R_j. For $i \neq j$,

$$U_{ij} = U_i \cap U_j = \operatorname{Spec} k\left[\frac{x_0}{x_i}, \ldots, \frac{x_{i-1}}{x_i}, \frac{x_{i+1}}{x_i}, \ldots, \frac{x_n}{x_i}, \frac{x_i}{x_j}\right].$$

Namely, F_j does not have terms containing x_j/x_i. Hence, F_j must be an element of k.

Problem 5. From the commutative diagram, we obtain the homomorphisms of additive groups $\varphi : \varinjlim_{\lambda \in \Lambda} L_\lambda \to \varinjlim_{\lambda \in \Lambda} M_\lambda$ and $\psi : \varinjlim_{\lambda \in \Lambda} M_\lambda \to \varinjlim_{\lambda \in \Lambda} N_\lambda$. We will prove the injectivity of φ. For the sake of simplicity, let $L = \varinjlim_{\lambda \in \Lambda} L_\lambda$. Suppose $\varphi(a) = 0$ for $a \in L$. Then a is an element determined by a certain $l_\lambda \in L_\lambda$. Then $\varphi(a) = 0$ means that $m_\lambda = \varphi_\lambda(l_\lambda)$ determines 0 in M. That is, there is $\mu > \lambda$ to satisfy $g_{\mu\lambda}(m_\mu) = 0$. On the other hand, $\varphi_\mu(f_{\mu\lambda}(l_\lambda)) = g_{\mu\lambda}(\varphi_\lambda(l_\lambda))$ implies $\varphi_\mu f_{\mu\lambda}(l_\lambda) = 0$. Since φ_μ is injective, $f_{\mu\lambda}(l_\lambda) = 0$. That is, $a = 0$. Next, assume $\psi(b) = 0$, $b \in M$. Then let b be the element in M determined by $m_\nu \in M_\nu$. Set $n_\nu = \psi_\nu(m_\nu)$, which determines 0 in N. Then for some ξ_ν, we have $h_{\xi_\nu}(n_\nu) = 0$. On the other hand, since $0 = h_{\xi_\nu}(\psi_\nu(m_\nu)) = \psi_\xi(g_{\xi_\nu}(m_\nu))$, the exact sequence of additive groups for ξ implies that there exists $l_\xi \in L_\xi$ to satisfy $\varphi_\xi(l_\xi) = g_{\xi_\nu}(m_\nu)$. Let a be the element of L determined by l_ξ. Then we get $\varphi(a) = b$. Consequently, we obtain $\operatorname{Ker}\psi = \operatorname{Im}\varphi$. Finally, we will prove the surjectivity of ψ. Choose $n_\lambda \in N_\lambda$ which determines $c \in N$. Since ψ_λ is surjective, there is $m_\lambda \in M_\lambda$ to satisfy $\psi_\lambda(m_\lambda) = n_\lambda$. Let b be the element of M determined by m_λ. Then we have $c = \psi(b)$. Namely, ψ is a surjection.

Problem 6. Since $0 = (\operatorname{Ker}\varphi)(U) = \operatorname{Ker}\{\varphi_U : \mathcal{F}(U) \to \mathcal{G}(U)\}$, φ_U is an injection. We will prove the surjectivity of φ. For an arbitrary point x in U, $\varphi_x : \mathcal{F}_x \to \mathcal{G}_x$ is an isomorphism. Therefore, for $t \in \mathcal{G}(U)$, there exists only one $s_x \in \mathcal{F}_x$ to satisfy $\varphi_x(s_x) = t_x$. Thus, one can choose an open covering $\{U_j\}_{j \in J}$ and $s_j \in \mathcal{F}(U_j)$ to satisfy $\varphi_{U_j}(s_j) = \rho_{U_j, U}(t)$. Since φ_{U_j} is injective, such an s_j is determined uniquely. For $U_{jk} = U_j \cap U_k \neq \varnothing$, we have $\rho_{U_{jk}, U_j}(s_j) = \rho_{U_{jk}, U_k}(s_k)$. Therefore, there exists $s \in \mathcal{F}(U)$ such that $\rho_{U_j, U}(s) = s_j$. Since $\rho_{U_j, U}(\varphi_U(s)) = \varphi_{U_j}(\rho_{U_j, U}(s)) = \varphi_{U_j}(s_j) = \rho_{U_j, U}(t)$, we obtain $\varphi_U(s) = t$.

Problem 7. Let $a \in \Gamma(X, \mathcal{O}_X)$ and $\varphi \in \operatorname{Hom}_{\mathcal{O}_X}(\mathcal{F}, \mathcal{G})$. Then, for each open set U, the $\mathcal{O}_X$-homomorphism $a\varphi$ is defined as

$$(a\varphi)(s) = \rho_{U,X}(a)\varphi_U(s), \quad s \in \mathcal{F}(U).$$

For φ and ψ in $\operatorname{Hom}_{\mathcal{O}_X}(\mathcal{F}, \mathcal{G})$, define $(\varphi + \psi)_U = \varphi_U + \psi_U$. Then $\varphi + \psi \in \operatorname{Hom}_{\mathcal{O}_X}(\mathcal{F}, \mathcal{G})$.

Problem 8. Note first that $\varphi \in \operatorname{Hom}_{\mathcal{O}_X|U}(\mathcal{O}_X|U, \mathcal{F})$ is determined uniquely by the element $\varphi_U(1_U) = a \in \mathcal{F}(U)$ for $1_U \in \mathcal{O}_X(U)$. That is because: for open sets $V \subset U$, we have $\varphi_V(1_V) = \varphi_V(\rho_{V,U}(1_U)) = \rho_{V,U}(\rho_U(1_U)) = \rho_{V,U}(a)$, and for any element $\alpha \in$

$\mathcal{O}_X(V)$, we have $\varphi_V(\alpha) = \alpha\varphi_V(1_U) = \alpha\rho_{V,U}(a)$. Conversely, for an arbitrary $a \in \mathcal{F}(U)$, define $\varphi_V(\alpha) = \alpha\rho_{V,U}(a)$, where $\alpha \in \mathcal{O}_X(V)$ and $V \subset U$. Then $\{\varphi_U\}$ is an element of $\mathrm{Hom}_{\mathcal{O}_X|U}(\mathcal{O}_X|U, \mathcal{F})$. This correspondence is an isomorphism of additive groups.

Problem 9. The first isomorphism of $\mathcal{O}_X(U)$-modules follows from

$$\mathrm{Hom}_{\mathcal{O}_X(U)}(\mathcal{F}(U) \oplus \mathcal{G}(U), \mathcal{H}(U))$$
$$\stackrel{\sim}{\to} \mathrm{Hom}_{\mathcal{O}_X(U)}(\mathcal{F}(U), \mathcal{H}(U)) \oplus \mathrm{Hom}_{\mathcal{O}_X(U)}(\mathcal{G}(U), \mathcal{H}(U)),$$

where U is an open set. The second isomorphism can be shown in a similar way.

Problem 10. For an exact sequence of R-modules $L \xrightarrow{\varphi} M \xrightarrow{\psi} N$, it is enough to show the exactness of $L_\mathfrak{p} \to M_\mathfrak{p} \to N_\mathfrak{p}$. For $\frac{m}{s} \in M_\mathfrak{p}$ ($m \in M, s \in R \setminus \mathfrak{p}$), if $\psi_\mathfrak{p}(\frac{m}{s}) = 0$, then $\psi_\mathfrak{p}(\frac{m}{s}) = \frac{\psi(m)}{s} = 0$. Namely, there exists $t \in R \setminus \mathfrak{p}$ such that $t\psi(m) = 0$. Since we have $0 = t\psi(m) = \psi(tm)$ and $\mathrm{Ker}\,\psi = \mathrm{Im}\,\varphi$, there exists $l \in L$ to satisfy $\varphi(l) = tm$. Since $st \in R \setminus \mathfrak{p}$, we get $\frac{l}{st} \in L_\mathfrak{p}$ and $\varphi_\mathfrak{p}(\frac{l}{st}) = \frac{\varphi(l)}{st} = \frac{tm}{st} = \frac{m}{s}$. That is, $\mathrm{Ker}\,\psi_\mathfrak{p} \subset \mathrm{Im}\,\varphi_\mathfrak{p}$. On the other hand, since $\psi \circ \varphi = 0$, we have $\mathrm{Im}\,\varphi_\mathfrak{p} \subset \mathrm{Ker}\,\psi_\mathfrak{p}$. Consequently, $\mathrm{Ker}\,\psi_\mathfrak{p} = \mathrm{Im}\,\varphi_\mathfrak{p}$.

Problem 11. Let $m \in M, n \in N$, and let a and b be positive integers. As an R_f-module, we have

$$\frac{m}{f^a} \otimes \frac{n}{f^b} = \frac{m}{f^{a+b}} \otimes \frac{n}{1} = \frac{1}{f^{a+b}}\left(\frac{m}{1} \otimes \frac{n}{1}\right)$$

in $M_f \otimes_{R_f} N_f$. Hence, we can define an R_f-homomorphism $\varphi : M_f \otimes_{R_f} N_f \to (M \otimes_R N)_f$ by $\varphi(\frac{m}{f^a} \otimes nf^b) = \frac{m \otimes n}{f^{a+b}}$. We also can define an R_f-homomorphism $\psi : (M \otimes_R N)_f \to M_f \otimes_{R_f} N_f$ by $\psi(\frac{m \otimes n}{f^a}) = \frac{m}{f^a} \otimes \frac{n}{1}$. Then we have $\psi \circ \varphi = \mathrm{id}$ and $\varphi \circ \psi = \mathrm{id}$. Namely, φ is an isomorphism.

Problem 12. For an exact sequence of $\mathcal{O}_X$-modules

$$\mathcal{F} \xrightarrow{\varphi} \mathcal{G} \xrightarrow{\psi} \mathcal{H},$$

we will show that

$$L = \Gamma(X, \mathcal{F}) \xrightarrow{\varphi_X} M = \Gamma(X, \mathcal{G}) \xrightarrow{\psi_X} N = \Gamma(X, \mathcal{H})$$

is exact. Since $\psi_X \circ \varphi_X = 0$, we have

$$\mathrm{Coker}\,\varphi_X \xrightarrow{\eta} N.$$

Then η is injective if and only if $\operatorname{Im}\varphi_X = \operatorname{Ker}\psi_X$. The map η induces the exact sequence

$$0 \to \operatorname{Ker}\eta \to \operatorname{Coker}\varphi_X \to \operatorname{Im}\eta \to 0.$$

Then, as the induced $\mathcal{O}_X$-modules, we get

$$\widetilde{\operatorname{Coker}\varphi_X} = \operatorname{Coker}\varphi, \quad \widetilde{\operatorname{Im}\eta} = \operatorname{Im}\psi.$$

On the other hand, since we have $\operatorname{Im}\varphi = \operatorname{Ker}\psi$, the homomorphism of $\mathcal{O}_X$-modules

$$\tilde{\eta} : \widetilde{\operatorname{Coker}\varphi_X} \to \widetilde{N} = \Gamma(X, \mathcal{H})$$

is injective. Namely, $\operatorname{Ker}\tilde{\eta} = 0$, and so $\operatorname{Ker}\eta = 0$. That is, η is injective.

Problem 13. It is clear that $\mathcal{F} \oplus \mathcal{G}$ is a finitely generated $\mathcal{O}_X$-module. For an open set U of X, consider an $\mathcal{O}_U$-homomorphism

$$\varphi : \mathcal{O}_U^{\oplus n} \to (\mathcal{F} \oplus \mathcal{G})|U.$$

Let p and q be the projections from $(\mathcal{F} \oplus \mathcal{G})|U = \mathcal{F}|U \oplus \mathcal{G}|U$ to $\mathcal{F}|U$ and $\mathcal{G}|U$, respectively. Then $\operatorname{Ker}(p \circ \varphi)$ and $\operatorname{Ker}(q \circ \varphi)$ are finitely generated $\mathcal{O}_U$-modules. As an $\mathcal{O}_U$-module, $\operatorname{Ker}\varphi$ is generated by $\operatorname{Ker}(p \circ \varphi)$ and $\operatorname{Ker}(q \circ \varphi)$. That is, $\operatorname{Ker}\varphi$ is also a finitely generated $\mathcal{O}_U$-module.

Problem 14. If for all $j \in J$, $\mathcal{F}|U_j$ is a finitely generated $\mathcal{O}_{U_j}$-module, then $\mathcal{F}$ is a finitely generated $\mathcal{O}_X$-module. For an open set V of X and for an $\mathcal{O}_V$-homomorphism

$$\xi : \mathcal{O}_V^{\oplus l} \to \mathcal{F}|V,$$

if $V_j = V \cap U_j \neq 0$, then we have an $\mathcal{O}_{V_j}$-homomorphism

$$\xi_j : \mathcal{O}_{V_j}^{\oplus l} \to \mathcal{F}|V_j$$

satisfying $\operatorname{Ker}\xi_j = \operatorname{Ker}\xi|V_j$. Since $\mathcal{F}|U_j$ is a coherent sheaf, $\operatorname{Ker}\xi_j$ is a finitely generated $\mathcal{O}_{V_j}$-module. Hence, $\operatorname{Ker}\xi$ is a finitely generated $\mathcal{O}_V$-module. Therefore, $\mathcal{F}$ is a coherent sheaf.

Problem 15. For an open set V in X, let U be an open set of Y satisfying $f(V) \subset U$. Then

$$0 \to \mathcal{F}_1(U) \to \mathcal{F}_2(U) \to \mathcal{F}_3(U) \to 0$$

is an exact sequence. From (4.14), we get the sequence of presheaves

$$0 \to \mathcal{F}_1^\bullet(U) \to \mathcal{F}_2^\bullet(U) \to \mathcal{F}_3^\bullet(U)$$

inducing the sequence of maps of sheaves
$$0 \to f^{-1}\mathcal{F}_1 \to f^{-1}\mathcal{F}_2 \to f^{-1}\mathcal{F}_3.$$
For $x \in X$, let $y = f(x)$. Then from (4.15), we have $(f^{-1}\mathcal{F}_j)_x = \mathcal{F}_{j,y}$ for $j = 1, 2, 3$. Since
$$0 \to \mathcal{F}_{1,y} \to \mathcal{F}_{2,y} \to \mathcal{F}_{3,y} \to 0$$
is exact, the sequence
$$0 \to (f^{-1}\mathcal{F}_1)_x \to (f^{-1}\mathcal{F}_2)_x \to (f^{-1}\mathcal{F}_3)_x \to 0$$
is exact.

Problem 16. Let $\mathcal{G} = f_*\mathcal{F}$ in Lemma 4.37. Then consider the homomorphism $\eta_\mathcal{F}$ corresponding to the identity map from $f_*\mathcal{F}$ to itself.

Problem 17 For an open set U of Y, we have the exact sequence
$$0 \to \mathcal{F}_1(f^{-1}(U)) \to \mathcal{F}_2(f^{-1}(U)) \to \mathcal{F}_3(f^{-1}(U)),$$
inducing the exact sequence for the direct images. On the other hand, for the inverse image, Problem 14 implies the exact sequence
$$0 \to f^{-1}\mathcal{G}_1 \to f^{-1}\mathcal{G}_2 \to f^{-1}\mathcal{G}_3 \to 0.$$
By tensoring this sequence with $\mathcal{O}_Y$ over $f^{-1}\mathcal{O}_Y$, we get the exact sequence of $\mathcal{O}_X$-modules
$$f^*\mathcal{G}_1 \to f^*\mathcal{G}_2 \to f^*\mathcal{G}_3 \to 0.$$

Problem 18 Choose an affine covering $\{U_j\}_{j \in J}$ of Y so that $f^{-1}(U_j)$ is an affine open set. Also choose an affine covering $\{V_i\}_{i \in I}$ of U to satisfy $V_i \subset U_{j_i}$, $j_i \in J$, and $V_i = D(g_i)$, $g_i \in \Gamma(U_{j_i}, \mathcal{O}_Y)$. We need to show that for each i, $f^{-1}(V_i)$ is an affine open set. Hence, consider $f^{-1}(U_{j_i}) \to U_{j_i}$. Let $A_i = \Gamma(f^{-1}(U_{j_i}), \mathcal{O}_X)$ and $B_i = \Gamma(U_{j_i}, \mathcal{O}_Y)$. Then we must show that the inverse image of $V = D(g)$, $g \in B_i$, under $f_i : \operatorname{Spec} A_i \to \operatorname{Spec} B_i$ is an affine open set. This is clear since $f_i^{-1}(D(g)) = D(\varphi(g)) \subset \operatorname{Spec} A_i$, where $\varphi : B_i \to A_i$ is the ring homomorphism induced by f_i.

Problem 19. It is enough to prove the case where Z and Y are affine schemes. The general case can be obtained by glueing the affine schemes. Let $Z = \operatorname{Spec} C, Y = \operatorname{Spec} B$ and $A = \Gamma(Y, \mathcal{A})$. Then

$\mathcal{A} = \widetilde{A}$, and $g^*\mathcal{A}$ is a commutative $\mathcal{O}_C$-algebra determined by $A \otimes_B C$. Since $\operatorname{Spec} \mathcal{A} = \operatorname{Spec} A \to \operatorname{Spec} B$ and

$$\operatorname{Spec} g^*\mathcal{A} = \operatorname{Spec} A \otimes_B C$$
$$\to \operatorname{Spec} C = \operatorname{Spec} A \times_{\operatorname{Spec} B} \operatorname{Spec} C \to \operatorname{Spec} C,$$

we conclude that $\operatorname{Spec} g^*\mathcal{A} \to Z$ is the base change of $\operatorname{Spec} \mathcal{A} \to Y$ under g.

Chapter 5

Problem 1. For base changes $f_W : X \times_Z W \to Y \times_Z W$ and $g_W : Y \times_Z W \to W$ by a morphism $h : W \to Z$, $g_W \circ f_W = (g \circ f)_W$ is universally closed. The map on underlying spaces induced by f_W is surjective. This is because a point (y, w) on $Y \times_Z W$ is precisely the points $y \in Y$ and $w \in W$ satisfying $g(f(y)) = h(w) = z$, and since there is $x \in X$ satisfying $f(x) = y$, we have $(x, w) \in X \times_Z W$ and $f_W((x,w)) = (y,w)$. For a closed set F of $Y \times_Z W$, $f_W^{-1}(F)$ is a closed set in $X \times_Z W$. The assumption implies that $(g \circ f)_W(f_W^{-1}(F))$ is a closed set in $X \times_Z W$. The assumption implies that $(g \circ f)_W(f_W^{-1}(F))$ is a closed set of W which is exactly $g_W(F)$. Therefore, g_W is a closed morphism.

Problem 2. Let $m_1, m_2, \ldots, m_l$ be generators for the R-module M. Then we have

$$m_i = \sum_{j=1}^{l} a_{ij} m_j, \quad a_{ij} \in I, \quad i = 1, 2, \ldots, l.$$

Using the Kronecker delta δ_{ij}, we have

$$\sum_{j=1}^{l}(\delta_{ij} - a_{ij})m_j = 0, \quad i = 1, 2, \ldots, l.$$

Let $f = \det(\delta_{ij} - a_{ij})$. Then $f \in 1 + I$. Solving this equation, we obtain $fm_i = 0$, $i = 1, 2, \ldots, l$.

Problem 3. For $x, y \in R$, from (V2) we have $v(x + y) \geq \min\{v(x), v(y)\} \geq 0$. Namely, $x + y \in R$. We also have $v(x, y) = v(x) + v(y) \geq 0$, i.e., $xy \in R$. Since $R \subset K$, R is a commutative ring. For $a, b \in R$ and $x, y \in \mathfrak{m}$, we have $v(ax + by) \geq \min\{v(ax), v(by)\} = \min\{v(a) + v(x), v(b) + v(y)\} > 0$, i.e., $ax + by \in \mathfrak{m}$. Therefore, $\mathfrak{m}$ is an ideal of R. For $a \in R \setminus \mathfrak{m}$, we have $v(a) = 0$. From

$0 = v(1) = v(a \cdot a^{-1}) = v(a) + v(a^{-1})$, we get $v(a^{-1}) = -v(a) = 0$; namely, $a^{-1} \in R$. Consequently, $\mathfrak{m}$ is a maximal ideal of R.

For $x \in K$, $x \neq 0$, we have either $v(x) \geq 0$ or $v(x) < 0$. If $v(x) > 0$, then $x \in R$, and if $v(x) < 0$, $v(x^{-1}) = -v(x) > 0$ implies $x^{-1} \in R$. That is, the quotient field of R is precisely K. Let $\mathfrak{A}$ be an ideal of R. For $a \in \mathfrak{A}$ we have $v(a) \geq 0$. If $v(a) = 0$, then $a^{-1} \in R$, and $1 = a^{-1} \cdot a \in \mathfrak{A}$. Therefore, if $\mathfrak{A} \neq R$, then for an arbitrary element a of $\mathfrak{A}$, we have $v(a) > 0$, i.e., $\mathfrak{A} \subset \mathfrak{m}$. We conclude that $(R, \mathfrak{m})$ is a local ring.

Problem 4. If R is the valuation ring of a valuation v of K, then for $x \neq 0$, we have either $v(x) \geq 0$ or $v(x^{-1}) \geq 0$. Namely, $x \in R$ or $x^{-1} \in R$.

Conversely, suppose that K has the properties as stated. Let $K^* = K \setminus \{0\}$, and $E = \{x \in K^* | x \in R, x^{-1} \in R\}$. Then, with respect to the product, K^* and E are abelian groups. Let $\Gamma = K^*/E$ be the additive group. For $x, y \in K^*$, write $x \geq y$ if $xy^{-1} \in R$. Then Γ is a totally ordered additive group. Let $v(x)$ be the element of Γ determined by $x \in K^*$. Then v is a valuation on K.

Problem 5. Let π be an element a of $\mathfrak{m}$ of the minimum $d = v(a)$. For an arbitrary $b \in \mathfrak{m}$, let $v(b) = md + e$, $0 \leq e < d$. If $e > 0$, then $v(\pi^{-m}b) = e > 0$, i.e., $\pi^{-m}b \in \mathfrak{m}$. This contradicts the choice of d. Hence, we get $e = 0$ and $v(\pi^{-m}b) = 0$. Let $\alpha = \pi^{-m}b$. Then $\alpha \in R$ and $\alpha^{-1} \in R$. We can write $b = \alpha\pi^m$. Namely, $\mathfrak{m} = (\pi)$. Note that $\beta\pi$ is a prime element if $v(\beta) = 0$.

For an ideal $\mathfrak{A}$ of R, let $d_1 = \min_{a \in \mathfrak{A}} v(a)$. For $v(a) = d_1$, let $d_1 = nd + e_1$, $0 \leq e_1 < d$. As in the above, we get $\pi^{-n}a \in \mathfrak{m}$, and so $e_1 = 0$. Therefore, we can write $a = \alpha\pi^n$ and $v(\alpha) = 0$. As shown above, we obtain $\mathfrak{A} = (\pi^n)$.

Problem 6. If X is an integral scheme, then X must be reduced and irreducible. Conversely, if X is reduced, then $\Gamma(U, \mathcal{O}_X)$ has no nilpotent elements. Therefore, X is an integral scheme. Let $U = \operatorname{Spec} A$ be an affine open set of X. Then A is an integral domain. Let x be the point in U corresponding to the ideal (0). Then x is an open point. If x is not a generic point, i.e., the closure $Z = \overline{\{x\}} \neq X$, then Z and $F = X \setminus U$ are proper closed subsets of X. That is, X would become reduced, contradicting the integralness of X. Hence $Z = X$, i.e., x is a generic point of X.

On the other hand, $\mathcal{O}_{X,x}$ coincides with the quotient field K of A. Since an arbitrary point y of U corresponds to a prime ideal $\mathfrak{p}$ of A, we have $\mathcal{O}_{X,y} = A_{\mathfrak{p}}$, and since A is an integral domain, we also have $A_{\mathfrak{p}} \subset K$. Furthermore, for an arbitrary point y of X, there exists an affine scheme $V = \operatorname{Spec} B$ containing y. Since $x \in V$, K is the quotient field of B as well. Therefore, we obtain $\mathcal{O}_{x,y} \subset K = \mathcal{O}_{X,x}$.

Problem 7. Since $F = \overline{\{x\}}$ is the minimum closed set of $X = \operatorname{Spec} A$ containing x, the complement F^c of F is the maximum open set without x. We can write $F^c = X \setminus F = \bigcup_{f \in \mathfrak{p}} D(f)$. A prime ideal $\mathfrak{q}$ of A satisfies $\mathfrak{q} \notin D(f)$ when $f \in \mathfrak{q}$. Hence, for $\mathfrak{q} \notin F^c$, an arbitrary element f of $\mathfrak{p}$ satisfies $f \in \mathfrak{q}$, i.e., $\mathfrak{p} \subset \mathfrak{q}$. Conversely, if $\mathfrak{p} \subset \mathfrak{q}$, we have $\mathfrak{q} \notin F^c$, i.e., $\mathfrak{q} \in F$. There is a one-to-one correspondence between $\operatorname{Spec} A/\mathfrak{p}$ and the totality of prime ideals of A containing $\mathfrak{p}$.

Problem 8. For a homogeneous element a of S_+, $D_+(a)$ forms an open covering of $X = \operatorname{Proj} S$. We have $D_+(a) = \operatorname{Spec} S_{(a)}$ and $D_+(a) \cap D_+(b) = D_+(ab) = \operatorname{Spec} S_{(ab)}$. The natural homomorphisms
$$\lambda_{ab,a} : S_{(a)} \to S_{(ab)}, \qquad \lambda_{ab,b} : S_{(b)} \to S_{(ab)}$$
induce the homomorphism
$$\psi_{ab} : S_{(a)} \otimes_R S_{(b)} \to S_{(ab)},$$
$$\alpha \otimes \beta \mapsto \lambda_{ab,a}(a)\lambda_{ab,b}(b),$$
which is a surjective homomorphism. Therefore
$$\Delta_{ab} : D_+(ab) \to D_+(a) \times_{\operatorname{Spec} R} D_+(b)$$
is a closed immersion.

Problem 9. It is well known that a finitely generated algebra over a Noetherian ring S_0 is Noetherian. Consider the case where S is a Noetherian ring. Let $\mathfrak{A}_1 \subset \mathfrak{A}_2 \subset \mathfrak{A}_3 \subset \cdots$ be a sequence of ideals of S_0 and let $\widetilde{\mathfrak{A}}_j$ be the ideal of S generated by $\mathfrak{A}_j$ and $S_+ = \bigoplus_{m=1}^{\infty} S_m$. We have $\widetilde{\mathfrak{A}}_j \cap S_0 = \mathfrak{A}_j$. Then we have $\widetilde{\mathfrak{A}}_1 \subset \widetilde{\mathfrak{A}}_2 \subset \widetilde{\mathfrak{A}}_3 \subset \cdots$. Since S is Noetherian, there exists a positive integer n so that we have $\widetilde{\mathfrak{A}}_n = \widetilde{\mathfrak{A}}_{n+1} = \cdots$. This implies $\mathfrak{A}_n = \mathfrak{A}_{n+1} = \cdots$, i.e., S_0 is a Noetherian ring. If S is not finitely generated as an S_0-algebra, then among generators there are elements $x_1, x_2, x_3, \ldots$ satisfying the infinite sequence of ideals $(x_1) \subsetneq (x_1, x_2) \subsetneq (x_1, x_2, x_3) \subsetneq \cdots$. This contradicts the Noetherianness of S.

Problem 10. If $\sqrt{J} \supset S_+$, then for $\mathfrak{p} \in \operatorname{Proj} S$, we have $\mathfrak{p} \not\subset J$. Then for an arbitrary $\lambda \in \Lambda$, if $f_\lambda \in \mathfrak{p}$, then $\mathfrak{p} \supset J$. Hence, there exists an element $f_\mu \notin \mathfrak{p}$, i.e., $\mathfrak{p} \in D_+(f_\mu)$. Conversely, if $\{D_+(f_\lambda)\}_{\lambda \in \Lambda}$ is an open covering of $\operatorname{Proj} S$, then for $\sqrt{J} \not\supset S_+$ there exists a homogeneous ideal $\mathfrak{p}$ of S that contains $\sqrt{J}$ but does not contain S_+. For example, choose a homogeneous element $g \in S_d$, $d \geq 1$, satisfying $g \notin \sqrt{J}$. For the localization S_g, choose a maximal ideal $\mathfrak{p}'$ containing the ideal generated by $\sqrt{J}$ in S_g. Let $\mathfrak{p}$ be the image of $\mathfrak{p}'$ under the natural homomorphism $S \to S_g$. Then $g \notin \mathfrak{p}$, and $\mathfrak{p}$ is a prime ideal. Since $f_\lambda \in \mathfrak{p}$, $\lambda \in \Lambda$, we have $\mathfrak{p} \notin \bigcup_{\lambda \in \Lambda} D_+(f_\lambda)$. But this would contradict our assumption that $\{D_+(f_\lambda)\}_{\lambda \in \Lambda}$ is an open covering of $\operatorname{Proj} S$.

Problem 11. Let $L' = \bigoplus_{n \geq n_0} L_n$. Then $\widetilde{L'} = \widetilde{L}, \widetilde{M'} = \widetilde{M}$ and $\widetilde{N'} = \widetilde{N}$. By the assumption, $0 \to \widetilde{L'} \to \widetilde{M'} \to \widetilde{N'} \to 0$ is an exact sequence.

Problem 12. From $\varphi : R \to A$, we get the homomorphism $\psi : S \to T$ of graded rings of degree 0. Since $\psi(S_+)$ generates T_+, we get $G(\psi) = \operatorname{Proj} T$. Then we can define a scheme morphism $\psi^a : \operatorname{Proj} T \to \operatorname{Proj} S$. For a homogeneous element $f \in S_d$, let $g = \psi(f)$. We have $T_{(g)} \xrightarrow{\sim} S_{(f)} \otimes_R A$. Hence, we get $D_+(g) = \operatorname{Spec} T_{(g)} = \operatorname{Spec} S_{(f)} \times_{\operatorname{Spec} R} \operatorname{Spec} A = D_+(f) \times_{\operatorname{Spec} R} \operatorname{Spec} A$. Consequently, $\operatorname{Proj} T = \operatorname{Proj} S \times_{\operatorname{Spec} R} \operatorname{Spec} A$.

Problem 13. (1) Let $U = \operatorname{Spec} R$ be an affine open set of X. Then for $f|U \in \Gamma(U, \mathcal{S})$, we have an open set $D_+(f|U)$ of $\pi^{-1}(U)$. For an affine covering $\{U_\lambda\}_{\lambda \in \Lambda}$ of X, Z_f is obtained by glueing $D_+(f|U_\lambda)$. Part (2) can be proved by restricting the assertion to $\pi^{-1}(U)$.

Problem 14. It is enough to consider the case where $f : X' = \operatorname{Spec} R' \to X = \operatorname{Spec} R$. Then the proof follows from Problem 11.

Problem 15. Choose an affine open set U to satisfy $\mathcal{L}|U = \mathcal{O}_X|U$. Then we get $\mathbb{S}(\mathcal{E})|U \xrightarrow{\sim} \mathbb{S}(\mathcal{E} \otimes \mathcal{L})|U$. Therefore, the restriction of $\mathbb{P}(\mathcal{E})$ to U is obtained. (Namely, for the structure morphism $\pi : \mathbb{P}(\mathcal{E}) \to X$, let $\mathbb{P}(\mathcal{E})|U$ and $\mathbb{P}(\mathcal{E} \otimes \mathcal{L})|U$ be the restriction over U determined by $\pi^{-1}(U) \to U$. We have $\mathbb{P}(\mathcal{E})|U \xrightarrow{\sim} \mathbb{P}(\mathcal{E} \otimes \mathcal{L})|U$.) For affine open sets $V \subset U$, the isomorphism $\mathbb{P}(\mathcal{E})|V \xrightarrow{\sim} \mathbb{P}(\mathcal{E} \otimes \mathcal{L})|V$ is induced by the isomorphism over U. Hence, by considering an affine covering of X, we obtain an isomorphism over X.

Problem 16. By letting $\mathcal{L}_1 = \mathcal{L}_2 = \mathcal{L}$ in Example 5.38, $\mathcal{L}^{\otimes 2}$ is very ample. Therefore, so is $\mathcal{L}^{\otimes 2} \otimes \mathcal{L} = \mathcal{L}^{\otimes 3}$, and so on.

Chapter 6

Problem 1. Break down the exact sequence into short exact sequences

$$0 \to \mathcal{F}_1 \xrightarrow{f} \mathcal{F}_2 \to \text{Coker}\, f_1 \to 0,$$
$$0 \to \text{Coker}\, f_1 \to \mathcal{F}_2 \to \text{Im}\, f_2 \to 0,$$
$$0 \to \text{Im}\, f_2 \to \mathcal{F}_3 \to \text{Coker}\, f_2 \to 0.$$

Then use Lemma 6.4 to prove the assertion.

Problem 2. For an open set U of X we have $\Gamma(U, \iota_*\mathcal{G}) = \Gamma(U \cap Y, \mathcal{G})$. Also we have $\Gamma(X, \iota_*\mathcal{G}) = \Gamma(Y, \mathcal{G})$. Since $\mathcal{G}$ is flabby, $\Gamma(Y, \mathcal{G}) \to \Gamma(U \cap Y, \mathcal{G})$ is surjective. Hence so is $\Gamma(X, \iota_*\mathcal{G}) \to \Gamma(U, \iota_*\mathcal{G})$.

Problem 3. Let

$$0 \to \mathcal{F} \to \mathcal{G}^0 \to \mathcal{G}^1 \to \mathcal{G}^2 \to \cdots$$

be a flabby resolution of $\mathcal{F}$ over Y. Then

$$0 \to \iota_*\mathcal{F} \to \iota_*\mathcal{G}^0 \to \iota_*\mathcal{G}^1 \to \iota_*\mathcal{G}^2 \to \cdots$$

is a flabby resolution of $\iota_*\mathcal{F}$. This is because, for a sheaf $\mathcal{H}$ over Y and for $x \in X$, we have

$$(\iota_*\mathcal{H})_x = \begin{cases} \mathcal{H}_x, & x \in Y, \\ 0, & x \notin Y. \end{cases}$$

The above exact sequence together with Problem 2 implies that $0 \to \iota_*\mathcal{F} \to \iota_*\mathcal{G}^0 \to \iota_*\mathcal{G}^1 \to \cdots$ is a flabby resolution of $\iota_*\mathcal{F}$. Furthermore, since $\Gamma(Y, \mathcal{G}^j) \xrightarrow{\sim} \Gamma(X, \iota_*\mathcal{G}^j)$, we get the isomorphism between cohomology groups.

Problem 4. See *Rings and Fields* 1 of this series.

Problem 5. Let $p = 0$. We have $\xi = \{f_i\}$, $(\delta^0\xi)_{ij} = \xi_j - \xi_i$, $(\delta^1\{\delta^0\xi\})_{ijk} = (\delta^0\xi)_{jk} - (\delta^0\xi)_{ik} + (\delta^0\xi)_{ij} = (\xi_k - \xi_j) - (\xi_k - \xi_i) + (\xi_j - \xi_i) = 0$. Namely, $\delta^1 \circ \delta^0 = 0$. One can prove the general case in a similar way.

Problem 6. Let

$$\xi = \{f_{i_0 \cdots i_p}\}, \qquad \eta = \{g_{i_0 \cdots i_p}\},$$
$$h_{i_0 i_1 \cdots i_{p+q}} = f_{i_0 \cdots i_p} \otimes g_{i_p \cdots i_{p+q}} | U_{i_0 i_1 \cdots i_{p+q}}.$$

In general, $\xi = \{h_{i_0 i_1 \cdots i_{p+q}}\}$ is not an alternating cochain. For $\delta^p \xi = 0$ and $\delta^q \eta = 0$, we will show that $\delta^{p+q}\xi = 0$. In fact,

$$(\delta^{p+q}\xi)_{j_0 j_1 \cdots j_{p+q+1}} = \sum_{k=0}^{p}(-1)^k \xi_{j_0 j_1 \cdots \check{j}_k \cdots j_{p+q+1}}$$
$$+ \sum_{k=p+1}^{p+q+1}(-1)^k \xi_{j_0 j_1 \cdots \check{j}_k \cdots j_{p+q+1}}$$
$$= \sum_{k=0}^{p}(-1)^k f_{j_0 \cdots \check{j}_k \cdots j_{p+1}} \otimes g_{j_{p+1} \cdots j_{p+q+1}}$$
$$+ \sum_{k=p+1}^{p+q+1}(-1)^k f_{j_0 \cdots j_p} \otimes g_{j_p \cdots \check{j}_k \cdots j_{p+q+1}}$$
$$= -(-1)^{p+1} f_{j_0 \cdots j_p} \otimes g_{j_{p+1} \cdots j_{p+q+1}}$$
$$- (-1)^p f_{j_0 \cdots j_p} \otimes g_{j_{p+1} \cdots j_{p+q+1}} = 0.$$

Hence $\delta^{p+q}\xi = 0$. If ξ is an element of $\operatorname{Im}\delta^{p-1}$, one can show that $\{h_{i_0 \cdots i_{p+q}}\} \in \operatorname{Im}\delta^{p+q-1}$ in the same way.

Problem 7. Consider $f_{i_0 i_1 \cdots i_p} \in S_{m+n(p+1)}$. Then the proof will follow as indicated.

Problem 8. For an exact sequence of vector spaces over a field k

$$0 \to U \to V \to W \to 0,$$

the isomorphism $W \xleftarrow{\sim} V/U$ implies $\dim_k W = \dim_k V - \dim_k U$. That is, $\dim_k U - \dim_k V + \dim_k W = 0$. From the given exact sequence, we see that the sequences

$$0 \to V_0 \to V_1 \to \operatorname{Im} f_1 \to 0,$$
$$0 \to \operatorname{Im} f_1 \to V_2 \to \operatorname{Im} f_2 \to 0,$$
$$\cdots,$$
$$0 \to \operatorname{Im} f_{m-3} \to V_{m-2} \to \operatorname{Im} f_{m-2} \to 0,$$
$$0 \to \operatorname{Im} f_{m-2} \to V_{m-1} \to V_m \to 0$$

are exact, where $\operatorname{Im} f_j = \operatorname{Ker} f_{j+1}$, $j = 0, 1, 2, \ldots, m-2$. From those exact sequences, we get

$$\dim_k V_1 = \dim_k V_0 + \dim_k \operatorname{Im} f_1,$$
$$\dim_k V_j = \dim_k \operatorname{Im} f_{j-1} + \dim \operatorname{Im} f_j, \quad 2 \leq j \leq m-2,$$
$$\dim_k V_{m-1} = \dim_k V_m + \dim_k \operatorname{Im} f_{m-2}.$$

Therefore,

$$\sum_{j=2}^{m-2}(-1)^j \dim_k V_j = \sum_{j=2}^{m-2}(-1)^j(\dim_k \operatorname{Im} f_{j-1} + \dim_k \operatorname{Im} f_j)$$
$$= \dim_k \operatorname{Im} f_1 + (-1)^{m-2}\dim_k \operatorname{Im} f_{m-2}$$
$$= \dim_k V_1 - \dim_k V_0 + (-1)^{m-1}(\dim_k V_{m-1} - \dim_k V_m).$$

Namely, $\sum_{j=2}^{m}(-1)^j \dim_k V_j = 0$.

Problem 9. By Corollary 6.20, for $m \geq 0$, $H^q(X, \mathcal{O}_X(m)) = 0$, $q \geq 1$, and $\dim_k H^0(X, \mathcal{O}_X(m)) = \binom{m+n}{n}$. On the other hand, for $m \leq -n-1$, we have $H^q(X, \mathcal{O}_X(m)) = 0$, $q \leq n-1$, and $\dim_k H^n(X, \mathcal{O}_X(m)) = \binom{-m-1}{n}$. For m not in the above cases, $H^q(X, \mathcal{O}_X(m)) = 0$ for all q.

Problem 10. For an arbitrary affine open set U of Y, $\iota^{-1}(U)$ is an affine open set of X. Therefore, we have $H^p(\iota^{-1}(U), \mathcal{F}) = 0$, $p \geq 1$. That is, we get $R^p f_* \mathcal{F} = 0$.

Problem 11. Choose a finite affine covering $\{U_i\}_{i=1}^l$, $U_i = \operatorname{Spec} R_i$, and consider

$$f_i = f|f^{-1}(U_i) : X_i = f^{-1}(U_i) \to U_i.$$

Then all the f_i's are projective morphisms. One can choose a positive integer n so that $\mathcal{L}^{\otimes n}$ is f-very ample (Theorem 5.41). Then from Theorems 5.39 and 6.21, for $k \geq k_0^{(i)}$ and for $p \geq 1$, there exists a $k_0^{(i)}$ such that

$$H^p(X_i, \mathcal{F} \otimes \mathcal{L}^{\otimes n}) = 0,$$
$$H^p(X_i, \mathcal{F}(1) \otimes \mathcal{L}^{\otimes n}) = 0,$$
$$\ldots,$$
$$H^p(X_i, \mathcal{F}(n-1) \otimes \mathcal{L}^{\otimes n}) = 0.$$

Let $k_0 = \max_i k_0^{(i)}$ and $m_0 = k_0 n$. For all i, if $m \geq m_0$, then we have $H^p(X_i, \mathcal{F}(m)) = 0$, $p \geq 1$. Therefore, we obtain $R^p f_* \mathcal{F}(m) = 0$, $p \geq 1$, $m \geq m_0$.

Solutions to Exercises

Chapter 4

4.1. Choose a small enough neighborhood U of $y \in \operatorname{supp} \mathcal{F}$ so that for $s_1, s_2, \ldots, s_l \in \mathcal{F}(U)$, the map

$$\mathcal{O}_U^{\oplus l} \to \mathcal{F},$$
$$(a_1, \ldots, a_l) \mapsto \sum a_j s_j,$$

is a surjection. By the assumption, $\mathcal{F}_y = 0$. Hence $s_{j,y} = 0$, $j = 1, 2, \ldots, l$. Let $V \subset U$ be a sufficiently small open neighborhood of y so that $s_{j,x} = 0$ for all $x \in V$. Namely, $\mathcal{F}_x = 0$. Thus, $X \setminus \operatorname{supp} \mathcal{F}$ is an open set.

4.2. We can write $\mathcal{F} = \widetilde{M}$, where M is an R-module. Then $\Gamma(D(f), \mathcal{F}) = M_f$. Therefore, we can express $t = \frac{m}{f^n}$, $m \in M$. That is, $f^n t = m \in M = \Gamma(X, \mathcal{F})$.

4.3. (1) For $X = \operatorname{Spec} R$, let $N = \sqrt{(0)}$. We show that $\mathcal{N} = \widetilde{N}$. For $f \in R$, if $t \in \Gamma(D(f), \mathcal{O}_X)$ is a nilpotent element, then one can find positive integers m and n to satisfy $f^n a^m = 0$, where $t = \frac{a}{f^l}$, $a \in R$. Therefore, we have $b = f^n a \in N$, i.e., $t \in N_f$. We obtain $\mathcal{N} = \widetilde{N}$. Hence, for a scheme X in general, $\mathcal{N}$ is a quasicoherent sheaf of ideals.

(2) From (1), X_{red} is a closed subscheme of X. For an affine scheme $X = \operatorname{Spec} R$, since all the prime ideals contain N, the underlying spaces of X and X_{red} coincide. The general scheme case follows from this result. Since $\widetilde{R} = R/\sqrt{(0)}$ does not contain nilpotent elements, for any prime ideal $\mathfrak{p}$ of $\widetilde{R}$, $\widetilde{R}_\mathfrak{p}$ has no nilpotent elements.

(3) For an open set U of Y, the image of a nilpotent element of $\mathcal{O}_Y(U)$ under the homomorphism $\theta_U : \mathcal{O}_Y(U) \to f_* \mathcal{O}_X(U) = \mathcal{O}_X(f^{-1}(U))$ is also a nilpotent element. Therefore, θ_U induces a homomorphism from $\mathcal{O}_Y/\mathcal{N}_Y(U)$ to $\mathcal{O}_X/\mathcal{N}_X(f^{-1}(U))$. Thus, a scheme

morphism $f_{\text{red}} : X_{\text{red}} \to Y_{\text{red}}$ is determined. When $\Delta_{X/Y} : X \to X \times_Y X$ is a closed immersion, similarly one can show that $\Delta_{X_{\text{red}}/Y_{\text{red}}} : X_{\text{red}} \to X_{\text{red}} \times_{Y_{\text{red}}} X_{\text{red}}$ is a closed immersion.

4.4. (1) For an arbitrary point x in X, one can choose an open neighborhood U of x and an n so that
$$\mathcal{A}^{\oplus n}|U \to \mathcal{B}|U$$
is a surjective $\mathcal{A}|U$-homomorphism. By the assumption, $f|U : \mathcal{B}|U \to \mathcal{E}|U$ is also a surjective $\mathcal{A}|U$-homomorphism. Hence, by composing those maps, we get a surjective $\mathcal{A}|U$-homomorphism
$$\mathcal{A}^{\oplus n}|U \to \mathcal{E}|U.$$

(2) From (1), $\operatorname{Im} g$ is a finitely generated $\mathcal{A}$-module. For an arbitrary open set U of X, consider the $\mathcal{A}|U$-homomorphism
$$\varphi : \mathcal{A}^{\oplus m}|U \to \operatorname{Im} g|U.$$
By composing φ with the natural injection $\iota : \operatorname{Im} g \to \mathcal{E}$, we get $\operatorname{Ker} \varphi = \operatorname{Ker} \iota \circ \varphi$. Since $\mathcal{E}$ is coherent, $\operatorname{Ker} \iota \circ \varphi$ is a finitely generated $\mathcal{A}|U$-module. Therefore, $\operatorname{Ker} \varphi$ is also a finitely generated $\mathcal{A}|U$-module. Hence, $\operatorname{Im} g$ is coherent.

(3) For an arbitrary point x in X, choose an open neighborhood U of x and a positive integer n so that $f : \mathcal{A}^{\oplus n}|U \to \mathcal{G}|U$ is a surjective $\mathcal{A}|U$-homomorphism. Then $\psi \circ f : \mathcal{A}^{\oplus n}|U \to \mathcal{H}|U$ is surjective, and $\operatorname{Ker}(\psi \circ f) \to \operatorname{Ker} \psi|U = \mathcal{F}|U$ is a surjective $\mathcal{A}|U$-homomorphism. Since $\mathcal{H}$ is a coherent sheaf, $\operatorname{Ker}(\psi \circ f)$ is a finitely generated $\mathcal{A}|U$-module. Hence, from (1), $\mathcal{F}|U$ is also a finitely generated $\mathcal{A}|U$-module. Since $\mathcal{F}$ and $\operatorname{Im} \varphi$ are isomorphic, (2) implies that $\mathcal{F}$ is a coherent sheaf.

(4) From (1), $\mathcal{H}$ is a finitely generated $\mathcal{A}$-module. For a homomorphism of $\mathcal{A}_U$-modules $h : \mathcal{A}_U^{\oplus n} \to \mathcal{H}|U$ over an open set U of X, let $a_j = h((0,\ldots,0,\overset{i}{1},0,\ldots,0)) \in \mathcal{H}(U)$. We will show that $\operatorname{Ker} h$ is a finitely generated $\mathcal{A}_U$-module. If needed, take a smaller U so that there may exist $b_j \in \mathcal{G}(U)$ to satisfy $\psi(b_j) = a_j$, $j = 1, 2, \ldots, n$. If needed, take an even smaller U so that there can exist a surjective $\mathcal{A}_U$-homomorphism $j : \mathcal{A}_U^{\oplus l} \to \mathcal{F}|U$. Then consider
$$\tilde{h} : \mathcal{A}_U^{\oplus l} \oplus \mathcal{A}_U^{\oplus n} \to \mathcal{G}|U,$$
$$(\alpha, \beta) \mapsto j(\alpha) + \sum \beta_j b_j, \quad \beta = (\beta_1, \ldots, \beta_n).$$

Since $\mathcal{G}$ is coherent, Ker $\tilde{h}$ is a finitely generated $\mathcal{A}_U$-module. By the definition of $\tilde{h}$, we have a surjective $\mathcal{A}_U$-homomorphism Ker $\tilde{h} \to$ Ker h. From (1), Ker h is a finitely generated $\mathcal{A}_U$-module. Therefore, $\mathcal{H}$ is coherent.

(5) Choose a neighborhood U of x so that $h : \mathcal{A}_U^{\oplus l} \to \mathcal{F}|U$ and $g : \mathcal{A}_U^{\oplus n} \to \mathcal{H}|U$ are surjective $\mathcal{A}_U$-homomorphisms. For $a_j = g((0, \ldots, 0, \overset{j}{1}, 0, \ldots, 0)) \in \mathcal{H}(U)$, choose a sufficiently small U so that there exist $b_j \in \mathcal{G}(U)$ to satisfy $\psi(b_j) = a_j$. Then
$$\mathcal{A}_U^{\oplus l} \oplus \mathcal{A}_U^{\oplus m} \to \mathcal{G}|U,$$
$$(\alpha_1, \ldots, \alpha_l, \beta_1, \ldots, \beta_m) \mapsto -\varphi(h(\alpha_1, \ldots, \alpha_l)) + \sum \beta_j b_j,$$
is a surjective $\mathcal{A}_U$-homomorphism. Hence, $\mathcal{G}$ is a finitely generated $\mathcal{A}_U$-module.

(6) Let $h : \mathcal{A}_U^{\oplus n} \to \mathcal{G}|U$ be an $\mathcal{A}_U$-homomorphism. In this case, $\mathrm{Ker}(\psi|U \circ h)$ is a finitely generated $\mathcal{A}_U$-module. For a sufficiently small open neighborhood V of $x \in U$, the homomorphism of $\mathcal{A}_V$-modules
$$g : \mathcal{A}_V^{\oplus m} \to \mathrm{Ker}(\psi|U \circ h)|V$$
is surjective. Let $a_i = g((0, \ldots, 0, \overset{i}{1}, 0, \ldots, 0)) \in \mathcal{A}_U^{\oplus n}(V)$ and $b_i = h(a_i) \in \mathcal{G}(V)$. Since $\psi_V(b_i) = 0$, there exists a unique $c_i \in \mathcal{F}(V)$ to satisfy $\varphi(c_i) = b_i$. Therefore, we obtain an $\mathcal{A}_V$-homomorphism
$$\tilde{h} : \mathcal{A}_V^{\oplus m} \to \mathcal{F}|V,$$
$$(\gamma_1, \ldots, \gamma_m) \mapsto \sum_{j=1}^{m} \gamma_j c_j.$$
Since $\mathcal{F}$ is coherent, Ker $\tilde{h}$ is a finitely generated $\mathcal{A}_V$-module. By the definition of $\tilde{h}$, there exists a surjective $\mathcal{A}_V$-homomorphism Ker $\tilde{h} \to$ Ker h. Hence Ker h is also a finitely generated $\mathcal{A}_V$-module.

4.5. (1) Let $[a_1 \otimes \cdots \otimes a_n]$ be the element of $\mathbb{S}(E)$ determined by $a_1 \otimes \cdots \otimes a_n \in T^n(E)$. For an R-linear map $\varphi : E \to A$, define
$$f([a_1 \otimes \cdots \otimes a_n]) = \varphi(a_1)\varphi(a_2)\cdots\varphi(a_n)$$
to obtain an R-homomorphism
$$f : \mathbb{S}(E) \to A.$$
Since $\mathbb{S}(E)$ is generated by $\sigma(E)$ as an R-algebra, the homomorphism f is uniquely determined. When an R-commutative algebra B and an

R-linear map $\tilde{\sigma} : E \to B$ have the same property, the $\sigma : E \to \mathbb{S}(E)$ can be decomposed as

$$E \xrightarrow{\tilde{\sigma}} B \xrightarrow{\tilde{f}} \mathbb{S}(E),$$

and furthermore, $\tilde{\sigma} : E \to B$ can be decomposed as

$$E \xrightarrow{\sigma} \mathbb{S}(E) \xrightarrow{\tilde{g}} B.$$

By the uniqueness of $\tilde{f}$ and $\tilde{g}$, we get

$$\tilde{f} \circ \tilde{g} = \mathrm{id} \quad \text{and} \quad \tilde{g} \circ \tilde{f} = \mathrm{id}.$$

(2) For the commutative diagram

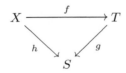

we have $f^* \mathcal{E}_{(T)} = f^*(g^* \mathcal{E}) = h^* \mathcal{E} = \mathcal{E}_{(W)}$ and $f^* \mathcal{O}_T = \mathcal{O}_W$. Therefore, f induces the following homomorphism of additive groups:

$$f^* : F_{\mathcal{E}}(T) = \mathrm{Hom}_{\mathcal{O}_T}(\mathcal{E}_{(T)}, \mathcal{O}_T) \to F_{\mathcal{E}}(W) = \mathrm{Hom}_{\mathcal{O}_W}(\mathcal{E}_{(W)}, \mathcal{O}_W).$$

From this, one can show that $F_{\mathcal{E}}$ is a contravariant functor.

(3) By Theorem 4.40, for an affine morphism $g : T \to S$,

$$\mathrm{Hom}_S(T, \mathbb{V}(\mathcal{E})) \xrightarrow{\sim} \mathrm{Hom}_{\mathcal{O}_S\text{-Alg}}(\mathbb{S}(\mathcal{E}), g_* \mathcal{O}_T)$$

is an isomorphism. From Proposition 3.4, since $f : \mathbb{V}(\mathcal{E}) \to S$ is an affine morphism, the above isomorphism holds for an arbitrary scheme $g : T \to S$ over S. On the other hand, the isomorphism $\mathrm{Hom}_R(E, A) \xrightarrow{\sim} \mathrm{Hom}_{R\text{-Alg}}(\mathbb{S}(E), A)$ implies

$$\mathrm{Hom}_{\mathcal{O}_S\text{-Alg}}(\mathbb{S}(\mathcal{E}), g_* \mathcal{O}_T) \xrightarrow{\sim} \mathrm{Hom}_{\mathcal{O}_S}(\mathcal{E}, g_* \mathcal{O}_T).$$

From Lemma 4.37, we have

$$\mathrm{Hom}_{\mathcal{O}_S}(\mathcal{E}, g_* \mathcal{O}_T) \xrightarrow{\sim} \mathrm{Hom}_{\mathcal{O}_T}(g^* \mathcal{E}, \mathcal{O}_T) = \mathrm{Hom}_{\mathcal{O}_T}(\mathcal{E}_{(T)}, \mathcal{O}_T) = F_{\mathcal{E}}(T).$$

Hence,

$$F_{\mathcal{E}}(T) \xrightarrow{\sim} \mathrm{Hom}_S(T, \mathbb{V}(\mathcal{E})).$$

Chapter 5

5.1. The natural injection $\mathcal{I}' \to \mathcal{S}$ induces the injection $\widetilde{\mathcal{I}'} \to \mathcal{O}_Z$. We need to show that the image of this injection is $\mathcal{J}$. By the definition of $\mathcal{J}$, we have the commutative diagram

$$\begin{array}{ccc} \mathcal{J} & \longrightarrow & \mathcal{S} \\ {\scriptstyle \alpha'}\downarrow & & \downarrow{\scriptstyle \alpha} \\ \Gamma_*(\mathcal{J}) & \longrightarrow & \Gamma_*(\mathcal{O}_Z). \end{array}$$

We have $\widetilde{\mathcal{S}} = \mathcal{O}_Z$ and $\beta : \widetilde{\Gamma_*(\mathcal{O}_Z)} \twoheadrightarrow \mathcal{O}_Z$. Hence, $\tilde{\alpha} : \widetilde{\mathcal{S}} \to \widetilde{\Gamma_*(\mathcal{O}_Z)}$ is also an isomorphism. Therefore, $\tilde{\alpha}' : \widetilde{\mathcal{I}'} \to \widetilde{\Gamma_*(\mathcal{J})}$ is an injection. Let $\mathcal{A} = \alpha(\mathcal{S}) \subset \Gamma_*(\mathcal{O}_Z)$ and $\mathcal{I}'' = \mathcal{A} \cap \Gamma_*(\mathcal{J})$. Since $\tilde{\alpha}$ is an isomorphism, we have $\widetilde{\mathcal{A}} = \widetilde{\Gamma_*(\mathcal{O}_Z)}$. Consequently, $\widetilde{\mathcal{I}''} = \widetilde{\Gamma_*(\mathcal{J})}$. Since $\alpha'(\mathcal{I}') = \mathcal{I}''$, $\tilde{\alpha}'$ is an isomorphism. Since $\beta_{\mathcal{J}} : \widetilde{\Gamma_*(\mathcal{J})} \to \mathcal{J}$ is an isomorphism, $\beta_{\mathcal{J}} \circ \tilde{\alpha}' : \widetilde{\mathcal{I}'} \to \mathcal{J}$ is an isomorphism.

5.2. (1) For a homogeneous polynomial $F(x_0, x_1, \ldots, x_n)$ of degree m, we have $\varphi_{\alpha A}(F(x_0, x_1, \ldots, x_n)) = \alpha^m \varphi_A(F(x_0, x_1, \ldots, x_n))$. Hence, for a homogeneous ideal I of R, we have $\varphi_A(I) = \varphi_{\alpha A}(I)$. Therefore, for a homogeneous ideal J of R, we have $\varphi_A^{-1}(J) = \varphi_{\alpha A}^{-1}(J)$. Namely, we get $f_A = f_{\alpha A}$.

(2) The defining ideal I of a point $(a_0 : a_1 : \cdots : a_n)$ is generated by the totality of all the minors of size 2 of the matrix $\begin{pmatrix} a_0 & a_1 & \cdots & a_n \\ x_0 & x_1 & \cdots & x_n \end{pmatrix}$. Therefore, $\varphi_A(I)$ is the ideal generated by the totality of all the minors of size 2 of the matrix

$$\begin{pmatrix} a_0 & a_1 & \cdots & a_n \\ \varphi_A(x_0) & \varphi_A(x_1) & \cdots & \varphi_A(x_n) \end{pmatrix}.$$

Let $(b_0, b_1, \ldots, b_n) = (a_0, a_1, \ldots, a_n){}^t A^{-1}$ and let J be the defining ideal of $(b_0 : b_1 : \cdots : b_n)$. Since we have

$$\begin{pmatrix} a_0 & a_1 & \cdots & a_n \\ \varphi_A(x_0) & \varphi_A(x_1) & \cdots & \varphi_A(x_n) \end{pmatrix} = \begin{pmatrix} b_0 & b_1 & \cdots & b_n \\ x_0 & x_1 & \cdots & x_n \end{pmatrix} {}^t A,$$

it follows that $\varphi_A(I) = J$, i.e., $J = \varphi_A^{-1}(I)$. That is,

$$f_A((a_0 : a_1 : \cdots : a_n)) = (b_0 : b_1 : \cdots : b_n).$$

(3) For $P_1 = (a_0 : a_1), P_2 = (b_0 : b_1)$ and $P_3 = (c_1 : c_2)$, choose α and β so that
$$\alpha a_0 + \beta b_0 = c_0,$$
$$\alpha a_1 + \beta b_1 = c_1.$$
Let
$$A = \begin{pmatrix} \alpha a_0 & \beta b_0 \\ \alpha a_1 & \beta b_1 \end{pmatrix}.$$
Then $f_{A^{-1}}((1:0)) = P_1, f_{A^{-1}}((0:1)) = P_2$, and $f_{A^{-1}}((1:1)) = P_3$. Let g be another projective transformation satisfying $g((1:0)) = P_1, g((0:1)) = P_2, g((1:1)) = P_3$. Then for $f = f_{A^{-1}}$, $f \circ g^{-1}$ is a projective transformation mapping $(1:0), (0:1), (1:1)$ onto themselves. There exists only one such map, i.e., the identity. As above, there exists a unique projective transformation h satisfying $h((1:0)) = Q_1, h((0:1)) = Q_2, h((1:1)) = Q_3$. Hence, $h \circ f_{A^{-1}}$ is the projective transformation that we want.

5.3. (1) For
$$l_1 : a_0 x_0 + a_1 x_1 + a_2 x_2 = 0,$$
$$l_2 : b_0 x_0 + b_1 x_1 + b_2 x_2 = 0,$$
choose c_1, c_2 and c_3 so that the 3×3 matrix
$$\begin{pmatrix} a_0 & a_1 & a_2 \\ b_0 & b_1 & b_2 \\ c_0 & c_1 & c_2 \end{pmatrix}$$
is regular. Let A be the inverse of that matrix. Then $\varphi_{A^{-1}}(x_0) = a_0 x_0 + a_1 x_1 + a_2 x_2$, $\varphi_{A^{-1}}(x_1) = b_0 x_0 + b_1 x_1 + b_2 x_2$. Hence, the desired projective transformation is given by f_A.

Next, for $Q_1 = (a_0 : a_1 : a_2), Q_2 = (b_0 : b_1 : b_2), Q_3 = (c_0 : c_1 : c_2), Q_4 = (d_0 : d_1 : d_2)$, choose (α, β, γ) so that
$$\begin{pmatrix} a_0 & a_1 & a_2 \\ b_0 & b_1 & b_2 \\ c_0 & c_1 & c_2 \end{pmatrix} \begin{pmatrix} \alpha \\ \beta \\ \gamma \end{pmatrix} = \begin{pmatrix} d_0 \\ d_1 \\ d_2 \end{pmatrix}.$$
Let
$$A^{-1} = \begin{pmatrix} \alpha a_0 & \beta b_0 & \gamma c_0 \\ \alpha a_1 & \beta b_1 & \gamma c_1 \\ \alpha a_2 & \beta b_2 & \gamma c_2 \end{pmatrix}.$$
Then, $f_A((1:0:0)) = Q_1, f_A((0:1:0)) = Q_2, f_A((0:0:1)) = Q_3, f_A((1:1:1)) = Q_4$. Repeat the argument as in 5.2(3).

(2) Let $\alpha x_0 + \beta x_1 + \gamma x_2 = 0$ be the line that we seek. Then solve
$$a_0\alpha + a_1\beta + a_2\gamma = 0,$$
$$b_0\alpha + b_1\beta + b_2\gamma = 0$$
to determine $\alpha : \beta : \gamma$.

(3) Solve
$$a_0 x_0 + a_1 x_1 + a_2 x_2 = 0,$$
$$b_0 x_0 + b_1 x_1 + b_2 x_2 = 0$$
to compute $x_0 : x_1 : x_2$.

(4) The assertion of the theorem is invariant under a projective transformation. Hence, one can assume $A_1 = (0:1:0), A_2 = (0:0:1)$, $B_1 = (1:0:0), B_2 = (1:1:1)$. Then, we have
$$l_1 = \overline{A_1 A_2} : x_0 = 0, \quad l_2 = \overline{B_1 B_2} : x_1 - x_2 = 0,$$
and $P = (1:0:1)$. Let $A_3 = (0:a_1:a_2)$ and $B = (b_0:b_1:b_2)$. Then
$$\overline{A_1 B_3} : b_1 x_0 - b_0 x_2 = 0,$$
$$\overline{A_3 B_1} : a_1 x_2 - a_2 x_1 = 0,$$
and $Q = (a_2 b_0 : a_1 b_1 : a_2 b_1)$. Furthermore, since
$$\overline{A_3 B_2} : (a_1 - a_2)x_0 + a_2 x_1 - a_2 x_2 = 0,$$
$$\overline{A_2 B_3} : b_1 x_0 - b_0 x_1 = 0,$$
we get $R = (a_1 b_0 : a_1 b_1 : a_1 b_0 - (b_0 - b_1)a_2)$. Then, P, Q and R are on the line
$$a_1 b_1 x_0 + a_2(b_1 - b_0)x_1 - a_1 b_1 x_1 = 0.$$

(5) See Figure 5.2.

5.4. (1) The scheme X is irreducible and reduced. If $S(X)$ is not an integral domain, then X is either reducible or non-reduced. Let K be the quotient field of S'. For $f \in \Gamma(X, \mathcal{O}_X(l))$ and $g \in \Gamma(X, \mathcal{O}_X(m))$, $g \neq 0$, the degree of $\frac{f}{g} \in K$ is defined by $l - m$. If $\frac{f}{g}$ satisfies
$$X^s + a_1 X^{s-1} + \cdots + a_s = 0, \quad a_i \in \Gamma(X, \mathcal{O}_X(m_i)),$$
then we have
$$f^s + a_1 f^{s-1} g + a_2 f^{s-2} g^2 + \cdots + a_s g^s = 0.$$

By comparing the degrees, we get $sl = sm + m_s$, i.e., $l - m \geq 0$. For each x in X, $\mathcal{O}_{X,x}$ is a normal ring. Therefore, $(\frac{f}{g})_x \in \mathcal{O}_{X,x}(l-m)$. That is, $\frac{f}{g} \in \Gamma(X, \mathcal{O}_X(l-m))$. We conclude that S' is a normal ring.

(2) By Theorem 5.21, $\alpha : S \to S'$ is injective. Consider the exact sequence of graded k-algebras
$$0 \to S \to S' \to T \to 0.$$
This gives the exact sequence of sheaves of $\mathcal{O}_X$-algebras over X
$$0 \to \widetilde{S} \to \widetilde{S'} \to \widetilde{T} \to 0.$$
Since we have $\widetilde{S} = \mathcal{O}_X$, by Theorem 5.21, $\widetilde{S'} = \mathcal{O}_X$ implies $\widetilde{T} = 0$. That is, for a sufficiently large d, we get $T_d = 0$. Therefore, $S_d = S'_d$.

(3) This is clear from (1) and (2).

5.5. (1) Since $\mathcal{S}$ is generated by $\mathcal{S}_1 = \mathcal{J}$ as an $\mathcal{O}_X$-algebra, the invertible sheaf $\mathcal{O}_{\widetilde{X}}(1)$ can be defined. Let V be an affine open set of X. Over $\pi^{-1}(V) = \operatorname{Proj} \mathcal{S}(V)$, $\mathcal{O}_{\widetilde{X}}(1)$ is the sheaf generated by the graded $\mathcal{S}(V)$-module $\mathcal{S}(V)(1) = \bigoplus_{n=-1}^{\infty} \mathcal{S}_{n+1}(V)$. From Example 5.23, one can regard $\mathcal{O}_{\widetilde{X}}(1)$ as the sheaf generated by $\bigoplus_{n=0}^{\infty} \mathcal{S}_{n+1}(V)$. Since we have $\mathcal{S}_{n+1}(V) = \mathcal{J}(V)^{n+1} = \mathcal{J}(V) \cdot \mathcal{J}(V)^n = \mathcal{J}(V)\mathcal{S}_n(V)$, we get $\bigoplus_{n=0}^{\infty} \mathcal{S}_{n+1}(V) = \mathcal{J}(V)\mathcal{S}(V)$. Therefore, $\widetilde{\mathcal{J}} = \mathcal{O}_{\widetilde{X}}(1)$, i.e., $\widetilde{\mathcal{J}}$ is an invertible sheaf.

(2) Since $\mathcal{J}|U = \mathcal{O}_U$ and $\mathcal{S}|U \overset{\sim}{\to} \mathcal{O}_U[T]$, we have
$$\pi^{-1}(U) \overset{\sim}{\to} \operatorname{Proj} \mathcal{O}_U[T] \overset{\sim}{\to} U.$$

(3) The assertion is of a local nature for X. Hence, we are allowed to assume that $X = \operatorname{Spec} R$ and that $\mathcal{J}$ is the sheaf generated by the ideal J of R. Let $S = \bigoplus_{n=0}^{\infty} J^n$. Then $\widetilde{X} = \operatorname{Proj} S$. For $J = (a_0, a_1, \ldots, a_m)$, the R-homomorphism of degree 0
$$\varphi : R[x_0, x_1, \ldots, x_m] \to S,$$
$$f(x_0, x_1, \ldots, x_m) \mapsto f(a_0, a_1, \ldots, a_n),$$
is surjective. Therefore, φ determines the closed immersion $^a\varphi : \widetilde{X} = \operatorname{Proj} S \to \mathbb{P}_R^n$, where $\widetilde{X}$ may be regarded as a closed subscheme of $\mathbb{P}_R^n$. Then $\operatorname{Ker} \varphi$ is a homogeneous ideal of $R[x_0, x_1, \ldots, x_n]$.

Since J is generated by $a_0, a_1, \ldots, a_m$, $\mathcal{J}$ is also generated as an $\mathcal{O}_X$-module by $a_j \in \Gamma(X, \mathcal{J})$, $j = 0, 1, \ldots, m$. Therefore, $\mathcal{L} = f^{-1}\mathcal{J} \cdot \mathcal{O}_Z$ is also generated by $s_j \in \Gamma(Z, \mathcal{L})$ corresponding to a_j as an $\mathcal{O}_Z$-module. Therefore, by Theorem 5.32, a morphism $g : Z \to \mathbb{P}_R^n$ satisfying $g^*\mathcal{O}_{\mathbb{P}_R^n}(1) \overset{\sim}{\to} \mathcal{L}$ is uniquely determined. Since, for a homogeneous polynomial $F \in \operatorname{Ker} \varphi$ of degree n, we have $F(a_0, a_1, \ldots, a_m) = 0$, it follows that $F(s_0, s_1, \ldots, s_m) = 0$ in $\Gamma(Z, \mathcal{L}^{\otimes n})$. Hence, the image of $g : Z \to \mathbb{P}_Z^n$ is contained in $\widetilde{X}$. Namely, one can define a morphism $g : Z \to \widetilde{X}$. Then we have $g^* \mathcal{O}_{\widetilde{X}}(1) \xrightarrow{\sim} \mathcal{L}$.

Chapter 6

6.1. (1) $\Rightarrow$ (2) is precisely Theorem 6.10 and (3) is a special case of (2). We will show that (3) $\Rightarrow$ (1). Choose an affine open neighborhood U of a closed point x. Let $Y = X \setminus U$. Let $\mathcal{J}_Y$ and $\mathcal{J}_{Y \cup \{x\}}$ be the defining ideals of Y and $Y \cup \{x\}$, respectively. Also let $A = \Gamma(X, \mathcal{O}_X)$. Then we have the exact sequence of $\mathcal{O}_X$-modules

$$0 \to \mathcal{J}_{Y \cup \{x\}} \to \mathcal{J}_Y \to k(x) \to 0,$$

where $k(x) = \mathcal{O}_{X,x}/\mathfrak{m}_x$, and $\mathfrak{m}_x$ is the maximal ideal of $\mathcal{O}_{X,x}$. From this exact sequence, we get the exact sequence

$$\Gamma(X, \mathcal{J}_Y) \to k(x) \to H^1(X, \mathcal{J}_{Y \cup \{x\}}) = 0.$$

Therefore, there exists $f \in \Gamma(X, \mathcal{J}_Y)$ to satisfy $f(x) = 1$. Since we can regard $\Gamma(X, \mathcal{J}_Y) \subset \Gamma(X, \mathcal{O}_X) = A$, we can consider $f \in A$. Let $X_f = \{y \in X | f(y) \neq 0\}$. Then X_f is an open set satisfying $x \in X_f \subset U$. This is because f is 0 on Y. Let $\overline{f}$ be the image of f in $\Gamma(U, \mathcal{O}_U)$. We have $X_f = U_{\overline{f}}$. Namely, X_f is an affine open set. Therefore, X can be covered by affine open sets of the type X_f. Since X is Noetherian, we can assume $X_{f_1}, X_{f_2}, \ldots, X_{f_n}$ cover X. Consider a homomorphism of $\mathcal{O}_X$-modules

$$\alpha : \mathcal{O}_X^{\oplus n} \to \mathcal{O}_X,$$

$$(a_1, \ldots, a_n) \mapsto \sum_{i=1}^n a_i f_i.$$

Since $\{X_{f_i}\}_{i=1}^n$ is an open covering of X, α is surjective. Let $\mathcal{F} = \operatorname{Ker} \alpha$. We have the exact sequence

$$0 \to \mathcal{F} \to \mathcal{O}_X^{\oplus n} \xrightarrow{\alpha} \mathcal{O}_X \to 0.$$

By the natural injections $\mathcal{O}_X \subset \mathcal{O}_X^{\oplus 2} \subset \cdots \subset \mathcal{O}_X^{\oplus(n-1)} \subset \mathcal{O}_X^{\oplus n}$, we get

$$\mathcal{F} = \mathcal{F} \cap \mathcal{O}_X^{\oplus n} \supset \mathcal{F} \cap \mathcal{O}_X^{\oplus(n-1)} \supset \cdots \supset \mathcal{F} \cap \mathcal{O}_X.$$

Let $\mathcal{F}_r = \mathcal{F} \cap \mathcal{O}_X^{\oplus r}$. Then $\mathcal{F}_r/\mathcal{F}_{r-1}$ is a coherent $\mathcal{O}_X$-ideal sheaf. By the assumption, we have $H^1(X, \mathcal{F}_r/\mathcal{F}_{r-1}) = 0$, and so $H^1(X, \mathcal{F}) = 0$.

Therefore
$$A^{\oplus n} = \Gamma(X, \mathcal{O}_X^{\oplus n}) \to A = \Gamma(X, \mathcal{O}_X)$$
is a surjection, and the ideal of A generated by $f_1, f_2, \ldots, f_n$ coincides with A. Therefore, we obtain $\operatorname{Spec} A = \bigcup_{i=1}^n \operatorname{Spec} A_{f_i}$. Consequently, there exists an isomorphism
$$\varphi_i : X_{f_i} \overset{\sim}{\to} \operatorname{Spec} A_{f_i}$$
such that $\varphi_i|X_{f_i} \cap X_{f_j} = \varphi_j|X_{f_i} \cap X_{f_j}$. By glueing the φ_i, we get the isomorphism $\varphi : X \overset{\sim}{\to} \operatorname{Spec} A$.

6.2. From Example 6.24,
$$\dim_k H^1(C, \mathcal{O}_C) = \binom{n-1}{2} = \frac{(n-1)(n-2)}{2}.$$

6.3. For an invertible sheaf $\mathcal{L}$, one can find an open covering $\mathcal{U} = \{U_i\}_{i \in I}$ satisfying $\varphi_i : \mathcal{L}|U_i \overset{\sim}{\to} \mathcal{O}_{U_i}$, $i \in I$. Then, over U_{ij}, $\varphi_{ij} = \varphi_i \circ \varphi_j^{-1}|U_{ij} : \mathcal{O}_{U_{ij}} \to \mathcal{O}_{U_{ij}}$ is uniquely determined by $f_{ij} = \varphi_{ij}(1) \in \Gamma(U_{ij}, \mathcal{O}_{U_{ij}})$. Since φ_{ij} is an isomorphism, we must have $f_{ij} \in \Gamma(U_{ij}, \mathcal{O}_{U_{ij}}^*)$. Furthermore, since $\varphi_{ik} = \varphi_{ij} \circ \varphi_{jk}$ holds over U_{ijk}, we get $f_{ik} = f_{ij} \circ f_{jk}$. Namely, $\{f_{ij}\}$ defines an element of $\check{H}^1(\mathcal{U}, \mathcal{O}_X^*)$, which induces an element of $\check{H}^1(X, \mathcal{O}_X^*)$. If there is an isomorphism of invertible sheaves $\Phi : \mathcal{L} \overset{\sim}{\to} \mathcal{M}$, choose an open covering $\mathcal{U} = \{U_i\}_{i \in I}$ so that we may have $\varphi_i : \mathcal{L}|U_i \overset{\sim}{\to} \mathcal{O}_{U_i}$ and $\psi_i : \mathcal{M}|U_i \overset{\sim}{\to} \mathcal{O}_{U_i}$. Let $\{f_{ij}\}$ and $\{g_{ij}\}$ be the elements of $\check{H}^1(\mathcal{U}, \mathcal{O}_X^*)$ induced by $\mathcal{L}$ and $\mathcal{M}$, respectively, as constructed above. Then, let $\Phi_i = \psi_i \circ \varphi_i^{-1} : \mathcal{O}_{U_i} \to \mathcal{O}_{U_i}$ and $h_i = \Phi_i(1)$. We have $h_i \in \Gamma(U_i, \mathcal{O}_{U_i}^*)$ and $h_i^{-1} f_{ij} h_j = g_{ij}$. That is, $\{f_{ij}\}$ and $\{g_{ij}\}$ define the same cohomologous class.

Conversely, a given element ξ of $\check{H}^1(X, \mathcal{O}_X^*)$ is the inductive limit of the cohomologous class $\{f_{ij}\} \in \check{H}^1(\mathcal{U}, \mathcal{O}_X^*)$ for a certain open covering $\mathcal{U}$ of X. By using $\{f_{ij}\}$, one can glue $\mathcal{O}_{U_i}$ and $\mathcal{O}_{U_j}$ over U_{ij} to obtain an invertible sheaf $\mathcal{L}$. This invertible sheaf $\mathcal{L}$ provides cohomologous classes $\{f_{ij}\}$.

Index

affine morphism, 46
alternating cochain, 133
ample invertible sheaf, 103
Artin-Rees lemma, 133

blowing up, 88, 110

canonical flabby resolution, 115
Čech cohomology group, 133, 134
Chern class, 148
closed immersion, 45
closed morphism, 54
closed subscheme, 45
cocycle, 7
coherent, 30
coherent sheaf, 30
cohomology, 7, 111, 118
cohomology group, 118
cokernel, 9
complete variety, 106
complex, 118
Cousin distribution, 15
Cousin problem, 15

direct sum, 18
discrete valuation, 61
discrete valuation ring, 61
dominate, 60
double complex, 122
dual sheaf, 52
duality principle, 109

Euler characteristics, 147
Euler-Poincaré characteristics, 147
exact, 11
exact sequence, 11

f-ample, 103

finitely generated, 24, 101, 103
finitely presented, 75, 89
finitely presented R-module, 23
flabby resolution, 113
flabby sheaf, 111
free module, 18
free $\mathcal{O}_X$-module, 18
f-very ample, 99

graded S-module, 72

higher direct image, 154
homogeneous coordinate ring, 109
homomorphism of degree 0, 70
homomorphism of sheaves, 2
hypersurface, 85

ideal sheaf, 44
image, 7
immersion, 99
injective, 11
injective resolution, 129
injective R-module, 129
integrally closed, 109
inverse image, 37, 40
invertible sheaf, 18
irrelevant ideal, 68
isomorphism, 3

kernel, 6

Leray's theorem, 136
line bundle, 19
line over k, 108
locally free $\mathcal{O}_X$-module, 18
  of rank n, 18
locally free resolution, 146
locally free sheaf, 18

long exact sequence, 125

monoidal transformation, 110

Nakayama's lemma, 59
nilpotent ideal sheaf, 50
normal ring, 109
normal scheme, 109

$\mathcal{O}_X$-flat sheaf, 23
$\mathcal{O}_X$-module, 17
$\mathcal{O}_X$-module homomorphism, 17
$\mathcal{O}_Y$-algebra, 47
$\mathcal{O}_Y$-commutative algebra, 47

pairing, 138
Pappus theorem, 109
perfect pairing, 138
Picard group, 23
prime element, 61
projective, 104
projective morphism, 103
projective scheme, 87
projective transformation, 108
projective variety, 106
projectively normal, 109
proper mapping theorem, 157
proper morphism, 56

quasicoherent, 24
quasicoherent ideal sheaf, 44
quasicoherent sheaf, 24
quasicompact morphism, 42
quasiprojective, 103
quasiprojective morphism, 103
quotient field, 15
quotient sheaf, 11

rank, 18
reduced, 50
refinement, 134
R-flat module, 23
Riemann-Roch theorem, 148
ring of total quotients, 15

S-homomorphism of degree n, 74
section, 19
Segre morphism, 98
sequence, 11
sheaf field of fractions, 15

sheaf obtained by extending by zero, 46
sheaf of local sections, 19
sheafification, 5
short exact sequence, 12
skyscraper sheaf, 11
specialization, 62
structure sheaf, 68
subsheaf, 6
support, 44
surjective, 11
symmetric algebra, 51
  over E, 51

tensor algebra, 51
  over E, 51
tensor product, 21
totally ordered module, 59

universally closed, 54
universally closed morphism, 54

valuation, 60
valuation ring, 60
valuative criterion, 59
  of properness, 63
  of separatedness, 66
vector bundle, 18, 19
vector fiber space, 52

weight, 70
weighted projective space, 70

Copying and reprinting. Individual readers of this publication, and nonprofit libraries acting for them, are permitted to make fair use of the material, such as to copy a chapter for use in teaching or research. Permission is granted to quote brief passages from this publication in reviews, provided the customary acknowledgment of the source is given.

Republication, systematic copying, or multiple reproduction of any material in this publication is permitted only under license from the American Mathematical Society. Requests for such permission should be addressed to the Assistant to the Publisher, American Mathematical Society, P.O. Box 6248, Providence, Rhode Island 02940-6248. Requests can also be made by e-mail to **reprint-permission@ams.org**.

Selected Titles in This Series

(*Continued from the front of this publication*)

172 **Ya. G. Berkovich and E. M. Zhmud',** Characters of finite groups. Part 1, 1998

171 **E. M. Landis,** Second order equations of elliptic and parabolic type, 1998

170 **Viktor Prasolov and Yuri Solovyev,** Elliptic functions and elliptic integrals, 1997

169 **S. K. Godunov,** Ordinary differential equations with constant coefficient, 1997

168 **Junjiro Noguchi,** Introduction to complex analysis, 1998

167 **Masaya Yamaguti, Masayoshi Hata, and Jun Kigami,** Mathematics of fractals, 1997

166 **Kenji Ueno,** An introduction to algebraic geometry, 1997

165 **V. V. Ishkhanov, B. B. Lur'e, and D. K. Faddeev,** The embedding problem in Galois theory, 1997

164 **E. I. Gordon,** Nonstandard methods in commutative harmonic analysis, 1997

163 **A. Ya. Dorogovtsev, D. S. Silvestrov, A. V. Skorokhod, and M. I. Yadrenko,** Probability theory: Collection of problems, 1997

162 **M. V. Boldin, G. I. Simonova, and Yu. N. Tyurin,** Sign-based methods in linear statistical models, 1997

161 **Michael Blank,** Discreteness and continuity in problems of chaotic dynamics, 1997

160 **V. G. Osmolovskiĭ,** Linear and nonlinear perturbations of the operator div, 1997

159 **S. Ya. Khavinson,** Best approximation by linear superpositions (approximate nomography), 1997

158 **Hideki Omori,** Infinite-dimensional Lie groups, 1997

157 **V. B. Kolmanovskiĭ and L. E. Shaĭkhet,** Control of systems with aftereffect, 1996

156 **V. N. Shevchenko,** Qualitative topics in integer linear programming, 1997

155 **Yu. Safarov and D. Vassiliev,** The asymptotic distribution of eigenvalues of partial differential operators, 1997

154 **V. V. Prasolov and A. B. Sossinsky,** Knots, links, braids and 3-manifolds. An introduction to the new invariants in low-dimensional topology, 1997

153 **S. Kh. Aranson, G. R. Belitsky, and E. V. Zhuzhoma,** Introduction to the qualitative theory of dynamical systems on surfaces, 1996

152 **R. S. Ismagilov,** Representations of infinite-dimensional groups, 1996

For a complete list of titles in this series, visit the AMS Bookstore at **www.ams.org/bookstore/**.